Lecture Notes in Mathematics

A collection of informal reports and seminars
Edited by A. Dold, Heidelberg and B. Eckmann, Zürich

228

Conference on Applications of Numerical Analysis

Held in Dundee/Scotland, March 23-26, 1971

Edited by
John Ll. Morris, University of Dundee, Dundee/Scotland

Springer-Verlag
Berlin · Heidelberg · New York 1971

AMS Subject Classifications (1970): 05 A 17, 30 A 08, 41 A 20, 45 A 55, 45 A 63, 45 B 05, 45 D 05, 45 E 10, 45 L 10, 47 B 45, 49 C 05, 49 D 10, 49 D 15, 49 D 99, 65 D 15, 65 D 30, 65 F 15, 65 H 05, 65 K 05, 65 L 05, 65 L 10, 65 L 15, 65 L 99, 15 M 05, 65 M 10, 65 N 05, 65 N 10, 65 N 15, 65 N 30, 65 N 99, 65 R 05, 90 C 20, 90 C 30, 90 C 50

ISBN 3-540-05656-4 Springer-Verlag Berlin · Heidelberg · New York
ISBN 0-387-05656-4 Springer-Verlag New York · Heidelberg · Berlin

Offsetdruck: Julius Beltz, Hemsbach.

FOREWORD

During the four days 23rd - 26th March, 1971 at the University of Dundee, Scotland, one hundred and seventy participants attended a conference on the Applications of Numerical Analysis. As the title suggests, the conference was intended to give an opportunity to researchers to present papers and hear the results of others' investigations in Numerical Analysis where methods and analyses were applied to particular problems or with some particular problem in mind. The theme of the conference grew out of the realization that Numerical Analysts, on the one side, and those research workers who meet real life physical problems, on the other side, have been in danger for some time of losing the ability to communicate with one another. We are only too familiar with the numerical analyst who produces a new algorithm and tests the method on a simple (well behaved) problem and claims, as the result of one or two tests, a fabulous new scheme which will solve all practical problems covered by the class of equations for which the scheme is designed. Similarly one is also only too familiar with the engineer (say) who when confronted with the reason for using a particular method states that he has discovered this method in a text book long out of date and uses the method therein simply because he found the notation easy to understand. Often as not the method does not work for the particular problem he has in mind or, perhaps there exist far better methods discovered over the last few years.

Are we to blame the numerical analyst and engineer? If the reason for the former not testing his algorithm rigourously is complacency then clearly the numerical analyst is at fault and he is consequently not producing what ought to be produced; i.e. a good working algorithm with a precise indication of the limitations on the method so that a non-specialist can see at a glance whether or not such a method is capable of solving his problem and in this context some indication of the new algorithm's merits relative to existing algorithms in the field. If, through bad communications with the people who have the problems, the numerical analyst is unable to gain access to real life problems then it is surely the system of communication which is at fault. This problem brings to mind the many papers one reads where the author either does not mention or glosses over those important auxiliary data (for example initial and boundary conditions for a physical problem involving partial differential equations) which make a solution possible or impossible.

Likewise, if our engineer does not use the best methods available simply because it is too much trouble for him to search out the requisite algorithm then there is little one can do to help him. However, as I think occurs in the majority of cases, the engineer is unable to understand what the numerical analyst is trying to say and, what is as bad, is often unable to gain access to the very man who could help him. In this case, once again, it is surely the system of communication which needs review.

The reader may be quick to point out that some groups do exist (for example the Oxford group) which endeavour to produce the necessary links between numerical analysts and 'those with the problems'. However, unfortunately, these instances are in the minority at present. Thus it is hoped that this conference (and others like it) will contribute something towards closing the communication gap. It was therefore gratifying to find at the conference many participants from establishments which work on the physical problems alluded to above. Furthermore, the British Theoretical Mechanics Colloquium (BTMC) was held in Dundee the week after the Applications of Numerical Analysis conference and it was hoped that these 'men of Applied Mathematics' would avail themselves of a numerical analyst colloquium directed towards the BTMC-like problems. Conversly it was hoped that participants at the first conference would attend some of the lectures of the second one. To a certain extent this was true but there is always room for improvement. Thus hopefully conferences such as the subject one of these proceedings will afford both sides the opportunity of communicating with one another and allow the workers on both sides to becomes more familiar with the other's point of view.

The conference took the form of eighteen one-hour lectures given by the invited speakers followed intermittently by half-hour lectures given by the authors of submitted papers, some of these latter talks being presented in parallel sessions. The full list of contributors follows this foreword. Because the theme of the conference was Applications rather than a particular branch of Numerical Analysis, the papers contained in the Volume cover a considerable breadth of interests. For example, function approximation, integral equations, ordinary and partial differential equations of inital and boundary value types, eigenvalue and inverse eigenvalue problems, optimization, and solution of polynomial equations.

As far as possible, we have endeavoured to secure all the papers given at the conference for inclusion in this Volume. However, unfortunately, five of the invited papers have not been included. These are the papers by Fox, Morton, Wilkinson, Golub and Powell. The latter two authors have produces short summaries which are included in the proceedings. The references to the full texts of the talks presented at the conference can be found at the end of this foreword.

The present Volume by Springer Verlag represents the second proceedings of conferences held at Dundee during the Numerical Analysis Year. This project has been financed by the Science Research Council and we acknowledge with gratitude their generous support. It is true to say that the academic year 1970/71 has been extremely exciting in Dundee and to a great extent this is due to the generous support of the Science Research Council. At the time of writing this foreword, a conference on Nonlinear Optimization has just finished when over two hundred participants attended a four day conference. The proceedings of this conference will be published by Academic Press, edited by F. Lootsma. During the next week another conference is to take place on the Ritz-Galerkin procedures and the Finite Element Method when other S.R.C. supported speakers will visit Dundee to present papers, namely Birkhoff, Douglas, Strang, Schoenberg, Thomée and Wachspress. The proceedings of this meeting will be the third this year to be published in the Springer Verlag series. This will be edited by R. Wait.

A considerable amount of the typing in the Volume has been undertaken by Yvonne Nédélec, in the Department of Mathematics. This work is gratefully acknowledged.

Finally, I wish to acknowledge the profound influence that Professor A. R. Mitchell has exercised over this and other proceedings during the Year. The success of the Numerical Analysis Year can be attributed largely to his organisation and guidance.

Dundee, July 1971

John Ll. Morris

References

Fox, L., Mayers, D.F., Taylor, A.B. and Ockendon, J.R. The Numerical Analysis of a Functional Differential Equation. Submitted to J.I.M.A., 1971.

Lewis, H.R., Morton, K.W., Sykes, A. and Wesson, J.H. Comparison of some particle-in-cell plasma simulation methods. To appear in J. Comp. Phys., 1971.

Wilkinson, J.H. Theoretical and practical properties of inverse interation. To appear in proceedings of National Institute of Rome 'Problems in Numerical Analysis', January 1972.

CONTENTS

Invited Papers

R.BELLMAN: On the Identification of Systems and Some Related Questions 1

R.BELLMAN: Invariant Imbedding: Semigroups in Time, Space and Structure 9

J.CEA, R.GLOWINSKI et J.C. NEDELEC: Minimisation de Fonctionnelles Non-Differentiables 19

L.COLLATZ: Nonlinear Integral Equations and Field Approximation Problems 39

G.H.GOLUB: Some Modified Eigenvalue Problems 56

R.GORENFLO: On Difference Schemes for Parabolic Differential Equations with Derivative Boundary Conditions 57

J.GREENSTADT: Cell Discretization 70

P.HENRICI: Circular Arithmetic and the Determination of Polynomial Zeros 86

R.HUSS and R.KALABA: Computation of the Moments of Solutions of Certain Random Two Point Boundary Value Problems 93

P.LASCAUX et P.A.RAVIART: Stabilité et Précision des Schemas DSN pour l'Equation de Transport en Géométrie Sphérique 103

B.NOBLE: Some Applications of the Numerical Solution of Integral Equations to Boundary Value Problems 137

M.R.OSBORNE: On the Inverse Eigenvalue Problem for Matrices and Related Problems for Difference and Differential Equations. 155

M.J.D.POWELL, I.BARRODALE and F.D.K.ROBERTS: The Differential Correction Algorithm for Rational L_∞ Approximation 169

P.A.RAVIART: Résolution Numérique de Certains Problèmes Hyperboliques Non Linéaires. Méthode de Pseudo-Viscosité 170

H.J.STETTER: Stability of Discretizations on Infinite Intervals 207

E.L.WACHSPRESS: A Rational Basis for Function Approximation . . 223

O.B.WIDLUND: Some Results on Best Possible Error Bounds for Finite Element Methods and Approximation with Piecewise Polynomial Functions 253

Submitted Papers

L.S.CARETTO, A.D.GOSMAN and D.B.SPALDING: Removal of an Instability in a Free Convection Problem 264

G.J.COOPER: Bounds for the Error in Approximate Solutions of Ordinary Differential Equations 270

D.J.EVANS: Numerical Solution of the Sturm Liouville Problem with Periodic Boundary Conditions 277

A.R.GOURLAY, G.McGUIRE and J.Ll.MORRIS: One Dimensional Methods for the Numerical Solution of Nonlinear Hyperbolic Systems . 290

A.JENNINGS: The Development and Application of Simultaneous Iteration for Eigenvalue Problems 297

J.LE FOLL: An Iterative Procedure for the Solution of Linear and Nonlinear Equations 310

J.J.H.MILLER: On Weak Stability, Stability, and the Type of a Polynomial 316

G.M.PHILLIPS: Error Estimates for Certain Integration Rules on the Triangle 321

S.SIGURDSSON: Linear Multistep Methods with Variable Matrix Coefficients 327

J.C.TAYLOR and J.V.TAYLOR: PARODE: A New Representational Method for the Numerical Solution of Partial Differential Equations 332

J.H.VERNER: On Deriving Explicit Runge-Kutta Methods 340

R.WAIT: A Finite Element for Three Dimensional Function Approximation 348

J.R.WHITEMAN, N.PAPAMICHAEL and Q.MARTIN: Conformal Transformation Methods for the Numerical Solution of Harmonic Mixed Boundary Value Problems 353

Papers presented whose proceedings do not appear here:

L.FOX: Numerical Analysis of a Functional Differential Equation

K.W.MORTON: Analysis of Particle Methods for Plasma Simulation

J.H.WILKINSON: Concerning Inverse Iteration

LIST OF SPEAKERS

Invited Speakers

Bellman, R.	Department of Mathematics, Electrical Engineering and Medicine, University of Southern California, Los Angeles, California 90007, U.S.A.
Cea, J.	Université de Nice, UERMST, Nice, France.
Collatz, L.	Institut für Angewandte Mathematik, Universität Hamburg, 2 Hamburg 13, Rothenbaumchaussee 67/69, W. Germany.
Fox, L.	Computing Laboratory, University of Oxford, 19 Parks Road, Oxford, England.
Golub, G.H.	Department of Mathematics, Imperial College, 52/53 Princes Gate, Exhibition Road, London. S.W.7.
Gorenflo, G.	Technische Hochschule, D-51 Aachen, W. Germany.
Greenstadt, J.	I.B.M. Scientific Center, 2670 Hanover Street, Palo Alto, California 94306. U.S.A.
Henrici, P.	Eidgenössische Technische Hochschule, Zürich Switzerland.
Kalaba, R.	Biomedical Engineering, Graduate Center for Engineering Sciences, University of Southern California, University Park, Los Angeles, California 90007. U.S.A.
Lascaux, P.	Commissariat à l'Energie Atomique, Paris, France.
Morton, K.W.	Culham Laboratory, Abingdon, Berkshire, England.
Noble, B.	Mathematics Research Center, University of Wisconsin, Madison 53706. U.S.A.
Osborne, M.R.	Computer Center, Australian National University, Canberra, N.S.W., Australia.
Powell, M.J.D.	Mathematics Branch, A.E.R.E., Harwell, England.
Raviart, P.	Université de Paris, Paris VI, France.
Stetter, H.J.	Technical University of Vienna, Vienna, Austria.
Wachspress, E.L.	General Electric Company, Schenectady, New York, U.S.A.
Widlund, O.B.	Courant Institute of Mathematical Sciences, AEC Computing and Applied Mathematics Center, New York University, 251 Mercer Street, New York, N.Y. 10012, U.S.A.
Wilkinson, J.	N.P.L., Teddington, Middlesex, England.

Other Speakers (submitted papers)

Caretto, L.S.	Mechanical Engineering Department, Imperial College of Science and Technology, London, S.W.7.
Cooper, G.T.	School of Mathematics and Physical Sciences, University of Sussex, Brighton. BN1 9QH.
Evans, D.J.	Computing Laboratory, Department of Applied Mathematics and Computing Science, University of Sheffield, Sheffield 10.
Jennings, A.	Civil Engineering Department, Queen's University, Belfast, N. Ireland.
Le Foll, J.	Research and Development Laboratories, C.A. Parsons and Company Ltd., Newcastle-upon-Tyne. NE6 2YL.
Miller, J.J.H.	School of Mathematics, Trinity College, Dublin, Ireland.
Morris, J.Ll.	Department of Mathematics, University of Dundee, Dundee, Scotland.
Phillips, G.M.	Department of Applied Mathematics, University of St. Andrews, St. Andrews, Scotland.
Sigurdsson, S.	Department of Mathematics, University of Dundee, Dundee, Scotland.
Taylor, J.C.	Department of Natural Philosophy, University of Glasgow, Glasgow, Scotland.
Verner, J.H.	Department of Mathematics, Queen's University, Kingston, Ontario, Canada.
Wait, R.	Department of Mathematics, University of Dundee, Dundee, Scotland.
Whiteman, J.R.	Department of Mathematics, Brunel University, Kingston Lane, Uxbridge, Middlesex, England.

ON THE IDENTIFICATION OF SYSTEMS AND SOME RELATED QUESTIONS

Richard Bellman

1. Introduction

A traditional mathematical problem is to find the answer to a specific question, which is to say, to find the solution to a particular equation. We are interested in the converse problem: Given the solution, what is the equation? This is the natural way that mathematical problems arise in science, in the form of inverse problems. Another way of describing this activity is that we are concerned with ascertaining the structure of a system on the basis of observation.

There are many different versions of the problem depending upon the mathematical model of the underlying physical process that is employed and the type of observation permitted. Here we wish to consider the problem in the following form: Given the differential equation

$$(1) \qquad \frac{dx}{dt} = g(x, a), \qquad x(0) = c,$$

where x is an n-dimensional vector and a an m-dimensional constant vector, a parameter, determine a and possibly c on the basis of the set of values $\{x(t_i)\}$, $i = 1, 2, \ldots, N$, the observations.

Many interesting and difficult questions arise in this investigation. We shall discuss some of them, as well as some methods for attacking this problem, and finally indicate how we are led to some new approaches to the numerical solution of partial differential equations and other types of functional equations.

2. Quasilinearization

We can formulate the identification problem in the following fashion. Consider the function

(1) $$f(a, c) = \sum_{i=1}^{N} \|x(t_i, a, c) - x(t_i)\|^2 \quad ,$$

where x(t, a, c) denotes the solution of (1) evaluated at t_i and $x(t_i)$ denotes the observed value of x at t_i. Our aim is to choose a and c to minimize f(a, c).

There are several kinds of problems here. The first arises when we suppose that $x(t_i)$ is obtained by observation of a system described by (1.1); the second when we wish to fit the observations by means of a model such as (1.1). This latter is a problem in differential approximation.

Let us simplify the notation by using a specific norm

(2) $$\| \ldots \|^2 = (\ldots, \ldots),$$

the usual inner product, and set $x(t_i) = c_i$. With the aid of linear and nonlinear programming, we can consider more general norms if we wish. It does not seem as if there is much to be gained in this way.

We shall employ a method of successive approximations based upon quasilinearization. Let $a^{(0)}$, $c^{(0)}$ be an initial guess of the values of a and c. These values are usually obtained from some knowledge of the underlying physical process and perhaps by the use of some search methods.

Let $x^{(0)}$ be determined by (1.1) using these values, i.e.

(3) $$\frac{dx^{(0)}}{dt} = g\left(x^{(0)}, a^{(0)}\right), \quad x^{(0)}(0) = c^{(0)} \quad .$$

The next step is to quasilinearize around $x^{(0)}$ and $a^{(0)}$. Vectors $y^{(1)}$, $a^{(1)}$ and $c^{(1)}$ are related by means of the equation

(4) $$\frac{dy^{(1)}}{dt} = g\left(x^{(0)}, a^{(0)}\right) + J_1\left(y^{(1)} - x^{(0)}\right) + J_2\left(a^{(1)} - a^{(0)}\right), \quad y^{(1)}(0) = c^{(1)} \quad .$$

Here J_1 and J_2 are the Jacobian matrices associated with the Taylor series around $x^{(0)}$ and $a^{(0)}$.

Solving the foregoing linear equation for $y^{(1)}$, we obtain $y^{(1)}$ as a linear function of $a^{(1)}$ and $c^{(1)}$,

(5) $$y^{(1)}(t) = z^{(1)}(t) + X_1(t)a^{(1)} + X_2(t)c^{(1)}.$$

The vectors $a^{(1)}$ and $c^{(1)}$ are now determined by minimizing the quadratic form

(6) $$\sum_{i=1}^{N}\left(z^{(1)}(t_i) + X_1(t_i)a^{(1)} + X_2(t_i)c^{(1)}, \quad \ldots\right)$$

with respect to these quantities. This requires the solution of linear algebraic equations.

With $a^{(1)}$ and $c^{(1)}$ fixed, $x^{(1)}$ is determined via

(7) $$\frac{dx^{(1)}}{dt} = g\left(x^{(1)}, a^{(1)}\right), \quad x^{(1)}(0) = c^{(1)},$$

and we proceed as above. A number of examples and further details may be found in [1].

3. Discussion

When convergence occurs, it is quadratic. The success of the method depends, of course, upon a judicious choice of $a^{(0)}$ and $c^{(0)}$. There are a number of questions connected with the choice of the t_i and N and the accuracy of the observations. These are stability matters, and quite difficult.

The method, nonetheless, is one of wide utility. It can be applied to very general classes of defining equations; differential difference equations and partial differential equations; to nonlinear boundary conditions and to the cases where the "observations" have the form

(1) (a) $(x(t_i), b_i) = c_i, \quad i = 1, 2, \ldots, N$, or

(b) $\int_0^T x(t)g_i(t)dt = c_i, \quad i = 1, 2, \ldots, N$.

The case (1a) is interesting since it corresponds to the frequent situation where the complete state cannot be measured at any time, or where there is not enough time to measure the complete state. The second corresponds to the case where only certain averages are available.

4. Approximation in Structure Space

A problem arising in many different fields is that of determining the parameters a_k and λ_k when a function u(t) is known to have the form

(1) $$u(t) = \sum_{k=1}^{N} a_k e^{\lambda_k t} .$$

This arises, for example, frequently in the field of pharmacokinetics, [2].

An immediate approach is to use the fact that u(t) satisfies an Nth order linear differential equation

(2) $$u^{(N)} + b_1 u^{(N-1)} + \dots + b_N u = 0, \quad u^{(i)}(0) = c_i,$$
$$i = 0, 1, \dots, N-1$$

and to proceed as above to calculate the b_i and c_i using values of u(t). Alternatively, we can use the finite difference version of (2). If u(t) is known for the entire interval $0 \leq t \leq T$, we can determine the b_i by minimizing the expression

(3) $$J(b) = \int_0^T \left(u^{(N)} + b_1 u^{(N-1)} + \dots + b_N u\right)^2 dt ,$$

This is the approach of differential approximation.

As we mentioned above, there is always the problem of obtaining a reasonable first approximation when using quasilinearization. We can overcome the problem to some extent by associating u(t) with a specific physical process. Suppose, for example, that u(t) is the concentration $x_1(t)$ in the first compartment of an N-compartmental model model governed by the equations.

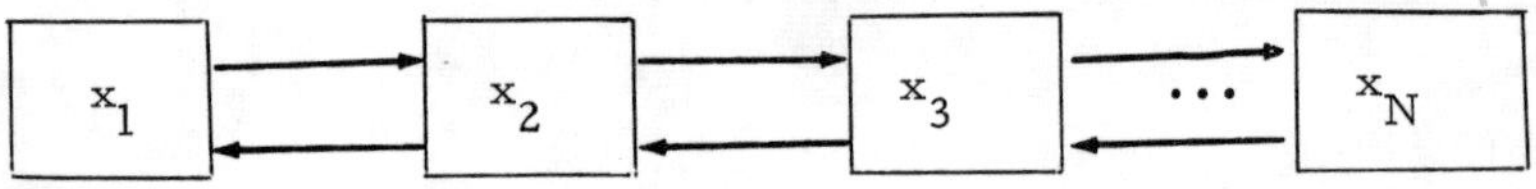

Fig. 1

$$(4) \qquad \dot{x}_1 = -k_{12}x_1 + k_{21}x_2, \quad x_1(0) = c_1,$$

$$\dot{x}_2 = k_{12}x_1 - (k_{21} + k_{23})x_2 + k_{32}x_3, \quad x_2(0) = c_2$$

$$\vdots$$

Here the k_{ij} are the rate constants and $c_1, c_2, \ldots,$ the initial concentrations.

The point is that every positive choice of the k_{ij} and c_i yields an admissible function $x_1(t)$ with values of a_k and λ_k and b_k, but not conversely. In other words, one parametrization of the a_k and λ_k is obtained in this fashion. Another parametrization is obtained using a different structure, say

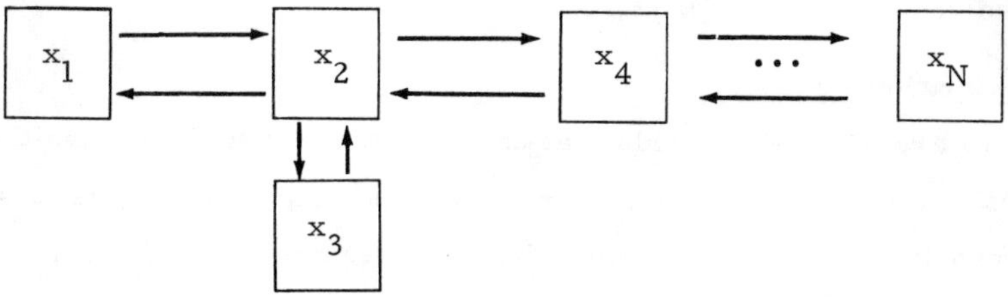

Fig. 2

Once we have decided on a structure, we can use observations of the function and quasilinearization to calculate the rate constants. The problem of determining which is the appropriate structure given $x_1(t)$ and the knowledge that one of the foregoing is the actual structure is unsolved at the present time, see [2].

There are many other linear processes apart from those pharmacokinetics which we can use for parametrization purposes.

5. A New Identification Method

Any technique for the identification of systems must involve some assumptions. Let us now present a method which hinges upon the nature of the solution.

Specifically, we suppose that the solution is sufficiently "smooth". By this we mean that $x'(t_i)$, $i = 1, 2, \ldots, N$, can be well approximated by a linear combination of the values $x(t_j)$, $j = 1,2,\ldots, N$,

$$(1) \qquad x'(t_i) = \sum_{j=1}^{N} a_{ij} x(t_j)$$

We shall discuss ways of choosing the coefficients a_{ij} below.

If (1) holds, the problem of minimizing $f(a, c)$, as given by (2.2) becomes that of minimizing

$$(2) \qquad \left(\sum_{j=1}^{N} a_{ij} x(t_j) - g(x(t_i), a) \ldots\right)$$

where $x(t_1) = c$. The case where c is known is easier than that where it is unknown. If $g(x, a)$ is linear in a, as is often the case, the minimization problem is readily resolved. See [3], [4], [5].

6. Approximation Procedure

There are several methods that can be employed to determine the a_{ij}. One, by analogy with the procedure used in Gaussian quadrature, is to suppose that the approximation is exact for a suitably chosen set of functions, e.g., $x(t)$ is a polynomial in t, or the sum of a set of orthonormal functions, or a spline function, see [7].

Many interesting questions arise in connection with the approximation of more general linear functions, e.g., $x'(t) + ax(t-1)$. In general, in using this method we face the task of approximating a functional of a function by means of observations involving the function.

7. A New Method for Calculating the Solution of Partial Differential Equations

We can use the foregoing approximation procedure to obtain a quick and easy algorithm for the solution of partial differential equations. Consider, for example the equation

$$(1) \qquad u_t = uu_x + u_{xx}, \quad u(x, 0) = g(x), \quad 0 < x < 1.$$

Choose N points $x_1, x_2, \ldots, x_N$ and write

(2) $$u(x_i, t) = v_i(t),$$

$$u_x(x,t)\Big|_{x=x_i} = \sum_{j=1}^{N} a_{ij}\, u(x_j, t)$$

$$u_{xx}(x,t)\Big|_{x=x_i} = \sum_{j=1}^{N} b_{ij}\, u(x_j, t)$$

Then (1) becomes a set of ordinary differential equations

(3) $$v_i'(t) = v_i(t) \sum_j a_{ij} v_j(t) + \sum_j b_{ij} v_j(t), \qquad v_i(0) = g_i,$$

$i = 1, 2, \ldots, N$. See [6] for some numerical results.

The author was supported by the National Institutes of Health under Grant No. GM 16197-03

REFERENCES

1. Bellman, R., and R. Kalaba, Quasilinearization and Nonlinear Boundary-Value Problems, American Elsevier, New York, 1965.

2. Bellman, R., and K. J. Astrom, "On Structural Identifiability," Mathematical Biosciences, Vol. 7, 1970, pp. 329-339.

3. Bellman, R., "A New Method for the Identification of Systems," Mathematical Biosciences, Vol. 5, 1969, pp. 201-204.

4. Bellman, R., and J. Casti, "Differential Quadrature and Long-Term Integration," University of Southern California, Tech. Rep. No. 70-14, March 1970.

5. Bellman, R., and J. Casti, "Differential Quadrature and the Identification of Chemical Rate Constants," University of Southern California, Tech. Rep. No. 70-59, December 1970.

6. Bellman, R., J. Casti, and B. Kashef, "Differential Quadrature: A Technique for the Solution of Nonlinear Partial Differential Equations," University of Southern California, Tech. Rep. No. 70-43, July 1970.

7. Galimberti, G., and V. Pereyra, "Numerical Differentiation and the Solution of Multidimensional Vandermonde Systems," Mathematics of Computation, Vol. 24, 1970, pp. 357-364.

INVARIANT IMBEDDING: SEMIGROUPS IN TIME, SPACE AND STRUCTURE

Richard Bellman

1. Introduction

Invariant imbedding is a mathematical theory devoted to the exploitation of structural features of processes. In consequence it is a loose confederation of ideas and techniques, of methods and methodology which can be employed in the analytic and computational study of large classes of mathematical and scientific questions. In what follows we will discuss some basic analytic aspects of the theory and provide references to more detailed analysis and numerical results.

We can begin with the observation that problem-solving is a principal occupation of the intellectual (see [2]). A powerful procedure widely employed in this pursuit is "imbedding." By this term we mean that the procedure whereby the resolution of a specific question is accomplished by consideration of a family of related questions. Rather remarkably, it turns out that it is often far easier to treat a set of problems in unison rather than a single problem in isolation. This is the essence of the comparative method familiar to so many disciplines: comparative linguistics, comparative anatomy, comparative religion, comparative anthropology, to name a few. Perhaps needless to say, it is not always an easy matter to discern the connecting links and thus an appropriate family. Banach is reputed to have said that brilliance consists of spotting analogies, and genius of seeing analogies between analogies.

The methods that are discussed below can be applied to many parts of mathematical physics and control theory, to mathematical economics, scheduling theory and operations research. They constitute an important part of the modern mathematical approach to the study of systems.

2. Imbedding in Time

Let us begin with a classical example of the method. Suppose that we are given the current state of a system and asked to predict the state at some subsequent time T. One way to go about this is to consider the general problem of predicting the state of the system at any subsequent time $t > 0$, where 0 denotes the present time.

We begin then by introducing a function $x(t)$, the state of the system at time t. For our present purposes we assume that this is a finite dimensional vector of dimension N. The next step is to obtain relations between functional values of x for different values of t. In many cases we can obtain an equation of the form

$$(1) \qquad x(t + \Delta) = x(t) + g(x(t))\Delta + \dots$$

for small positive Δ. In the limit as $\Delta \to 0$ this yields the differential equation

$$(2) \qquad x'(t) = g(x(t)) ,$$

with an initial condition $x(0) = c$.

The prediction problem has been transformed into the task of solving a functional equation.

3. Advantages and Disadvantages

This is a very powerful and flexible method which has had widespread success in science. It can be made the basis of numerous computational algorithms, algorithms which can be quickly and accurately carried out with the aid of digital computers. There are, however, some drawbacks as, of course, there must be to every method. In a number of cases too much data is calculated at too high a cost in both time and accuracy.

One way to circumvent these difficulties is to use some semigroup properties of the process. From the physical point of view this means taking advantage of the law of causality; from the mathematical point of view it means exploiting existence and uniqueness of solution. The impetus to this approach is due to Hadamard; see Hille-Phillips [10].

If (2.2) is a linear equation,

$$(1) \qquad x'(t) = Ax , \quad x(0) = c ,$$

the semigroup property is made apparent using the exponential form of the solution,

$$(2) \qquad x = e^{At} c .$$

We see that

(3) $$e^{A(s+t)} = e^{As}(e^{At}).$$

This allows us to use doubling techniques,

(4) $$e^{2At} = (e^{At})^2, \quad e^{2^N t} = (e^t)^{2^N}.$$

Thus N successive squarings will yield $e^{2^N t}$ starting with e^t. This is a considerable acceleration time. Nonetheless, there remain many interesting questions connected with the calculation of e^{NAt} given e^{At}.

4. Iteration

If (2.2) is nonlinear, we must use the more general approach of iteration to illustrate the underlying semigroup properties. Write

(1) $$x(t) = x(t, c) = f(c, t).$$

Then uniqueness of solution (assuming that g(x) is well-behaved, e.g., analytic) yields the basic semigroup relation

(2) $$f(c, s+t) = f(f(c, s), t), \quad s, t > 0,$$

with $f(c, 0) = 0$. In some cases this approach, together with the concept of relative invariants [5] can be used to accelerate the calculation of $f(c, T)$.

The semigroup determined by the linear equation of (3.1) generalizes in several ways. One generalization is afforded by a nonlinear differential equation as the basic equation; one is provided by two-point boundary-value problems of the type discussed subsequently in place of an initial value problem; one is provided by the theory of multistage decision processes, which is to say dynamic programming, in place of a descriptive process. There the equation is quasilinear, namely

(3) $$x'(t) = \max_q \, [A(q)x(t) + b(q)];$$

See [3, 4]. In this fashion the calculus of variations is imbedded in semigroup theory.

5. Imbedding in Space

Let us consider a further, and equally important, example of the imbedding method. Consider a steady-state transport process in a one-dimensional rod with an incident flux c at one end point T,

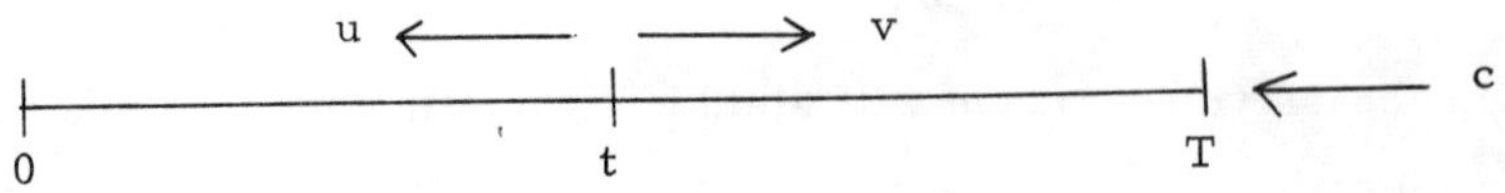

Fig. 1

We are asked to determine the reflected and transmitted fluxes, under various assumptions concerning the interaction of the flux with the medium and with itself. To answer this question using the technique of imbedding, we enlarge the investigation by asking for the values of the left-hand and right-hand fluxes, u(t) and v(t), at any interior point t. The quantity v(t) is the desired reflected flux; u(0) is the required transmitted flux.

Examination of the relations between u(t), $u(t \pm \Delta)$, v(t) and $v(t \pm \Delta)$ (local conservation relations) yields a pair of differential equations

$$(1) \qquad u'(t) = g(u(t), \; v(t)), \quad u(T) = c \, ,$$

$$v'(t) = h(u(t), \; v(t)), \quad v(0) = 0 \, .$$

Observe that this is a two-point boundary-value problem. We have insufficient information at t = 0 and t = T to resolve the equation as an initial value problem; see [11].

6. Advantages and Disadvantages

If we can solve (5.1), we will have obtained a solution of the original problem as well as a good deal of additional information of interest. However, a serious drawback to this approach lies in the fact that this equation cannot be used to provide a guaranteed algorithm for a digital computer the way an initial value equation can. Two-point boundary value problems are notoriously difficult, both analytically and computationally.

This obstacle, as well as the time barrier, may be turned by the use of the hybrid computer, analog plus digital. Since these questions, however, have not been investigated to any extent, we shall say no more at this point.

7. An Imbedding in Structure

Let us now imbed the original questions, the determination of the reflected and transmitted fluxes in a different family of problems. Let us seek to obtain these desired fluxes as functions of the initial intensity and the thickness of the rod.

To this end we write the reflected and transmitted fluxes.

(1) $$v(T) = r(c, T),$$

$$u(0) = t(c, T),$$

as functions of these parameters.

To obtain equations for these functions, we use some semigroup ideas. It turns out that for this purpose it is convenient to introduce an additional variable of physical significance. Suppose that a flux d is incident from the left at 0, as indicated below.

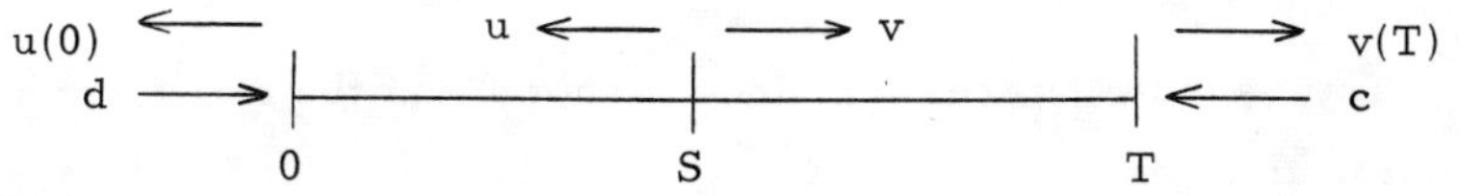

Fig. 2

Let then

(2) $$v(T)= r(c, d, T),$$

$$u(0) = t(c, d, T),$$

denote the respective "reflected" and "transmitted" fluxes, which is to say the fluxes emergent from the right and left, as indicated. Thus

(3) $$v(S) = r(u(S), d, S).$$

Similarly, the interval [S, T] yields the relation

(4) $u(S) = t(c, v(s), T-S)$.

We also have, using the entire interval,

(5) $r(c, d, T) = r(c, v(S), T-S)$,

$t(c, d, T) = t(u(S), d, T)$.

Elimination of $u(s)$ and $v(S)$ yield the basic semigroup relations for the function $r(c, d, T)$ and $t(c, d, T)$.

8. Associated Partial Differential Equation

The limiting form of the foregoing relations as $S \to T$ are partial differential equations for the functions $r(c, d, T)$ and $t(c, d, T)$. These are of initial value type since the reflection and transmission functions are simply specified for $T = 0$, namely

(1) $r(c, d, 0) = d$,

$t(c, d, 0) = c$.

This is quite interesting since it shows that the two-point boundary-value problem for an ordinary differential equation can be solved in terms of an initial value problem for a partial differential equation, and conversely. This of course is connected with characteristic theory. This flexibility is important for both analytic and computational reasons.

If (5.1) is linear, then

(2) $r(c, d, T) = R_1(T)c + T_1(T)d$,

$t(c, d, T) = T_1(T)d + R_1(T)d$,

where R_1 and T_1 are reflection and transmission matrices. Here we are assuming both homogeneity and isotropy. This need not be the case, requiring the introduction of additional matrices. The functional equations of (7.3), yield addition formulas for these matrices, and the partial differential equations then yield the familiar Riccati equations for the reflection and transmission matrices.

9. Imbedding in Structure

Let us give another example of imbedding in structure associated with the potential equation. Let R be a region with boundary B.

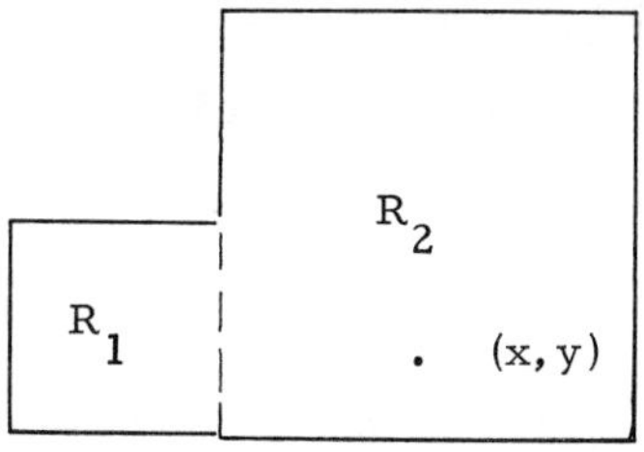

Fig. 3

Consider the equation

$$(1) \qquad u_{xx} + u_{yy} = 0, \quad (x, y) \in R ,$$

$$u = g(x, y), \quad (x, y) \in B .$$

The usual determination of u at a particular point (x_1, y_1) depends upon a simultaneous determination of u for all (x, y) in R. We can approximate to this problem by using a suitable subset of points and thus reduce the problem to a finite one.

Can we use a "building block" method to resolve the problem for $R = R_1 + R_2$ based upon the subregions R_1 and R_2 of simpler nature? It turns out that this can be done using ideas of invariant imbedding or of dynamic programming; [1].

10. Applications in Mathematical Physics

We have mentioned above an application of invariant imbedding to transport theory, the origin of all of the subsequent investigations. It was the pioneering work of Ambarzumian in the theory of radiative transfer and then the deeper and more extensive research of Chandrasekhar, his "principles of invariance," which made it clear that a vast domain of mathematical analysis awaited for exploration.

It turns out that there are many processes in mathematical physics whose structures are ideally suited to invariant imbedding. Let us mention random walk, wave propagation and analytical mechanics; see [6].

11. Application to Combinatories

In the foregoing, a very simple partitioning, or stratification, technique was used, being applied to both intervals and regions. We can also employ a partitioning of sets. This enables us to treat a number of combinatorial problems by means of the functional equation of dynamic programming; see [7], where many additional references to applications to routing and scheduling problems will be found.

12. Discussion

The foregoing studies lead to the posing of a large number of important, interesting and difficult questions. Some of these are

1. In how many ways may we meaningfully imbed a particular process in a family of processes and how do we determine these imbeddings?
2. How do we calculate desired data with as little extraneous calculation as possible?
3. How do we minimize the time required to calculate desired data?
4. How does one obtain the desired data using algorithms of specified type? This is a generalized Mascheroni construction.
5. What is the minimum data set required to obtain a specified value?
6. What are the simplest possible analytic structures which will yield the desired results?

Let us give one example of what we have in mind. Suppose we throw a stone straight up with velocity v, and wish to determine the maximum altitude attained. One approach is immediate. We write

$$(1) \qquad x'' = -g, \qquad x(0) = 0, \; x'(0) = v ,$$

leading to

(2) $x' = v - gt$.

Setting $x' = 0$, we find

(3) $t_{max} = v/g$,

$x_{max} = v^2/2g$.

However, we can also proceed as follows, using a quite different imbedding. Let

(4) $f(v)$ = the desired maximum altitude, defined for $v \geq 0$.

Then, considering what occurs in an initial time Δ, we have

(5) $f(v) = v\Delta + f(v - g\Delta) + o(\Delta)$,

Thus

(6) $f'(v) = v/g, \quad f(0) = 0$,

leading to

(7) $f(v) = v^2/2g$

as before.

This example, as well as those previously discussed, shows that we need a careful study of mathematical modelling in the physical sciences. This will certainly involve invariant theory, group theory, and an examination of the analogue of sufficient statistics.

One outgrowth of this study will be a modern theory of experimentation.

REFERENCES

1. Angel, E., and R. Bellman, Partial Differential Equations and Dynamic Programming, Academic Press, New York (to appear).

2. Bellman, R., and P. Brock, "On the Concepts of a Problem and Problem-Solving," Amer. Math. Monthly, Vol. 67, 1960, pp. 119-134.

3. Bellman, R., Dynamic Programming, Princeton University Press, Princeton, New Jersey, 1957.

4. Bellman, R., Adaptive Control Processes: A Guided Tour, Princeton University Press, Princeton, New Jersey, 1961.

5. Bellman, R., Perturbation Techniques in Mathematics, Physics and Engineering, Holt, Rinehart and Winston, New York, 1964.

6. Bellman, R., R. Kalaba, and G. M. Wing, "Invariant Imbedding and Mathematical Physics--I: Particle Processes," J. Math. Phys., Vol. 1, 1960, pp. 280-308.

7. Bellman, R., K. L. Cooke, and J. Lockett, Algorithms, Graphs and Computers, Academic Press, New York, 1970.

Some computational results will be found in

8. Bellman, R., R. Kalaba, and M. Prestrud, Invariant Imbedding and Radiative Transfer in Slabs of Finite Thickness, Amer. Elsevier Publ. Co., Inc., New York, 1963.

9. Bellman, R., H. Kagiwada, R. Kalaba, and M. Prestrud, Invariant Imbedding and Time-Dependent Processes, Amer. Elsevier Publ. Co., Inc., New York, 1964.

10. Hille, E., and R. Phillips, "Functional Analysis and Semigroups," Anal., Math. Soc. Colloq., Vol. 31, 1948.

A discussion of transport theory and invariant imbedding will be found in

11. Wing, G. M., An Introduction to Transport Theory, John Wiley and Sons, Inc., New York, 1962.

The author was supported by the National Science Foundation under Grant No. GP 20423 and the Atomic Energy Commission, Division of Research under Contract No. AT-11-1-1113, Project 19

MINIMISATION DE FONCTIONNELLES NON-DIFFERENTIABLES

J.Cea[1], R.Glowinski[2] et J.C.Nedelec[3]

On étudie le problème de la minimisation d'une famille très vaste de fonctionnelles non différentiables : La méthode utilisée est la suivante : à un problème de minimisation d'une fonctionnelle non différentiable sans contrainte (le problème primal), on associe un problème de minimisation d'une fonctionnelle différentiable avec des contraintes très simples ; il se trouve que la transformation de la difficulté due à la non différentiabilité en une difficulté due à l'introduction de contraintes simplifie considérablement l'étude du problème initial. La dualité étant le point essentiel de ce transfert de difficultés, nous donnerons tout d'abord le Lagrangien et nous expliciterons ensuite les problèmes primal et dual. Après avoir donné quelques exemples de non différentiabilité, nous étudierons trois méthodes numériques très adaptées à ce genre de problème.

1 - LA DUALITE , LE PROBLEME PRIMAL ET LE PROBLEME DUAL. LES DONNEES:

Tous les éléments sont réels ; on donne deux espaces de Hilbert V et L, de produits scalaires respectifs $(.,.)$ et $[.,.]$ et de normes respectives

$\|\cdot\|$ et $[\![\cdot]\!]$; de plus on donne $F \in V$ et $S \in \mathcal{L}(V,L)$.

---

1) Université de Nice, U E R M S T
2) Université de Paris VI et IRIA, P A R I S
3) Université de Rennes

On pose

$$J_o(v) = \frac{1}{2} \|v\|^2 - (F,v) \qquad v \in V$$

$$\mathcal{L}(v,\mu) = J_o(v) + [\mu , Sv] \qquad v \in V \ , \ \mu \in L$$

On donne enfin un ensemble Λ convexe, fermé et borné dans L. On suppose que :

$$0 \in \Lambda \ .$$

1.1. LA DUALITE :

On va démontrer le

Théorème 1.1 : Sous les données et hypothèses précédentes, il existe un couple u , λ tel que

$$(1.1) \qquad \begin{cases} u , \lambda \in V \times \Lambda \\ \mathcal{L}(u,\mu) \leq \mathcal{L}(u,\lambda) \leq \mathcal{L}(v,\lambda) \\ \forall v, \mu \in V \times \Lambda \end{cases}$$

Démonstration : On utilise le théorème de Ky Fan [8] et Sion [9] avec les données suivantes : V muni de la topologie faible,

$\mathcal{U}_p = \{ v \mid v \in V, \|v\| \leq p \}$, L muni de la topologie faible, Λ et $\mathcal{L}(v,\mu)$: il en résulte l'existence d'un couple u_p , λ_p vérifiant :

$$(1.2) \qquad \begin{cases} u_p , \lambda_p \in \mathcal{U}_p \times \Lambda \\ \mathcal{L}(u_p ,\mu) \leq \mathcal{L}(u_p , \lambda_p) \leq \mathcal{L}(v , \lambda_p) \\ \forall v , \mu \in \mathcal{U}_p \times \Lambda \end{cases}$$

ou encore

$$(1.2)' \qquad \begin{cases} u_p , \lambda_p \in \mathcal{U}_p \times \Lambda \\ J_o(u_p) + [\mu , Su_p] \leq J_o(u_p) + [\lambda_p , Su_p] \leq J_o(v) + [\lambda_p , Sv] \\ \forall v , \mu \in \mathcal{U}_p \times \Lambda \end{cases}$$

Si on choisit $\mu = 0$, il vient

$$[\lambda_p , Su_p] \geq 0$$

d'où

$$J_o(u_p) \leq J_o(v) + [\lambda_p , Sv]$$

et, en particulier,lorsque $v = 0$

$$J_o(u_p) \leq J_o(0) = 0 \quad .$$

De là, il vient facilement

$$\| u_p \| \leq c < +\infty \qquad \forall p.$$

Si nous choisissons $p > c$, alors u_p est un minimum local pour la fonctionnelle convexe $v \longrightarrow \mathcal{L}(v, \lambda_p)$, par suite u_p est un minimum global et dans (1.2) nous pouvons remplacer $\forall v \in \mathcal{U}_p$ par $\forall v \in V$; on obtient donc (1.1) à partir de (1.2), en choisissant $p > c$ et en posant $u = u_p$, $\lambda = \lambda_p$. ■

<u>Corollaire</u> 1.1. :

Le couple u, λ solution de (1.1) est aussi solution de (1.3) et réciproquement :

$$(1.3) \qquad \begin{cases} u - F + S^* \lambda = 0 \\ [Su, \mu - \lambda] \leq 0 \qquad \forall \mu \in \Lambda \end{cases} .$$

<u>Démonstration</u> :

La 1ère relation de (1.3) exprime que le gradient de la fonctionnelle $v \longrightarrow \mathcal{L}(v, \lambda)$ est nul (ce qui est équivalent à l'inégalité de droite de (1.1)) la 2ème relation de (1.3) n'est autre que l'inégalité de gauche de (1.1).

<u>Corollaire</u> 1.2. :

Si u_1 , λ_1 et u_2 , λ_2 vérifient 1.1. alors $u_1 = u_2$.

Nous utilisons les relations (1.3) équivalentes à (1.1) :

$$u_1 - F + S^* \lambda_1 = 0$$

$$[Su_1 , \mu - \lambda_1] \leq 0 \qquad \forall \mu \in \Lambda$$

$$u_2 - F + S^* \lambda_2 = 0$$

$$[Su_2 , \mu - \lambda_2] \leq 0 \qquad \forall \mu \in \Lambda \quad .$$

Nous choisissons $\mu = \lambda_2$ dans la 1ère inégalité et $\mu = \lambda_1$ dans la $2^{\text{ième}}$ et par addition, il vient :

$$[Su_1 - Su_2 , \lambda_2 - \lambda_1] \leq 0$$

ou encore

$$(u_1 - u_2 , S^*\lambda_2 - S^*\lambda_1) \leq 0$$

ou encore

$$(u_1 - u_2 , u_1 - u_2) \leq 0$$

d'où $u_1 = u_2$. ■

Corollaire 1.3 :

Si le couple u, λ vérifie (1.1), alors :

(1.4) $$\mathcal{L}(u,\lambda) = \operatorname*{Inf}_{v \in V} \sup_{\mu \in \Lambda} \mathcal{L}(v,\mu) = \operatorname*{Sup}_{\mu \in \Lambda} \operatorname*{Inf}_{v \in V} \mathcal{L}(v,\mu) \; .$$

Démonstration :

C'est une conséquence classique de (1.1) . ■

En fait, nous démontrerons plus loin qu'on peut remplacer Inf. par Min. et Sup. par Max dans (1.4). Au vu de (1.4) il est naturel de poser les définitions suivantes :

Définition 1.1 :

i) On désigne par problème $\mathcal{P}$ ou problème primal le problème suivant:

$$\operatorname*{Inf}_{v \in V} \operatorname*{Sup}_{\mu \in \Lambda} \mathcal{L}(v,\mu) \; .$$

ii) On désigne par problème $\mathcal{P}^*$ ou problème dual le problème suivant :

$$\operatorname*{Sup}_{\mu \in \Lambda} \operatorname*{Inf}_{v \in V} \mathcal{L}(v,\mu) \; .$$

1.2. LE PROBLEME PRIMAL :

Désignons par h la fonction d'appui du convexe Λ de L :

$$h(\nu) = \operatorname*{Sup}_{\mu \in \Lambda} [\mu, \nu]$$

Dans le cas présent, Λ étant un ensemble convexe, fermé et borné, nous avons

$$h(\nu) = \operatorname{Max}_{\mu \in \Lambda} [\mu, \nu]$$

et puisque $0 \in \Lambda$, on a

(1.5) $$h(\nu) \geq 0 \qquad \forall \nu \in L .$$

Nous savons que h est une fonction convexe et semi-continue inférieurement.

Nous avons

$$\operatorname{Sup}_{\mu \in \Lambda} \mathcal{L}(v, \mu) = J_0(v) + \operatorname{Sup}_{\mu \in \Lambda} [\mu, Sv] = J_0(v) + h(Sv) .$$

Posons

(1.6) $$\begin{cases} J_1(v) = h(Sv) \\ J(v) = J_0(v) + J_1(v) \end{cases} .$$

Le problème primal est donc : Minimiser J(v)

La fonctionnelle J est strictement convexe, faiblement semi-continue inférieurement et vérifie

$$\lim_{\|v\| \to +\infty} J(v) = +\infty .$$

Nous savons que, dans ces conditions, le problème primal admet une solution et une seule.

Proposition 1.1. Si u, λ vérifie (1.1) alors u est la solution du problème primal.

Démonstration :

L'inégalité de gauche de (1.1) s'écrit :

$$J_1(u) = h(Su) = [\lambda, Su]$$

et l'inégalité de droite devient

$$J(u) \leq J_0(v) + [\lambda, Sv]$$

mais

$$[\lambda, Sv] \leq h(Sv) = J_1(v)$$

d'où

$$J(u) \leq J(v) \qquad \forall v \in V . \qquad \blacksquare$$

La proposition 1.1 et l'unicité de la solution du problème $\mathcal{P}$ permettent de retrouver le corollaire 1.2.

1.3. LE PROBLEME DUAL (OU $\mathcal{P}^*$) :

Il s'agit du problème $\sup_{\mu \in \Lambda} \inf_{v \in V} \{ J_o(v) + [\mu, Sv] \}$.

On vérifie facilement que

(1.7) $$-\frac{1}{2} \| u_\mu \|^2 = J_o(u_\mu) + [\mu, Su_\mu] \leq J_o(v) + [\mu, Sv] \qquad \forall v$$

lorsque u_μ est défini par

(1.8) $$u_\mu - F + S^*\mu = 0$$

une autre formulation du problème dual est donc la suivante

(1.9) $$\begin{cases} \sup_{\mu \in \Lambda} & -\frac{1}{2} \| u_\mu \|^2 \\ u_\mu & - F + S^*\mu = 0 \end{cases}$$

ou encore

(1.9)' $$\begin{cases} + \inf_{\mu \in \Lambda} & \frac{1}{2} \| u_\mu \|^2 \\ u_\mu - F + S^*\mu = 0 . \end{cases}$$

Nous pouvons éliminer u_μ : il vient

(1.9)" $$\inf_{\mu \in \Lambda} \frac{1}{2} \| S^*\mu - F \|^2$$

et en posant

(1.10) $$J^*(\mu) = \frac{1}{2} \| S^*\mu - F \|^2 - \frac{1}{2} \| F \|^2 = \frac{1}{2} [SS^*\mu, \mu] - [SF, \mu] =$$

$$= \frac{1}{2} \| u_\mu \|^2 - \frac{1}{2} \| F \|^2$$

le problème dual peut encore s'écrire sous la forme

(1.11) $$\inf_{\mu \in \Lambda} J^*(\mu)$$

(au signe près et à une constante additive près).

Notons qu'en posant

$$\mathcal{A} = SS^*$$

$$\mathcal{F} = SF$$

on a :

(1.12)
$$\begin{cases} J^*(\mu) = \frac{1}{2}[\mathcal{A}\mu, \mu] - [\mathcal{F}, \mu] \\ \mathcal{A} \in \mathcal{L}(L,L) \\ \mathcal{A} = \mathcal{A}^*, \quad [\mathcal{A}\mu, \mu] \geq 0 \qquad \forall \mu \in L . \end{cases}$$

Comme Λ est un ensemble convexe fermé et <u>borné</u>, on sait que le <u>problème</u> $\mathcal{P}^*$ <u>a au moins une solution</u>.

<u>Proposition</u> 1.2 : Si λ est une solution du problème $\mathcal{P}^*$, si $u = F - S^*\lambda$, alors le couple u, λ vérifie (1.1) et u est la solution du problème $\mathcal{P}$.

<u>Démonstration</u> : Nous savons que tout λ qui minimise

$$J^*(\mu) = \frac{1}{2}[\mathcal{A}\mu, \mu] - [\mathcal{F}, \mu]$$

vérifie aussi :

$$[\mathcal{A}\lambda - \mathcal{F}, \mu - \lambda] \leq 0 \qquad \forall \mu \in \Lambda$$

c'est à dire

$$[SS^*\lambda - SF, \mu - \lambda] \leq 0 \qquad \forall \mu \in \Lambda$$

ou encore

$$[Su, \mu - \lambda] \leq 0 \qquad \forall \mu \in \Lambda .$$

On utilise alors le corollaire 1.1. et la proposition est démontrée. ■

Remarquons que

(1.13)
$$\begin{cases} \text{grad } J^*(\mu) = \mathcal{A}\mu - \mathcal{F} = SS^*\mu - SF = -Su_\mu \\ u_\mu - F + S^*\mu = 0 \end{cases}$$

Compte tenu de ce qui a été fait jusqu'à présent, il est clair que (1.4) s'écrit de façon plus précise :

$$\mathcal{L}(u,\lambda) = \underset{v\in V}{\text{Min}}\ \underset{\mu\in\Lambda}{\text{Max}}\ \mathcal{L}(v,\mu) = \underset{\mu\in\Lambda}{\text{Max}}\ \underset{v\in V}{\text{Min}}\ \mathcal{L}(v,\mu)$$

2 - EXEMPLES

Nous choisirons des exemples liés à la théorie des équations aux dérivées partielles. Naturellement le cadre du n°1 est suffisamment vaste pour pouvoir l'utiliser dans d'autres cas.

Soit Ω un ouvert borné de $\mathbb{R}^n$. Rappelons que

$$v \in H^1(\Omega) \Longleftrightarrow v \in L^2(\Omega),\ D_i v = \frac{\partial}{\partial x_i} v \in L^2(\Omega), \quad i = 1,\dots,n$$

et que

$$((u,v))_{H^1(\Omega)} = \sum_{i=1}^{n} (D_i u, D_i v)_{L^2(\Omega)} + (u,v)_{L^2(\Omega)}$$

$$||u||_{H^1(\Omega)} = ((u,u))^{1/2}_{H^1(\Omega)} \quad .$$

Soit $f \in L^2(\Omega)$; dans tous les exemples nous choisirons

$V = H^1(\Omega)$, $J_o(v) = \frac{1}{2}||v||^2_{H^1(\Omega)} - (f,v)_{L^2(\Omega)}$; Notons qu'il existe $F \in H^1(\Omega)$ tel que $((F,v)) = (f,v)\ \forall\ v \in V$, et que par suite J_o est bien du type indiqué dans le n°1.

Dans tout ce qui va suivre, nous donnerons les éléments L, S, Λ et nous expliciterons J_1. Rappelons que

$$J_1(v) = \operatorname*{Max}_{\mu \in \Lambda} \left[\mu, Sv\right] = h(Sv)$$

et que le problème primal consite en la minimisation de

$$J(v) = J_o(v) + J_1(v) \quad .$$

Comme les démonstrations conduisant à l'expression de $J_1(v)$ sont toujours du même type, nous ne donnerons qu'une démonstration (voir exemple 2.3.1.);

2.1. UNE PREMIERE FAMILLE

$$L = \mathbb{R} \ , \quad Sv = (g,v)_{L^2(\Omega)} \quad .$$

Exemple 2.1.1.

$\Lambda = \left[-1, +1\right]$, g donnée dans $L^2(\Omega)$

$$J_1(v) = |(g,v)_{L^2(\Omega)}| \quad .$$

Exemple 2.1.2

$\Lambda = [0, +1]$, g donnée dans $L^2(\Omega)$

$J_1(v) = (g,v)^+_{L^2(\Omega)}$

(Rappelons que $X^+ = \max\{X,0\}$).

2.2. UNE DEUXIEME FAMILLE

$L = L^2(\Omega)$, $S = I$.

Exemple 2.2.1

$\Lambda = \{\mu | \mu \in L^2(\Omega),\ |\mu(x)| \leq 1 \quad \text{pp } x \in \Omega\}$

$$J_1(v) = \underset{\mu \in \Lambda}{\text{Max}} \int_\Omega \mu(x)\, v(x) dx = \int_\Omega |v(x)| dx \quad .$$

Exemple 2.2.2.

$\Lambda = \{\mu | \mu \in L^2(\Omega) \ ,\ 0 \leq \mu(x) \leq 1 \quad \text{pp } x \in \Omega\}$

$$J_1(v) = \underset{\mu \in \Lambda}{\text{Max}} \int_\Omega \mu(x)\, v(x) dx = \int_\Omega v(x)^+ dx \quad .$$

2.3. UNE TROISIEME FAMILLE

$L = (L^2(\Omega))^n$, $Sv = (D_1 v, \ldots, D_n v) \in L$ quand $v \in H^1(\Omega)$.

Notons que $Sv(x) = \text{grad } v(x)$ et que $[\mu, Sv] = \sum_{i=1}^{n} (\mu_i, D_i v)_{L^2(\Omega)}$.

Exemple 2.3.1.

$\Lambda = \{\mu | \mu \in L,\ |\mu(x)|_{\mathbb{R}^n} \leq 1 \quad \text{pp } x \in \Omega\}$

où $|\mu(x)|_{R^n} = (\sum_{i=1}^{n} |\mu_i(x)|^2)^{1/2}$.

On a $[\mu, Sv] = \sum_{i=1}^{n} (\mu_i, D_i v)_{L^2(\Omega)} = \int_\Omega \sum_{i=1}^{n} \mu_i(x) . D_i v(x) dx$

d'où en utilisant l'inégalité de Cauchy Schwarz dans $\mathbb{R}^n$

$$\begin{cases} [\mu, Sv] \leq \int_\Omega |\mu(x)|_{\mathbb{R}^n} \, |\text{grad } v(x)|_{\mathbb{R}^n} dx \leq \int_\Omega |\text{grad } v(x)| dx \\ \forall \mu \in \Lambda \end{cases} .$$

D'autre part, si λ_ε est défini par :

$$\lambda_{\varepsilon,i}(x) = \frac{D_i v(x)}{|\text{grad } v(x)| + \varepsilon} \quad , \varepsilon \text{ donné} > 0 \quad .$$

On a alors :

$$\left[\lambda_\varepsilon, S_v\right] = \int_\Omega \frac{|\text{grad } v(x)|^2}{|\text{grad } v(x)| + \varepsilon}\, dx \geq \int_\Omega \frac{|\text{grad } v(x)|^2 - \varepsilon^2}{|\text{grad } v(x)| + \varepsilon}\, dx$$

$$\left[\lambda_\varepsilon, Sv\right] \geq \int_\Omega |\text{grad } v(x)|\, dx - \varepsilon \int_\Omega dx.$$

Compte tenu des relations précédentes, il vient :

$$\int_\Omega |\text{grad } v(x)|\, dx + \varepsilon \int_\Omega dx \leq h(Sv) = \underset{\mu \in \Lambda}{\text{Max}} \left[\mu, Sv\right] \leq \int_\Omega |\text{grad } v(x)|\, dx$$

par suite

$$J_1(v) = h(Sv) = \int_\Omega |\text{grad } v(x)|\, dx$$

ou encore

$$J_1(v) = \int_\Omega \left(\sum_{i=1}^{n} |D_i\, v(x)|^2 \right)^{1/2} dx.$$

Cet exemple a été étudié par J.CEA, R.GLOWINSKI dans [2] .

Exemple 2.3.2.

$$\Lambda = \{\mu | \mu \in L,\ |\mu_i(x)| \leq 1 \quad \text{pp } x \in \Omega,\ i = 1,\ldots,n\}$$

$$J_1(v) = \underset{\mu \in \Lambda}{\text{Max}} \sum_{i=1}^{n} \int_\Omega \mu_i(x)\, D_i\, v(x) dx = \sum_{i=1}^{n} \int_\Omega |D_i\, v(x)| dx .$$

Exemple 2.3.3.

$$\Lambda = \{\mu | \mu \in L,\ \sum_{i=1}^{n} |\mu_i(x)| \leq 1 \quad \text{pp } x \in \Omega\}$$

$$J_1(v) = \int_\Omega \underset{i=1,\ldots,n}{\text{Max}} |D_i\, v(x)|\, dx .$$

On peut construire de nouveaux exemples à partir des précédents en ajoutant la contrainte $\mu_i(x) \geq 0$ pp $x \in \Omega$, $i = 1,\ldots,n$.

Nous donnerons un seul exemple.

Exemple 2.3.4.

$$\Lambda = \{\mu | \mu \in L,\ \sum_{i=1}^{n} |\mu_i(x)| \leq 1 ,\ \mu_i(x) \geq 0 \quad \text{p.p. } x \in \Omega,\ i=1,\ldots,n\}.$$

Dans ce cas, on montre que

$$J_1(v) = \int_\Omega \underset{i=1,\ldots,n}{\text{Max}} (D_i\, v(x))^+ dx.$$

3 - METHODES NUMERIQUES

Nous nous proposons d'approcher la solution u du problème primal en employant une méthode itérative ; l'idée directrice est la suivante : nous introduisons une suite λ_m destinée à approcher une solution du problème dual, nous construisons une suite u_m par la relation $u_m = F - S^*\lambda_m$ et enfin nous montrons que u_m converge fortement vers la solution u du problème primal.

3.1. UNE PREMIERE METHODE

Il s'agit de l'adaptation de la méthode d'UZAWA. [10] qui n'est autre que la méthode du gradient avec projection appliquée au problème dual. La méthode du gradient avec projection est exposée dans A.A. GOLDSTEIN[7]. De façon plus précise, si P désigne la projection de L sur Λ, on définit λ_m par :

(3.1)
$$\begin{cases} \lambda_o \text{ donné dans } \Lambda \\ \lambda_{m+1} = P(\lambda_m - \rho \operatorname{grad} J^* (\lambda_m)) \end{cases}$$

ou en utilisant (1.13)

(3.2)
$$\begin{cases} \lambda_o \text{ donné dans } \Lambda \\ u_m = F - S^*\lambda_m \\ \lambda_{m+1} = P(\lambda_m + \rho\, Su_m) \end{cases} .$$

Dans ces formules ρ désigne un nombre positif fixe qui sera précisé ultérieurement. Nous allons montrer que la suite $J^*(\lambda_m)$ est décroissante ; pour cela, commençons par utiliser une propriété de la projection : on a :

(3.3)
$$[\lambda_m + \rho\, Su_m - \lambda_{m+1}, \mu - \lambda_{m+1}] \leqslant 0 \qquad \forall \mu \in \Lambda$$

et en choisissant $\mu = \lambda_m$, il vient :

(3.4)
$$[\![\lambda_m - \lambda_{m+1}]\!]^2 \leq \rho\left[Su_m, \lambda_{m+1} - \lambda_m\right] .$$

Le développement de Taylor de J^* conduit à :

$$J^*(\lambda_{m+1}) = J^*(\lambda_m) - [Su_m, \lambda_{m+1} - \lambda_m] + \frac{1}{2}\left[SS^*(\lambda_{m+1}-\lambda_m), \lambda_{m+1}-\lambda_m\right]$$

d'où

$$J^*(\lambda_{m+1}) \leq J^*(\lambda_m) - [Su_m, \lambda_{m+1} - \lambda_m] + \frac{1}{2} ||S^*||^2 . [\![\lambda_{m+1} - \lambda_m]\!]^2$$

où $||S^*||$ est la norme de S^* dans $\mathcal{L}(L,V)$.

En utilisant (3.4) il vient :

$$J^*(\lambda_{m+1}) \leq J^*(\lambda_m) - \frac{1}{\rho} [\![\lambda_{m+1} - \lambda_m]\!]^2 + \frac{1}{2} ||S^*||^2 . [\![\lambda_{m+1} - \lambda_m]\!]^2$$

ou encore

(3.5) $$J^*(\lambda_{m+1}) \leq J^*(\lambda_m) - \left(\frac{1}{\rho} - \frac{||S^*||^2}{2}\right) . [\![\lambda_{m+1} - \lambda_m{}^2]\!] .$$

Si nous choisissons ρ tel que

(3.6) $$0 < \rho < \frac{2}{||S^*||^2}$$

alors la suite $J^*(\lambda_m)$ est décroissante. Comme Λ est borné, la suite $J^*(\lambda_m)$ est bornée ; par suite $\lim_{m\to\infty} J^*(\lambda_m) - J^*(\lambda_{m+1}) = 0$ d'où avec (3.5)

(3.7) $$\lim_{m\to\infty} [\![\lambda_m - \lambda_{m+1}]\!]^2 = 0$$

Etablissons maintenant la convergence faible de la suite u_m vers u. Puisque λ_m est bornée, la suite u_m est aussi bornée ; soit u^* un point adhérent faible de la suite u_m ; par une éventuelle extraction d'une sous-suite toujours notée u_m, on peut se ramener au cas où :

(3.7) $$\begin{cases} \lim u_m = u^* & \text{dans } V \text{ faible} \\ \lim \lambda_m = \lambda^* & \text{dans } L \text{ faible} \\ u_m - F + S^*\lambda_m = 0 \\ u^* - F + S^*\lambda^* = 0 \end{cases}$$

On peut écrire (3.3) sous la forme :

(3.3)' $$\begin{aligned} &[Su_m , \mu - \lambda_m] \leq \varepsilon_m(\mu) \\ &\varepsilon_m(\mu) = \frac{1}{\rho} [\lambda_{m+1} - \lambda_m , \rho Su_m - \mu + \lambda_{m+1}] . \end{aligned}$$

Grâce à (3.7) et au fait que λ_m et u_m sont bornées, il vient

(3.8) $$\lim_{m\to\infty} \varepsilon_m(\mu) = 0 \qquad \forall\mu$$

or

$$[Su_m, \mu - \lambda_m] = [Su_m, \mu] - [Su_m, \lambda_m]$$

$$= (u_m, S^*\mu) - (u_m, F - u_m)$$

$$= ||u_m||^2 - (u_m, S^*\mu + F)$$

et puisque $u_m \rightarrow u^*$ dans V faible, il vient :

$$||u^*||^2 - (u^*, S^*\mu + F) \leqslant \underline{\lim}_{m\rightarrow+\infty} ||u_m||^2 - (u_m, S^*\mu + F)$$

ce qui entraine, en utilisant (3.3') et (3.8)

$$||u^*||^2 - (u^*, S^*\mu + F) \leqslant 0 \qquad \forall\mu \in \Lambda$$

ou encore, puisque $u^* - F + S^*\lambda = 0$

$$[Su^*, \mu - \lambda^*] \leqslant 0 \qquad \forall\mu \in \Lambda \ .$$

En résumé, on a la situation suivante :

$$\begin{cases} \lambda^* \in \Lambda \\ [Su^*, \mu - \lambda^*] \leqslant 0 \qquad \forall\mu \in \Lambda \\ u^* - F + S^*\lambda^* = 0 \ . \end{cases}$$

Nous savons, grâce au corollaire 1.1 que u^* est la solution du problème primal et que λ^* est une solution du problème dual. Comme le problème $\mathcal{P}$ a une solution unique u^*, on a $u = u^*$ et la suite u_m n'ayant qu'un seul point adhérent faible u, elle converge vers ce point.

Etablissons maintenant la convergence forte de u_m vers u ; un développement de Taylor conduit à :

$$J^*(\lambda) = J^*(\lambda_m) - [Su_m, \lambda - \lambda_m] + \frac{1}{2}[SS^*(\lambda - \lambda_m), \lambda - \lambda_m]$$

d'où

$$J^*(\lambda) \geqslant J^*(\lambda_m) - [Su_m, \lambda - \lambda_m]$$

et avec (3.3)' il vient

$$J^*(\lambda) \geqslant J^*(\lambda_m) - \varepsilon_m(\lambda) \ ,$$

Si λ est une solution du problème dual, alors $u - F + S^*\lambda = 0$, et compte-tenu de (1.10), il vient :

$$\frac{1}{2}\ ||u||^2 \geqslant \frac{1}{2}||u_m||^2 - \varepsilon_m(\lambda)\ .$$

A partir de la situation suivante

$$\begin{cases} ||u_m||^2 \leqslant ||u||^2 + 2\ \varepsilon_m(\lambda) \\ \lim \varepsilon_m(\lambda) = 0 \\ \lim u_m = u \text{ dans } V \text{ faible ,} \end{cases}$$

il vient facilement

$$\lim_{m\to+\infty} ||u_m - u||_V = 0\ . \qquad \blacksquare$$

On a donc démontré la

Théorème 3.1.

Sous les données et hypothèses du n°1, si la suite u_m est construite par (3.2), si ρ vérifie (3.6) alors

$$\lim ||u_m - u|| = 0$$

où u désigne la solution du problème primal.

Autre démonstration

Si (u,λ) est le point selle cherché, on a vu que

$$\left[Su,\ \mu - \lambda\right] \leqslant 0 \qquad \forall\ \mu \in \Lambda$$

ce qui équivaut à

(3.9) $$\lambda = P(\lambda + \rho\, Su) \qquad \forall \rho \geqslant 0$$

d'où compte tenu de (3.2) et la projection sur Λ étant contractante on a

(3.10) $$\begin{cases} u_m - u = -\ S^*(\lambda_m - \lambda) \\ [\![\lambda_{m+1} - \lambda]\!] \leqslant [\![\lambda_m - \lambda + \rho S(u_m - u)]\!]\ . \end{cases}$$

Posons $u_m - u = \tilde{u}_m$ et $\lambda_m - \lambda = \tilde{\lambda}_m$ alors :

$$(3.11)\quad \begin{cases} \tilde{u}_m = - S^* \tilde{\lambda}_m \\ [\![\tilde{\lambda}_{m+1}]\!]^2 \leq ||\tilde{\lambda}_m||^2 + 2\rho [\![\tilde{\lambda}_m, Su_m]\!] + \rho^2 [\![S]\!]^2 \, ||\tilde{u}_m||^2 . \end{cases}$$

De la première équation (3.11) on déduit :

$$(3.12)\quad ||\tilde{u}_m||^2 = -[\![\tilde{\lambda}_m, S\tilde{u}_m]\!]$$

d'où

$$(3.13)\quad [\![\tilde{\lambda}_m]\!]^2 - [\![\tilde{\lambda}_{m+1}]\!]^2 \geq ||\tilde{u}_m||^2 (2 - \rho ||S||^2) .$$

On déduit de (3.13) que si ρ vérifie :

$$(3.14)\quad 0 < \rho < \frac{2}{||S||^2}$$

alors

$$(3.15)\quad u_m \to u \quad \text{dans } V \text{ fort} .$$

3.2. - UNE DEUXIEME METHODE

On adapte la méthode de Frank et Wolfe [6] au problème dual. Décrivons cet algorithme : soit λ_o donnée dans Λ ; supposons construit l'itéré λ_m : le passage de λ_m à λ_{m+1} se fait en introduisant un point auxiliaire ν_m :

Soit ν_m un point tel que

$$(3.16)\quad \begin{cases} [\text{grad } J^*(\lambda_m), \nu_m] \leq [\text{grad } J^*(\lambda_m), \mu] & \forall \mu \in \Lambda \\ \nu_m \in \Lambda \end{cases}$$

ou encore

$$(3.16)' \quad \begin{cases} [-Su_m, \nu_m] \leqslant [-Su_m, \mu] & \forall\mu \in \Lambda \\ \nu_m \in \Lambda \\ u_m - F + S^*\lambda_m = 0 \end{cases}$$

ce qui s'écrit aussi, sous la forme

$$(3.16)'' \quad \begin{cases} [Su_m, \mu - \lambda_m] \leqslant \varepsilon_m & \forall\mu \in \Lambda \\ \varepsilon_m = [Su_m, \nu_m - \lambda_m] \\ \nu_m \in \Lambda \\ u_m - F + S^*\lambda_m = 0 \quad . \end{cases}$$

Notons que ν_m existe car Λ est un convexe fermé et borné (par contre, il n'y a pas nécessairement unicité de ν_m). Soit c un nombre positif fixe dont la valeur sera fixé plus loin ; posons

$$(3.17) \quad \begin{cases} \rho_m = \min \{ 1, \frac{\varepsilon m}{c} \} \\ \lambda_{m+1} = \lambda_m + \rho_m(\nu_m - \lambda_m) \quad . \end{cases}$$

Nous allons démontrer le

Théorème 3.2

Si $c > \frac{1}{2} ||S^*||^2 d^2$, où d désigne le diamètre de Λ, tout point adhérent faible à la suite λ_m est une solution du problème dual, et la suite u_m converge fortement vers u quand $m \to +\infty$.

Démonstration

En utilisant le développement de Taylor, il vient

$$J^*(\lambda_{m+1}) = J^*(\lambda_{m}) - [Su_m, \lambda_{m+1} - \lambda_m] + \frac{1}{2}[SS^*(\lambda_{m+1} - \lambda_m), \lambda_{m+1} - \lambda_m]$$

et en remplaçant $\lambda_{m+1} - \lambda_m$ par $\rho_m(\nu_m - \lambda_m)$, on obtient

$$J^*(\lambda_{m+1}) = J^*(\lambda_m) - \rho_m\varepsilon_m + \frac{1}{2}\rho_m^2[SS(\nu_m - \lambda_m), \nu_m - \lambda_m]$$

d'où il vient

$$(3.18) \quad J^*(\lambda_{m+1}) \leqslant J^*(\lambda_m) - \rho_m . \varepsilon_m + \frac{1}{2}||S^*||^2 . d^2 . \rho_m^2 \quad .$$

Si dans (3.9)" on fait $\mu = \lambda_m$, il vient

(3.19) $$\varepsilon_m \geqslant 0 \;.$$

Etudions le cas $\varepsilon_m = 0$; alors (3.9)" s'écrit

$$\begin{cases} [Su_m , \mu - \lambda_m] \leqslant 0 & \forall \mu \in \Lambda \\ \lambda_m \in \Lambda \\ u_m - F + S^*\lambda_m = 0 \end{cases}$$

d'après le corollaire 1.1, nous savons que u_m est la solution du problème primal et que λ_m est une solution du problème dual.

Seul reste donc à étudier le cas $\varepsilon_m > 0$. Si $\rho_m = 1$ nous avons, d'une part

$$1 \leqslant \frac{\varepsilon m}{c}$$

et d'autre part

(3.20) $$J^*(\lambda_{m+1}) \leqslant J^*(\lambda_m) - c + \frac{1}{2}\,||S^*||^2\, d^2$$

et puisque $J^*(\lambda_m)$ est bornée inférieurement et que

(3.21) $$c > \frac{1}{2}\,||S^*||^2\, d^2$$

la relation (3.20) ne peut avoir lieu que pour un nombre fini d'indices m; autrement dit, pour m assez grand, nous avons

$$\rho_m = \frac{\varepsilon m}{c} \;.$$

En utilisant (3.18) on obtient

$$J^*(\lambda_{m+1}) < J^*(\lambda_m) - \frac{1}{2}(c - \frac{1}{2}||S^*||^2\, d^2)\, \varepsilon_m^2 \quad ;$$

la suite $J^*(\lambda_m)$ est donc décroissante; de plus

$$\lim_{m\to+\infty} J^*(\lambda_m) - J^*(\lambda_{m+1}) = 0 \text{ ; par suite}$$

(3.22) $$\lim_{m\to+\infty} \varepsilon_m = 0 \;.$$

La situation est donc la suivante pour m assez grand

$$\begin{cases} u_m - F + S^* \lambda_m = 0 \\ [Su_m, \mu - \lambda_m] \lesssim \varepsilon_m \\ \lim\limits_{m\to+\infty} \varepsilon_m = 0 \\ \lambda_m \in \Lambda \end{cases}$$

la suite de la démonstration est identique à celle du théorème 3.1 après (3.3)' et (3.8).

3.3 - UNE METHODE DU TYPE ARROW-HURWICZ

Lorsque V est de dimension finie, J_o est de la forme

$$J_o(v) = \frac{1}{2}(Av,v) - (f,v)$$

où A est une matrice symétrique définie positive et où $f \in \mathbf{V}$.

Tout point selle (u,λ) est solution du système :

$$\text{(3.23)} \quad \begin{cases} Au + S^*\lambda = f \\ \lambda = P(\lambda + \rho Su) \qquad (\rho \geqslant 0) \end{cases}$$

l'opérateur P étant l'opérateur de projection sur Λ.

Pour résoudre (3.23) on va utiliser l'algorithme :

$$\text{(3.24)} \quad \begin{cases} \lambda_o \text{ donné} \\ u_{m+1} = u_m - \rho_1(Au_m + S^*\lambda_m - f) \qquad (\rho_1 > 0) \\ \lambda_{m+1} = P(\lambda_m + \rho_2 \, Su_{m+1}) \qquad (\rho_2 > 0) \end{cases}$$

qui est du type ARROW-HURWICZ (cf [11]) ; on a alors le

Théorème 3.3

Lorsque ρ_1 et ρ_2 sont suffisamment petit, tout point adhérent à la suite $(\lambda_m)_m$ est solution du problème dual et $\lim\limits_{m\to+\infty} u_m = u$

Démonstration

On pose $\rho_1 = \rho$, $\rho_2 = \rho c$ $(c > 0)$, $\tilde{\lambda}_m = \lambda_m - \lambda$, $\tilde{u}_m = u_m - u$ où (u,λ) est solution de (3.23).

La projection étant contractante, on a

(3.25) $$[\![\tilde{\lambda}_{m+1}]\!]^2 \leq [\![\tilde{\lambda}_m]\!]^2 + 2\rho c[\tilde{\lambda}_m, S\,\tilde{u}_{m+1}] + \rho^2 c^2 [\![S\,\tilde{u}_{m+1}]\!]^2$$

par ailleurs

(3.26) $$\tilde{u}_{m+1} = \tilde{u}_m - \rho_1(A\,\tilde{u}_m + S^*\tilde{\lambda}_m)$$

et en multipliant scalairement les deux membres de (3.26) par $\tilde{u}_{m+1}$ on a :

(3.27) $$||\tilde{u}_{m+1}||^2 = (\tilde{u}_m,\tilde{u}_{m+1}) - \rho(A\tilde{u}_m, \tilde{u}_{m+1}) - \rho(S^*\tilde{\lambda}_m, u_{m+1})$$

$$= (\tilde{u}_m,(I-\rho A)\,\tilde{u}_{m+1}) - \rho[\tilde{\lambda}_m, S\tilde{u}_{m+1}]$$

on pose α = rayon spectral de $I - \rho A$, A étant symétrique définie positive, si ρ est suffisamment petit on a $0 \leq \alpha < 1$ d'où

(3.28) $$||\tilde{u}_{m+1}||^2 \leq \alpha\,||\tilde{u}_m||\;||\tilde{u}_{m+1}|| - \rho[\tilde{\lambda}_m, S\tilde{u}_{m+1}]\,.$$

En utilisant $ab \leq \frac{a^2+b^2}{2}$ et la continuité de S, on déduit de (3.25) et (3.28)

(3.29) $$\frac{1}{2c}[\![\tilde{\lambda}_m]\!]^2 + \frac{\alpha}{2}||\tilde{u}_m||^2 - \left(\frac{1}{2c}[\![\tilde{\lambda}_{m+1}]\!]^2 + \frac{\alpha}{2}||\tilde{u}_{m+1}||^2\right)$$

$$\geq \left(1 - \alpha - \rho^2\frac{c}{2}||S||\right)||\tilde{u}_{m+1}||^2$$

Si $\rho^2 c$ est suffisamment petit on a $1 - \alpha - \rho^2\frac{c}{2}||S|| > 0$ la suite $\frac{1}{2c}[\![\tilde{\lambda}_m]\!]^2 + \frac{\alpha}{2}||\tilde{u}_m||^2$ est donc décroissante et étant bornée inférieurement par zéro, elle est convergente, le premier membre de (3.29) tend donc vers zéro, d'où $||\tilde{u}_m|| \to 0$ i.e. $u_m \to u$ ce qui démontre la convergence.

4 - GENERALISATION

On peut utiliser le théorème de Ky Fan - Sion dans le cas où $S \notin \mathcal{L}(V,L)$: Il suffit que $v \to [Sv,\mu]$ soit convexe et faible ment semi-continue inférieurement. Notons aussi qu'il n'est pas nécessaire que V et L soient des espaces de Hilbert ; on pourra trouver deux exemples dans J.CEA - R.GLOWINSKI - J.C.NEDELEC [3] et dans J.CEA - K.MALANOVSKI [4] .

BIBLIOGRAPHIE

[1] CEA J. Optimisation, théorie et algorithmes. Dunod, 1971

[2] CEA J. - GLOWINSKI R. Méthodes Numériques pour l'écoulement dans une conduite cylindrique d'un fluide rigide visco-plastique incompressible. - A paraitre.

[3] CEA J. - GLOWINSKI R. - NEDELEC J.C. - Méthodes Numériques pour la torsion elasto-plastique d'une barre cylindrique. A paraitre.

[4] CEA J. - MALANDYSKI K.- An example of a max-min problem in partial differential equations. SIAM CONTROL, Vol 8, n°3, August 1970.

[5] DUVAUT G. - LIONS J.L. - Ecoulement d'un fluide rigide visco-plastique incompressible. C.R. Acad. Sc. Paris, 270 Série A, 58-60, 1970.

[6] FRANK M. - WOLFE P. - An Algorithm for quadratic programming. Naval res. Log. quart. 3, 95-110, 1956.

[7] GOLDSTEIN A.A - Real contructive analysis, Harper and Row, New-York, 1967.

[8] KY FAN - Sur un théorème minimax. C.R. Acad. Sc. Paris, 259, 3925-3928, 1964.

[9] SION M. On general minimax theorems. Pacific J. of Math. 8, 171-176, 1958.

[10] UZAWA H. Cf. Livre Arrow K.J - Hurwicz L. - Uzawa H. : Studies in linear and non linear programming. Stan ford University Press, 1958.

[11] ARROW K.J - HURWICZ Cf Livre Arrow K.J - Hurwicz L - Uzawa H. Studies in Linear and non Linear Programming. Stan ford University Press, 1958.

NONLINEAR INTEGRAL EQUATIONS AND FIELD APPROXIMATION PROBLEMS

L. Collatz

Summary: Some inclusion theorems for the solutions of linear and nonlinear integral equations can be used for the numerical calculation of the solutions. To obtain good error bounds, one has to solve unusual types of approximation problems e.g. in particular field type approximation problems as one sided approximations with infinitely many restrictions.

0. Introduction: To solve integral equations, one can use optimization-principles with finitely many parameters $a_1,\dots,a_p$. If one wishes to get the best values of these parameters, i.e. those values which give the best possible error bounds, one has to solve in many cases approximation problems of field approximation type. In another paper (Collatz [71]) I described the application of some fixed point theorems to integral equations. Here I shall consider the connection of the fixed point theorems with the approximations.

1. Some types of integral equations. A list of examples of different types of integral equations is given in Collatz [71]. (There B is a domain in the n-dimensional real space R^n of points $x = \{x_1,\dots,x_n\}$; $u:B \longrightarrow R^1$ is the unknown function, $f:B \longrightarrow R^1$, $\varphi:R^1 \longrightarrow R^1$, and all listed functions K are given functions. If B is not a fixed domain, but depends on x, we get the corresponding Volterra-integral-equation.) It will suffice, therefore, to give some further examples for the following types:

Type	Example
Linear equation (1.1) $u(x) = f(x) + \mu \int_B K(x,t)u(t)dt$	Renewal equation $u(x) = f(x) + \int_0^x h(x-t)u(t)dt$
Hammerstein-Equation (1.2) $u(x) = f(x) + \int_B K(x,t)\varphi(u(t))dt$	branching process $u(x) = p_o f(x) + \sum_{v=1}^{R} p_v \int_0^x [u(x-s)]^v df(s)$
Urysohn-Equation (1.3) $u(x) = \int_B K(x,t,u(t))dt$	equivalent boundary value problem $-\Delta u(x,y) = \sum_{v=1}^{R} p_v(x,y)[u(x,y)]^v$ + boundary conditions
Biargument-Equation (1.4) $\int_B K(x,t,u(x),u(t))dt = 0$	Communication (Transmission-Signals) Arthurs [70] $\int_0^{\frac{\pi}{2}} \frac{\sin(x-t)}{\pi (x-t)} u(x)u(t)dt = 1$
Integrodifferential-Equation (1.5) $\int_B K(x,t,u(x),u(t),\frac{\partial u(t)}{\partial x_j},\ldots)dt = 0$	Plasmaphysics (Boltzmann-Vlasow-Equation) $\frac{\partial u}{\partial t} + y\frac{\partial u}{\partial x} + c\int_a^x \left[\int_{-\infty}^{+\infty} u(t,\xi,y)dy-n \right] d\xi \, \frac{\partial u}{\partial y} = 0$ (c,n constants)
Integrofunctional-Equation (1.6) $\int_B K(x,t,u(x),u(t),\frac{\partial u(t)}{\partial x_j},\ldots,u(\psi(t)),\ldots)dt = 0$	Regulation (equivalent differential equation): For instance $u'(t) = -a\, u(t-\varphi(t,u))$

2. The contraction Mapping Theorem and Approximation

Let us consider the equation $u = Tu$ in a metric space R with a given linear or nonlinear operator T which maps a complete set $D \subset R$ into R and which satisfies a Lipschitz condition

$$(2.1) \qquad \rho(Tv, Tw) \leq K\rho(v,w) \quad \forall v,w \in D$$

with a constant $K < 1$. We start the iterative procedure

$$(2.2) \qquad u_{n+1} = T\,u_n \quad (n=0,1,2,\ldots)$$

with an element $u_o \in D$ and suppose, that the sphere

$$(2.3) \qquad S = S(u_o,u_1) = \{\, z \mid \rho(z,u_1) \leq \frac{K}{1-K}\, \rho(u_o,u_1)\}$$

belongs to D. Then a solution u to the equation $u = Tu$ exists in S.

In applying this theorem to integral equations, where T, for example, is an integral operator and $u(x)$ is a function defined on a domain B as in No.1, one usually chooses for u_o a function w belonging to a class $W = \{w(x,a)\}$, $a = (a_1,\ldots,a_q)$ being a vector of the parameters $a_1,\ldots,a_q$. Then u_1 depends upon a as well. If one wishes now to obtain a best possible error estimate using $u_o \in W$ as the initial element in the iteration procedure, one has to solve the approximation problem

$$(2.4) \qquad \rho(u_o,u_1) = \text{Min.}$$

If we take as R the Banachspace $C(B)$ of the continuous-functions $h(x)$ in B with the supremum norm

$$(2.5) \qquad \|h\| = \sup_{x \in B} p(x)\,|h(x)|$$

where p(x) is a fixed, continuous, positive function p(x) in B, we have the Tchebycheff Approximation problem (abbreviated T.A.)

$$\|u_1 - u_o\| = \text{Min.} \tag{2.6}$$

For the linear equation (1.1) one can work with a linear manifold

$$w(x,a) = \sum_{v=1}^{q} a_v \varphi_v(x) \tag{2.7}$$

with given functions $\varphi_v(x) \in C(B)$; setting $u_o = \sum_v a_v \varphi_v$ and calculating

$$z_v(x) = \int_B K(x,t)\varphi_v(t)dt \tag{2.8}$$

one must solve the classical T.A.

$$\rho(f(x), \sum_{v=1}^{q} a_v(\mu z_v(x) - \varphi_v(x))) = \text{Min.} \tag{2.9}$$

In the other cases of equations (1.2)-(1.6), nonlinear T.A. occur.

Numerical example. Let us consider the Urysohn-Equation

$$u(x) = \mu \int_0^1 \frac{dt}{1+xtu(t)} = Tu. \tag{2.10}$$

We use the Banachspace R = C(B) with B = [0,1] and for the domain D, the set of functions $h \in R$ with $h(x) \geq 0$ for $x \in B$. With u_o = constant = c as starting element for the iteration (2.2), we get $u_1 = Tu_o = \frac{\mu}{cx} \ln(1+cx)$.

A possibility for determining c is solving the T.A.

$$\operatorname*{Max}_x |u_1 - u_o| = \operatorname*{Max}_x \left| \frac{\mu}{cx} \ln(1+cx) - c \right| \overset{!}{=} \text{Min.}$$

We calculate the Lipschitz-constant K of the operator T in (2.1), using the norm (2.5):

$$|Tv-Tw| \leq \mu \int_0^1 \frac{xt|w(t)-v(t)|dt}{[1+xtv(t)][1+xtw(t)]}$$

or by using $1+xtv(t) \geq 1$, $1+xtw(t) \geq 1$:

$$K = \mu \operatorname{Max}_{x \in B} (xp(x)) \cdot \int_0^1 \frac{tdt}{p(t)} \tag{2.11}$$

Numerical results are given in the table (in which „(2.3) fails" means, that the sphere condition $S \subset D$ is violated):

μ	c	$u_1(1)$	p(x)	K	$\frac{K}{1-K}$	error estimation $\lvert u-u_1\rvert \leq$
1	0.86	0.72	1	0.5	1	0.14
1.6	1.31	1.02	1	0.8	4	(2.3) fails
			2-x	0.618	1.62	0.935/(2-x)
2	1.597	1.194	1	1	∞	(2.3) fails
			2-x	0.772	3.38	(2.3) fails

For small values of μ one can use as weight function $p(x) \equiv 1$; for greater values of μ (for instance $\mu = 1.6$ in the table), one obtains existence of a solution and a rough error estimate with the weight function $p(x) = 2-x$, for large values of μ, the contraction mapping theorem is not applicable; but the MDO-method of No.3 is applicable for all $\mu > 0$., See No.5.

3. Monotonically Decomposible Operators (M.D.O.) and Approximation

Now let R be a partially ordered Banach-space. An operator T with domain of definition D is called syntone (resp. antitone) if $v \prec w$ implies $Tv \prec Tw$ (resp. $Tv \succ Tw$) for all $v, w \in D$. An operator T is called monotonically decomposible or, for short an MDO, if one can write $T = T_1 + T_2$, where T_1 is a syntone operator and T_2 is an antitone operator.

Theorem (J. Schröder [56]): Consider the equation

$$u = \hat{T}u = T_1 u + T_2 u + r \tag{3.1}$$

in a partially ordered Banachspace R. Let T_1 and T_2 be given completely continuous operators which are defined in a convex domain D, T_1 be syntone, T_2 antitone and r a given element of R. Let us use the iteration procedure

$$\left\{ \begin{array}{l} v_{n+1} = T_1 v_n + T_2 w_n + r \\ w_{n+1} = T_1 w_n + T_2 v_n + r \end{array} \right\}, \quad n=0,1,2,\ldots, \tag{3.2}$$

where the elements $v_o, w_o \in D$ satisfy the initial conditions

$$v_o \prec v_1 \prec w_1 \prec w_o \tag{3.3}$$

Then (as consequence of the Schauder fixed point theorem [30]) there exists at least one solution to the equation (3.1) with $v_1 \prec u \prec w_1$.

This theorem includes two important special cases:

A) $\hat{T}$ is syntone, $T_2 = \mathcal{O}$ = zero-operator, $u = T_1 u + r$; then (3.2) consists of two separate sequences:

(3.4) $$v_{n+1} = T_1 v_n + r , \quad w_{n+1} = T_1 w_n + r .$$

v_n is monotonically non-decreasing, w_n non-increasing.

B) $\hat{T}$ is antitone, $T_1 = \mathcal{O}$. It is possible to use only v_o, w_1, v_2 with

(3.5) $$w_1 = T_2 v_o + r , \quad \tilde{v}_2 = T_2 \tilde{w}_1 + r \quad \text{with } \tilde{w}_1 = w_1 \text{ or } \tilde{w}_1 \succ w_1$$

If, then,

(3.6) $$v_o \prec \tilde{v}_2 \prec w_1 ,$$

one knows that

(3.7) $$v_2 \prec u \prec w_1 .$$

Let us begin with the simplest case of a linear equation (1.1). Every real kernel has the form

(3.8) $$K(x,t) = K_1(x,t) + K_2(x,t) \quad \text{where } K_1 \geq 0 \text{ and } K_2 \leq 0;$$

Then the operator

(3.9) $$T_i h = \int_B K_i(x,t) h(t) dt$$

is syntone for $i = 1$ and antitone for $i = 2$.

a) Let $K_2 = 0$, and $K(x,t) = K_1(x,t) \geq 0$ in BxB. If one can find a domain D in which the operator T_1 has a Lipschitz-constant $K < 1$, one gets a „onesided Tchebycheff Approximation problem" (abbreviated T_1-A.), which may be written as optimization problem in the form

(3.10) $$0 \leq v_1 - v_o \leq \delta , \quad \delta = \text{Min}$$

for an approximate solution

(3.11) $$v_o = \sum_{v=1}^{q} a_v \varphi_v(x).$$

Here φ_v has the same meaning as in (2.7). Furthermore one has the field condition, that v_o, v_1 and the sphere $S(v_o, v_1)$ [see (2.3)] belong to D.

b) In the case $K_1 = 0$, we obtain field-approximation problems.

J. Schröder [60] gave the following generalization of the M.D.O. method. Suppose there exist two operators $H_j(\xi,\eta)$ $(j = 1,2)$ defined in DxD which are syntone in ξ and antitone in η. Suppose further that

(3.12) $$H_j(\xi,\eta) \prec H_j(\hat{\xi},\hat{\eta}) \quad \text{for } \xi \prec \hat{\xi}, \ \eta \succ \hat{\eta} \ \text{ and } \xi, \eta, \hat{\xi}, \hat{\eta} \in D.$$

(3.13) $$H_1(\xi,\xi) \prec T\xi \prec H_2(\xi,\xi) \quad \text{for all } \xi \in D.$$

To solve the equation $u = Tu$ one can use the iteration procedure

(3.14) $$\left\{ \begin{aligned} v_{n+1} &= H_1(v_n, w_n) \\ w_{n+1} &= H_2(w_n, v_n) \end{aligned} \right\} \qquad n = 0,1,2,\ldots$$

starting with the elements $v_o, w_o \in D$. Again we suppose that the initial conditions (3.3) are satisfied and that the operator T is completely continuous. Then exists at least one solution u of $u = Tu$ with

$$(3.15) \qquad v_1 < u < w_1 \ .$$

Numerical example: For the biargument-equation

$$(3.16) \qquad u(x) = Tu(x) = \int_0^1 \frac{1+t}{1+x}\frac{u(x)}{u(t)}\,dt$$

we can choose $H_j(\xi,\eta) = H(\xi,\eta) = \int_0^1 \frac{1+t}{1+x}\frac{\xi(x)}{\eta(t)}\,dt$.

For $v_o = 0$, $w_o = 2$ we get $v_1 = \frac{1}{1+2x}$ and $w_1 = 2$. The initial conditions (3.3) are satisfied and therefore equation (3.16) has at least one solution u(x) with $\frac{1}{1+2x} \leq u(x) \leq 2$.

4. Field-Approximation

Let R again be the partially ordered Banachspace C(B) and $\phi(x,a)$ and $G_j(x,a)$ for j=1,...,m, given functions defined on BxA, where A is given subset of the q-dimensional space R^q of the parameters $a_1,\dots,a_q$. Then we define as Field-Approximation (abbreviated Field A.) the problem of finding a parameter vector a, which solves the optimization problem (onesided T.A. with restrictions).

$$(4.1) \qquad 0 \leq \phi(x,a) \leq \delta \qquad \delta = \text{Min.} \quad \text{with the field restrictions:}$$

$$(4.2) \qquad 0 \leq G_j(x,a) \ , \qquad j = 1,\dots,m \ ,$$

for $x \in B$, $a \in A$.

This is a generalization of the approximation problems with restricted ranges (Taylor [69], Schumaker [69]

5. MDO and Field-Approximation

a) Let $\mathcal{K}(x,t) = \mathcal{K}_2(x,t) \leq 0$ in the linear equation (1) and let us use for v_o and w_o an expression as in (3.11)

$$(5.1) \qquad v_o = \sum_{v=1}^{q} a_v \varphi_v(x) \;, \quad w_o = \sum_{v=1}^{q} b_v \varphi_v(x) \;, \quad a \in A,\; b \in A,$$

and calculate the functions $z_v(x)$ from (2.8).

We obtain the field-approximation

$$(5.2) \qquad 0 \leq \sum_{v=1}^{q} (a_v - b_v)\, z_v(x) \leq \delta, \quad \delta = \text{Min}$$

with the field restrictions

$$(5.3) \qquad \begin{cases} 0 \leq f - \sum_{v=1}^{q} a_v \varphi_v(x) + \sum_{v=1}^{q} b_v z_v(x) \\ 0 \leq -f + \sum_{v=1}^{q} b_v \varphi_v(x) - \sum_{v=1}^{q} a_v z_v(x) \text{ and of course } v_o, w_o \in D. \end{cases}$$

b) In the general case of a kernel $\mathcal{K}(x,t)$ which does not necessarily have a fixed sign, we calculate with (3.8)

$$(5.4) \qquad z_v^{(i)}(x) = \int_B \mathcal{K}_i(x,t) \varphi_v(t)\,dt$$

Then (5.1) gives the Field-A.

$$(5.5) \qquad 0 \leq w_1 - v_1 = \sum_{v=1}^{q} (b_v - a_v)(z_v^{(1)}(x) - z_v^{(2)}(x)) \leq \delta, \quad \delta = \text{Min}$$

with the field restrictions $v_o, w_o \in D$ and

$$(5.6) \qquad \begin{cases} 0 \leq v_1 - v_o = f(x) + \sum_{v=1}^{q} a_v (z_v^{(1)}(x) - \varphi_v(x)) + \sum_{v=1}^{q} b_v z_v^{(2)}(x) \\ 0 \leq w_o - w_1 = -f(x) - \sum_{v=1}^{q} b_v (z_v^{(1)}(x) - \varphi_v(x)) - \sum_{v=1}^{q} a_v z_v^{(2)}(x) \end{cases}$$

In the case of a syntone operator $K = K_1$, all $z_v^{(2)}(x) = 0$, and (5.5) and (5.6) reduce to simpler forms; one has comparing with (3.10) the advantage that here no calculation of a Lipschitz-constant is necessary.

Numerical example. Let us consider again the Urysohn equation (2.10)

$$(5.7) \qquad u(x) = Tu(x) \text{ with } Tz(x) = \mu \int_0^1 \frac{dt}{1+xtz(t)} \text{ with } \mu > 0.$$

The operator T here is antitone and completely continuous in the domain $z(t) \geq 0$.

The iteration (3.2) reduces to

$$(5.8) \qquad v_{n+1} = T\, w_n, w_{n+1} = T\, v_n \quad (n = 0,1,2,\ldots)$$

Choosing $v_0 = 0$ we get $w_1 = Tv_0 = T(0) = \mu$

and $w_0 = \mu$ gives $v_1 = T(\mu) = \frac{1}{x} \ln(1+\mu x)$.

We have $v_1 < \mu$ for $x > 0$ and therefore the initial conditions (3.3) are satisfied for every $\mu > 0$. This gives the existence of at least one solution of (5.7) for any given $\mu > 0$ and the inclusion, fig.1 (see page 54)

$$(5.9) \qquad \frac{1}{x} \ln(1+\mu x) \leq u(x) \leq \mu .$$

One sees, that in this example the MDO-method is much superior to the method of contraction mappings of No.2.

A slight improvement is the following: (5.9) shows that $u(x) \geq \gamma = \ln(1+\mu)$. Taking $\hat{v}_0 = \gamma$ instead of $v_0 = 0$, one arrives at $\hat{w}_1 = T(\gamma) = \frac{\mu}{\gamma x} \ln(1+\gamma x)$ and the better bounds

(5.10) $$\frac{1}{x}\ln(1+\mu x) \leq u(x) \leq \frac{\mu}{\gamma x}\ln(1+\gamma x) \quad \text{with } \gamma = \ln(1+\mu) \ .$$

For the generalization of Schröder [60] of the M.D.O. Method [see (3.12) to (3.15)] we have the same type of field-approximation because the formulas (3.3) and (3.15) are common to both methods, the M.D.O.Method as well as to the generalization.

6. Theory of H-sets for field-approximation

The Theory of H-sets for approximation problems [see e.g. Meinardus [64], Bredendiek [69], Taylor [69], Collatz [70a] and others] can be applied also to the field-approximation. In the linear case one has to test whether the corresponding systems of linear inequalities are solvable or not [compare Collatz [66] p.420-431]; for this decision, algorithms such as the Gaussian elimination procedure are available; in the nonlinear case one has systems of nonlinear inequalities which have only been discussed in special cases [Collatz [69]].

Let us illustrate these methods in the special case of a linear integral equation

(6.1) $$u(x) = Tu(x) \text{ with } Tz(x) = 2 - \frac{1}{2}\int_{-1}^{1} |x-t|\, z(t)dt$$

A. The simplest trial uses constants as functions v_o, w_o, say $v_o = a$, $w_o = b$.
Using (with $x^2 = s$) the result

$$T(c_o + c_1 s) = 2 - \frac{c_o}{2}(1+s) - \frac{c_1}{12}(3+s^2),$$

one gets

$$v_1 = 2 - \frac{b}{2}(1+s), \quad w_1 = 2 - \frac{a}{2}(1+s)$$

With $a=0$, $b=2$ and $v_o = 0$, $v_1 = 1-s$, $w_1 = w_o = 2$, the initial conditions (3.3) are satisfied and there exists therefore at least one solution $u(x)$ of (6.1) with $1-x^2 \leq u(x) \leq 2$. These bounds are very rough, but the best one can possibly obtain using constant functions v_o, w_o. We observe; fig.2

(considering all functions as functions of s)

$$w_o(0)-w_1(0)=0, \quad v_1(1)-v_o(1)=0, \quad \max_s[w_1(s)-v_1(1)]=w_1(1)-v_1(1)=\delta = 2.$$

For if there were a better result using constants $\hat{v}_o = \hat{a}$, $\hat{w}_o = \hat{b}$, $\hat{v}_1 = 2 - \frac{\hat{b}}{2}(1+s)$, $\hat{w}_1 = 2 - \frac{\hat{a}}{2}(1+s)$, $\max_s[\hat{w}_1(s)-\hat{v}_1(s)] = \hat{\delta} < 2$, then the differences $\hat{a}-a=\bar{a}$, $\hat{b}-b=\bar{b}$, $\hat{v}_j - v_j = \bar{v}_j$, $\hat{w}_j - w_j = \bar{w}_j$ $(j=1,2)$ would satisfy the inequalities:

$$\bar{w}_o(0)-\bar{w}_1(0)=\hat{w}_o(0)-w_o(0)-\hat{w}_1(0)+w_1(0)=\hat{b}-2+\frac{\hat{a}}{2}-[b-2+\frac{a}{2}]=\bar{b}-\frac{1}{2}\bar{a} \geq 0$$

$$\bar{v}_1(1)-\bar{v}_o(1)=\hat{v}_1(1)-v_1(1)-\hat{v}_o(1)+v_o(1)=2-\hat{b}-\hat{a}-[2-b-a]=-\bar{b}-\bar{a} \geq 0$$

$$\bar{w}_1(1)-\bar{v}_1(1)=\hat{w}_1(1)-w_1(1)-\hat{v}_1(1)+v_1(1)=-\hat{a}+\hat{b}-[-a+b] = \bar{b}-\bar{a} < 0$$

This gives $-\bar{a}+2\bar{b} \geq 0$, $-\bar{a}-\bar{b} \geq 0$, $\bar{a}-\bar{b} > 0$ or the contradiction $\bar{b} > 0$, $-2\bar{b} > 0$ and therefore, no better approximation with real constants a,b exists.

B. Let us take $v_o = a+\alpha s$, $w_o = b+\beta s$ as starting elements with

$$a = \frac{10}{7}, \quad \alpha = -\frac{36}{49}, \quad b = \frac{72}{49}, \quad \beta = -\frac{32}{49},$$

$$v_1 = Tw_o = \frac{1}{49}[70-36s+\frac{8}{3}s^2], \qquad w_1 = Tv_o = \frac{1}{49}[72-35s+3s^2] .$$

We obtain the inclusion $v_1 \leq u \leq w_1$, $0 \leq w_1 - v_1 \leq \delta = \frac{10}{147} \approx 0.0680$.

We have the symbolic sketch, fig.3, with

$$v_1(0)=v_o(0)=v_1'(0)-v_o'(0) = w_1(0)-w_o(0)=w_1(1)-w_o(1) = 0,$$

$$\max_s [w_1(s)-v_1(s)] = w_1(1)-v_1(1) = \delta$$

If we compare this with any other values $\hat{a},\hat{\alpha},\hat{b},\hat{\beta}$ where $v_1'(0)-v_0'(0)=0$, then we have achieved the best possible approximation (with the smallest value of δ), since a calculation analogous to A. yields here the system of linear inequalities

$$+\bar{a}+\frac{1}{3}\bar{\alpha}-\bar{b}-\frac{1}{3}\bar{\beta} > 0,\quad \bar{a}+\frac{1}{3}\bar{\alpha}+\bar{b}+\frac{1}{3}\bar{\beta} \geqslant 0,\quad \frac{1}{2}\bar{a}+\frac{1}{4}\bar{\alpha}+\bar{b} \geqslant 0,\quad -\bar{a}-\frac{1}{2}\bar{b}-\frac{1}{4}\bar{\beta} \geqslant 0,$$

which has with $v_1'(0)-v_0'(0) = 0$ or $2\bar{\alpha}+\bar{b} = 0$ no solution.

7. Nonlinear elliptic boundary value problems and field approximation

Many other problems in analysis are connected with field approximation. We select elliptic boundary value problems of following type, which are equivalent to nonlinear integral equations. For simplicity we consider the equation

$$(7.1) \qquad Tz = -\Delta z+h(x,z,z_j) \quad \text{in } B$$

for a function $z(x) = z(x_1,\dots,x_n)$. Here Δ is the Laplacean operator, B a given domain with piecewise smooth boundary Γ and z_j means $\frac{\partial z}{\partial x_j}$. The function $h(x,z,z_j)$ should not „vary too strongly" with z_j, see in detail Collatz [66] p.385. In the simplest case which we now discuss h is independent of z_j. Let us suppose that the given function $h(x,z)$ is non decreasing in z. Then for two functions $v,w \in C^2(B)$ the following monotonicity theorem holds (in fact in a much more general form, see Redheffer [62], Collatz [66] p.389).

$$(7.2) \qquad Tv \leq 0 \leq Tw \text{ in } B \text{ and } v \leq w \text{ on } \Gamma \text{ implies } v \leq w \text{ in } B.$$

Suppose there exists a solution $u(x)$ of

$$(7.3) \qquad Tu = 0 \text{ in } B, \quad u = g(x) \text{ on } \Gamma$$

with given boundary values $g(x)$; then we have the inclusion

(7.4) $v(x) \leq u(x) \leq w(x)$ in B for $Tv \leq 0 \leq Tw$ in B, $v \leq g \leq w$ on Γ.

We try to choose, v,w from classes V,W of functions

$$v(x,a) \in V, \quad w(x,b) \in W$$

where a,b are parameter-vectors as above.

(Suppose that $v \leq g \leq w$ on Γ is satisfied, otherwise there are further conditions the parameter a,b must satisfy; then we have the field approximation

$$(7.5) \quad \begin{aligned} &Tv(x,a) \leq 0 \leq Tw(x,b) \quad \text{for } x \in B \\ &w(x,b) - v(x,a) \leq \delta, \quad \delta = \text{Min.} \end{aligned}$$

(Comparing with (4.1): The condition $0 \leq w(x,b)-v(x,a)$ is not necessary, because this is true by the theorem of monotonicity). Although the following example is slightly more general than the theory just discussed, the results of Redheffer do apply.

Example:

(7.6) $Tu = -u''+1+xuu' = 0$ with the boundary conditions

(7.7) $u(\pm 1) = 1$;

We try to take

$$v(x,a) = 1+(1-x^2)a_1+(1-x^4)a_2, \quad w(x,b) = 1+(1-x^2)b_1+(1-x^4)b_2$$

and we get with $x^2 = s$

$$Tv = 1+2a_1+12a_2s-2s(a_1+2a_2s)[1+a_1(1-s)+a_2(1-s^2)]$$

and analogously Tw.

We have

$$Tv \leq 0 \leq Tw, \quad (1-s)[b_1-a_1)+(1+s)(b_2-a_2)] \leq \delta \text{ for } 0\leq s\leq 1, \qquad \delta = \text{Min}.$$

A. Working only with a_1, b_1 that means $a_2 = b_2 = 0$, the problem has no solution.

With $a_1 = b_1 = -\frac{1}{2}$, $a_2 = -\frac{1}{8}$, $b_2 = 0$ one gets

$$Tv = \frac{1}{16}s\ (s-1)(s^2+7s+18) \leq 0 \leq Tw = \frac{1}{2}\ (s+s^2)$$

and therefore one has

$$\frac{1+x^2}{2} - \frac{1-x^4}{8} \leq u(x) \leq \frac{1+x^2}{2} \quad .$$

B. Working with a_1, a_2, b_1, b_2 one gets

$$a_1 = -0.5\ , \quad a_2 = -0.125\ , \quad b_1 = -0.4955\ , \quad b_2 = -0.0497$$

with the error bound $\delta = -0.0797$.

I wish to thank Mr.Lorenz and Mr.Sprekels, Hamburg, for numerical calculation carried out on a computer.

Figures:

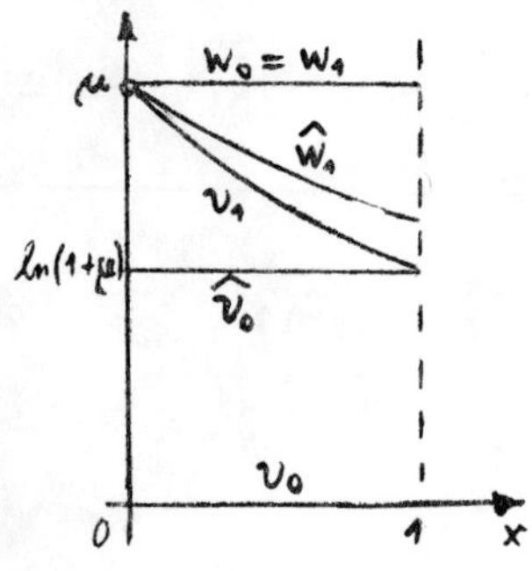

Figure 1

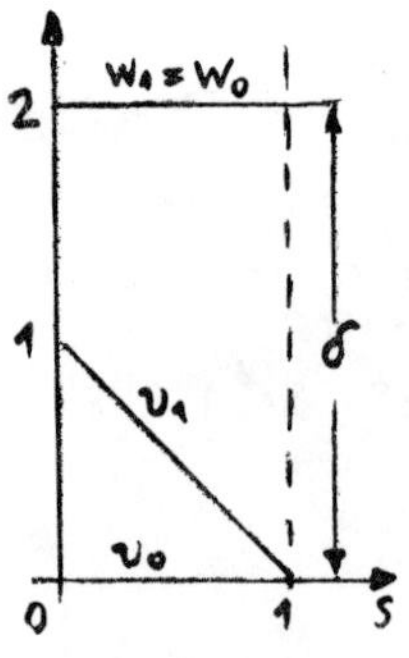

Figure 2

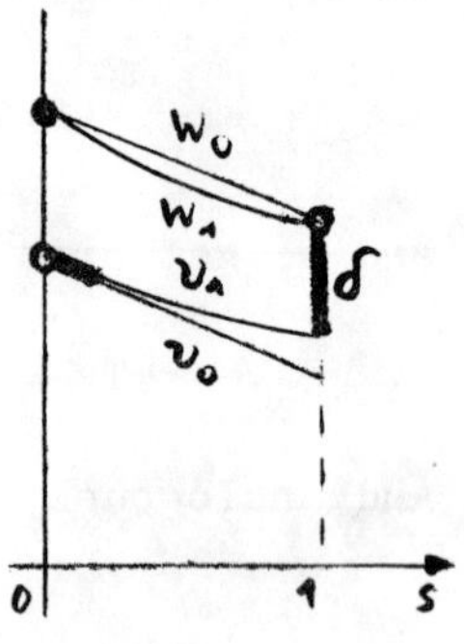

Figure 3

LITERATURE

Arthurs, A.M. [70]: Complementary Variational Principles, Oxford 1970, 95 S.

Bredendiek, E. [69]: Simultan-Approximation. Arch.Rat.Mech.Anal.33 1969, 307-330.

Collatz, L. [66]: Functional Analysis and Numerical Mathematics, Acad. Press 1966, 473 p.

Collatz, L. [69]: The determination of H-sets for the inclusion theorem in nonlinear Tschebyscheff-Approximation, Proc.Symp. on Approx. Theory and its Applications, Lancaster/England, 22.-26. July 1969, p. 179-189.

Collatz, L. [70]: Applications of nonlinear optimization to approximation problems, in Abadie: Nonlinear and integer programming, North Holland Publ. Company (1970).285-308.

Collatz, L. [71]: Some applications of functional analysis to analysis, particularly to nonlinear integral equations, Proc. Advanced Seminar, University of Wisconsin Press, Madison 1971.

Hammerstein, A. [30]: Nichtlineare Integralgleichungen nebst Anwendungen, Acta Math.54(1930),117-176.

Meinardus, G. [64]: Approximation von Funktionen und ihre numerische Behandlung, Springer 1964.

Michlin, S.G. - Smolizki, Ch.L. [69]: Näherungsmethoden zur Lösung von Differential- und Integralgleichungen, Leipzig 1969, 284 S.

Kantorowitsch, L.W. - Akilow, G.P. [64]: Funktionalanalysis in normierten Räumen, Berlin 1964, 622 p.

Redheffer, R.M. [62]: An Extension of certain maximum principles, Mh.Math.Phys.66(1962),32-42.

Schröder, J. [56]: Das Iterationsverfahren bei allgemeinerem Abstandsbegriff, Math.Z.66(1956),111-116.

Schröder, J. [60]: Funktionalanalytische Herleitung von Fehlerabschätzungen und ihre praktische Durchführung auf Rechenanlagen; Z.angew.Math.Mech.40(1960),T27-37.

Schumaker, L.L. - Taylor, G.D. [69]: An Approximation by polynomials having restricted ranges. SIAM J.Numer.Anal.6(1969),31-36.

Taylor, G.D. [69]: An Approximation by polynomials having restricted ranges I., SIAM J.Numer.Anal.

SOME MODIFIED EIGENVALUE PROBLEMS

G.H. Golub

In many applications, various eigenvalue problems arise which are slightly nonstandard. That is, the usual algorithms for computing eigensystems are not directly applicable. For instance, in various statistical data-fitting problems, it is desirable to find $\min_{x \neq 0} \frac{x'Ax}{x'Bx}$ subject to the constraint $C'x = 0$. In this talk we shall present an algorithm for reducing this problem to the usual eigenvalue problem. In addition, we shall consider methods for solving the eigenvalue problem $Ax = \lambda Bx$ when A and B are singular and we shall consider the problem of determining the eigensystem of a matrix which has been modified by a matrix of rank one.

ON DIFFERENCE SCHEMES FOR PARABOLIC DIFFERENTIAL EQUATIONS WITH DERIVATIVE BOUNDARY CONDITIONS

Rudolf Gorenflo

§ 1 Introduction

On $D = [0,1] \times [0,T]$ we consider the differential equation

(1.1) $$\frac{\partial u}{\partial t} - a\frac{\partial^2 u}{\partial x^2} - b\frac{\partial u}{\partial x} - cu = r(x,t), \qquad a > 0 ,$$

for $u = u(x,t)$ with initial condition

(1.2) $$u(x,0) = g(x)$$

and lateral boundary conditions

(1.3) $$-\frac{\partial u(0,t)}{\partial x} + p\, u(0,t) = \phi(t),$$

(1.4) $$\frac{\partial u(1,t)}{\partial x} + q\, u(1,t) = \psi(t).$$

In the interests of notational simplicity we assume the coefficients a,b,c,p,q to be real constants. $T > 0$ is finite.

In treating difference schemes for this problem (and its generalizations to problems with variable coefficients or even with nonlinearities) it is usually supposed that p and q are non-negative and that c is non-positive. See, for example, Babuška, Práger and Vitásek [2], Batten [3], Isaacson [12], Rose [20], and Varga [22]. In recent years, however, there has been done some research on what happens when these assumptions are dispensed with. See Anderssen [1], Campbell and Keast [4], Gorenflo [8], [9], [10], [11], Mitchell [16], Osborne [17], and Taylor [21].

We must carefully discern between the various kinds of stability involved, because the corresponding homogeneous problem ($r(x,t) = \phi(t) = \psi(t) \equiv 0$) may have exponentially increasing solutions as $t \to \infty$ for particular choices of $g(x)$. For the heat equation ($a=1$, $b=c=0$) Copson and Keast [6] show that this kind of instability arises if $p+q < 0$ or $p + q + pq < 0$. See also Campbell and Keast [4]. In such cases a difference scheme cannot be very accurate over a long intervall $[0,T]$.

Let us agree to consider stability and convergence of difference schemes in a finite region $D = [0,1] \times [0,T]$ in the sense of Lax and Richtmyer (see [18]).

It is often said that in applications p and q are always non-negative. There are, however, diffusion processes with drift to a reflecting barrier which correspond to a problem of the kind described with p or q negative (see [8]).

Let $\Theta \in [0,1]$ be a parameter, $\bar{\Theta} = 1 - \Theta$, $Jh = 1$, $J \geq 2$ an integer, $\tau = \mu h^2$ where $\mu > 0$ may depend on h, and $N = [T/\tau]$. Let

$U^n = (U_o^n, U_1^n, \ldots, U_J^n)'$, U_j^n being considered as an approximation to $u_j^n = u(jh, n\tau)$. We shall investigate the standard class of two-level-schemes.

(1.5) $U^{n+1} - U^n - \mu M(\Theta U^{n+1} + \bar{\Theta} U^n) = S^n$, $n = 0,1,\ldots,N-1$,

(1.6) $U_j^o = g(jh)$, $j = 0,1,\ldots,J$,

which one obtains by central difference approximation of (1.1), (1.2), (1.3), (1.4) in x-direction and elimination of the fictitious values U_{-1}^m and U_{J+1}^m. Here M is a tridiagonal matrix, and S^n is a vector:

$$M = \begin{pmatrix} -2a-ph(2a-bh)+ch^2, & 2a & & & \\ a-\frac{bh}{2}, & -2a+ch^2, & a+\frac{bh}{2} & & \\ & \cdots & \cdots & \cdots & \\ & & a-\frac{bh}{2}, & -2a+ch^2, & a+\frac{bh}{2} \\ & & & 2a, & -2a-qh(2a+bh)+ch^2 \end{pmatrix},$$

$$S^n = (\tau r_o^{n+\Theta} + \mu(2a-bh)h\phi^{n+\Theta}, \tau r_1^{n+\Theta}, \ldots, \tau r_{J-1}^{n+\Theta}, \tau r_J^{n+\Theta} + \mu(2a+bh)h\psi^{n+\Theta})' .$$

We may take $\phi^{n+\Theta} = \phi((n+\Theta)\tau)$ or $\Theta\phi((n+1)\tau) + \bar{\Theta}\phi(n\tau)$ and analogously $r_j^{n+\Theta}$ and $\psi^{n+\Theta}$. We are interested in the passage to the limit $h \to 0$, $\tau \to 0$.

If the solution u exists and belongs to $C^{4,2}(D)$, then we find

(1.7) $u^{n+1} - u^n - \mu M(\Theta u^{n+1} + \bar{\Theta} u^n) = \tau\varepsilon^n$

where $u^m = (u(0,m\tau), u(h,m\tau),\dots,u(1,m\tau))'$,

$$\varepsilon^n = \begin{cases} (O(h+\tau), O(h^2+\tau),\dots, O(h^2+\tau), O(h+\tau))' & \text{if } \Theta \neq 1/2 \\ (O(h+\tau^2), O(h^2+\tau^2),\dots, O(h^2+\tau^2), O(h+\tau^2))' & \text{if } \Theta = 1/2 . \end{cases}$$

There are three methods by which the problem can be treated: (i) eigenvalue techniques, (ii) the monotonicity method,(iii) energy methods.

Using eigenvalue techniques Osborne [17] has shown this scheme to be stable and convergent in the L_2-norm under the additional assumptions $\Theta \geq 1/2$ and $h = O(\tau)$. The order of convergence is $O(h^2+\tau)$ for $\Theta > 1/2$, $O(h^2+\tau^2)$ for $\Theta = 1/2$. There is no restriction for μ. Eigenvalue techniques are also used by Campbell, Keast and Mitchell; see [4] and [16].

We shall concentrate on the monotonicity method which directly yields convergence and stability in the maximum norm but requires a boundedness condition for μ if $\Theta < 1$. We shall sketch how this method can also be applied to weakly coupled parabolic systems where, for the class of simple schemes considered, the case $\Theta = 1$ is the most important in practical computations. The monotonicity method is a generalization of the method of discrete maximum principles (see Rose [20] , Isaacson [12] , Batten [3]) and has been described by Krawczyk [14] and Kolar [13] for the first boundary value problem, by the author [8],[9],[10],[11] for the third boundary value problem. It works with discrete analogues of the Nagumo-Westphal monotonicity lemma for parabolic equations(see Collatz [5] and Walter [23]). The advantage of this method lies in its simplicity (and suitedness for introductory courses) and in its easy generalizability to more general problems ([9], [10], [11]). Its drawback lies in the fact that for $\Theta = 1/2$, because of the boundedness condition for μ, the higher accuracy of the Crank-Nicholson scheme cannot be fully exploited.

The author thinks that by energy methods (see, for example, Lees [15], and Babuška, Práger, Vitásek [2]) it should be possible to prove stability and convergence in the maximum norm without any restrictions for μ in the Crank-Nicholson case $\Theta = 1/2$.

§ 2 The monotonicity method

Solving (1.5) for U^{n+1} yields

$$(2.1)\quad U^{n+1} = (I-\mu\Theta M)^{-1}(I+\mu\bar{\Theta}M)U^n + (I-\mu\Theta M)^{-1} S^n,$$

with I = identity matrix. By induction we see that all U_j^n for $(j,n) \in \{0,1,\ldots,J\} \times \{0,1,\ldots,N\}$ exist and depend isotonically on the inhomogeneities r,g,ϕ,ψ, if $I-\mu\Theta M$ is invertible and all elements of $(I-\mu\Theta M)^{-1}$ and $I+\mu\bar{\Theta}M$ and the coefficients $2a \pm bh$ in S^n are non-negative.

With regard to the matrix $I-\mu\Theta M$ a well-known sufficient condition (see Collatz [5], p. 297) is that its non-diagonal elements are non-positive and that all its row-sums are positive. We obtain a smallness condition

$$(2.2)\qquad |b|h \leq 2a$$

for h and a rather weak restriction

$$(2.3)\qquad \mu\Theta h\{ch - \min(p(2a-bh),\ q(2a+bh),0)\} < 1$$

for the mesh-width ratio $\mu = \tau/h^2$. (2.2) also warrants the isotonic dependence of the U_j^n on ϕ and ψ. All entries of $I+\mu\bar{\Theta}M$ are non-negative if (2.2) and

$$(2.4)\quad \mu\bar{\Theta}\{2a-ch^2 + h\max(p(2a-bh),\ q(2a+bh),0)\} \leq 1$$

are satisfied. Thus the monotonicity conditions (2.2),(2.3),(2.4) are sufficient for U to depend isotonically on the inhomogeneities. The problem of determining U from r,g,ϕ,ψ as data is a problem of monotonic type in the sense of Collatz [5].

To establish convergence and stability let us assume $u \in C^{4,2}(D)$ and (2.2), (2.3), (2.4) to be satisfied. Let V be the solution of a disturbed scheme

$$(2.5)\quad V^{n+1} - V^n - \mu M(\Theta V^{n+1} + \bar{\Theta}V^n) = S^n + \tau\rho^n,\quad n = 0, 1,\ldots, N-1,$$

$$(2.6)\quad V_j^o = g(jh) + \hat{\rho}_j,\quad j = 0,1,\ldots,J,$$

the ρ_j^m and $\hat{\rho}_j$ representing numerical disturbances (for example rounding errors) obeying

$$(2.7)\quad \rho^n = (O(h+\tau),\ O(h^2+\tau),\ldots,O(h^2+\tau),\ O(h+\tau))',$$

(2.8) $\hat{\rho}_j = O(h^2+\tau)$.

For the errors $w_j^m = V_j^m - u_j^m$ we find

(2.9) $(Pw)^n := w^{n+1} - w^n - \mu M(\Theta w^{n+1} + \bar{\Theta} w^n) = \tau(\rho^n - \varepsilon^n)$

(2.10) $w_j^o = \hat{\rho}_j$.

For linearity and symmetry reasons and because our problem is of monotonic type it suffices now to find a grid function W with $W_j^n = O(h^2+\tau)$ and $W_j^n \geq |w_j^n|$ for sufficiently small h and τ.

Any grid function W will do which has the properties

(2.11) $$(PW)_j^n \geq \begin{cases} R\tau(h+\tau) & \text{for } j=0, \quad j=J \\ R\tau(h^2+\tau) & \text{for } j=1,2,\ldots,J-1 \end{cases},$$

$n = 0,1,\ldots,N-1$,

(2.12) $W_j^o \geq R(h^2 + \tau), \quad j = 0, 1,\ldots, J,$

where $R > 0$ is a suitably chosen constant.

It is important to use a sufficiently powerful error majorizing function W. We achieve our goal with

(2.13) $W_j^n = R(h^2+\tau) \cosh\ (\Omega(jh - \frac{1}{2}))\ \exp\ (Sn\tau)$.

For h and τ sufficiently small (2.11) is satisfied for j=0 and j=J by choosing the constant Ω large enough, and then for $1 \leq j \leq J-1$ by choosing the constant S large enough. We obtain the

Theorem: Let (1.1), (1.2), (1.3), (1.4) have a solution $u \in C^{4,2}(D)$ and let the monotonicity conditions (2.2), (2.3), (2.4) be fulfilled. Let V be the solution of a disturbed scheme (2.5), (2.6) with (2.7) and (2.8). Then

$$|V_j^n - u(jh, n\tau)| \leq L(h^2 + \tau) \text{ for}$$

$$(j,n) \in \{0,1,\ldots,J\} \times \{0,1,\ldots,N\}$$

with a constant L independent of the mesh-widths h and τ.

<u>Remarks:</u> <u>1.</u> If $c \leq 0$ and $p \geq 0$, $q > 0$, we can use

$$W_j^n = (h^2 + \tau) \{S - \exp (\Omega jh)\}$$

by choosing the constants Ω and S sufficiently large. See Isaacson [12] and Wendroff [24], pp. 223 - 228. The restrictions for p and q look artificial; their asymmetry is caused by the asymmetry of W. The restriction for c can be dispensed with and those for p and q can be symmetrically relaxed to $p > -4$, $q > -4$ by taking

$$W_j^n = \{R + \Omega(jh - \tfrac{1}{2})^2\}(h^2 + \tau) \exp (Sn\tau)$$

with sufficiently large constants Ω and S. See Batten [3] who, however, discretizes the lateral boundary conditions as we do in § 3 and supposes $p \geq 0$, $q \geq 0$.

<u>2.</u> It is not difficult to modify the scheme and the described theory to the case of Dirichlet boundary conditions at $x = 0$ and/or $x = 1$. If, for example, $u(0,t)$ is prescribed, then in the vectors U^m the first component and in the matrix M the first row and the first column are to be deleted, and the vector S^n is to be correspondingly adjusted.

<u>Comment:</u> Whereas for μ (2.3) and (2.4) roughly give a bound $1/(2a)$ in the explicit case $\Theta = 0$ and at worst a bound $O(1/h)$ in the fully implicit case $\Theta = 1$ (this bound is satisfactory in practice), they give a bound about $1/a$ in the Crank-Nicholson case $\Theta = 1/2$. Nevertheless, it is known that in the case $\Theta = 1/2$ there is convergence in the maximum norm of the order $O(h^2 + \tau^2)$ for any $\mu > 0$ if $c \leq 0$ and $p \geq 0$, $q \geq 0$, and if $u \in C^{4,3}(D)$. See Babuška, Práger, Vitásek [2] for a proof by an energy method. It is the author's opinion that these restrictions on c, p and q can be dispensed with. By use of the following idea a proof might be found. By a substitution $u(x,t) = v(x,t) \times \cosh (\Omega(x-\frac{1}{2})) \exp (St)$ the problem (1.1), (1.2), (1.3), (1.4) is transformed into a similar problem for v (but with some of the coefficients being variable) in which v has a negative coefficient in the differential equation and positive coefficients in the lateral boundary conditions.

To achieve this one has only to choose the constants Ω and S sufficiently large. Analogously one may transform the system of error equations (2.9), (2.10) into a new system which should be amenable to energy methods. Because there arise some technical difficulties the author has not yet carried out this programme. See Rjabenki and Filippow [19], p. 58, for a simple example of such an error transformation (with S = 0).

§ 3. A modified discretization of the lateral boundary conditions

Replace (1.3), (1.4) by

$$(3.1) \qquad -\beta \frac{\partial u(0,t)}{\partial x} + p\, u(0,t) = \phi(t),$$

$$(3.2) \qquad \gamma \frac{\partial u(1,t)}{\partial x} + q\, u(1,t) = \psi(t),$$

where

$$(3.3) \qquad \beta \geq 0,\ \gamma \geq 0, \qquad \beta + |p| > 0, \quad \gamma + |q| > 0,$$

$$(3.4) \qquad p > 0 \text{ if } \beta = 0, \qquad q > 0 \text{ if } \gamma = 0.$$

β, γ, p, q are assumed to be real constants (see [10] for a detailed treatment of the case of variable coefficients).

(3.1) and (3.2) with (3.3) and (3.4) comprise the three standard boundary value problems. By one - sided discretization it is possible to avoid treating the first boundary value problem separately from the second and third ones.

Replace in (3.1) $u(0, m\tau)$ by U_o^m and $\frac{\partial u(0, m\tau)}{\partial x}$ by the second order discretization $(-3U_o^m + 4U_1^m - U_2^m)/(2h)$, and in (3.2) $u(1, m\tau)$ by U_J^m and $\frac{\partial u(1, m\tau)}{\partial x}$ by $(3U_J^m - 4U_{J-1}^m + U_{J-2}^m)/(2h)$. For $1 \leq j \leq J - 1$ discretize (1.1) as in § 2. Note that we have the standard discretization of the Dirichlet problem for (1.1) if $\beta = \gamma = 0$.

Eliminating all U_o^m and U_J^m again yields a matrix-vector scheme, now for the vectors $U^m = (U_1^m, U_2^m, \ldots, U_{J-1}^m)'$. After some manipulations we find as sufficient monotonicity conditions

$$(3.5) \qquad h(|b| + a\eta) \leq a,$$

$$(3.6) \qquad \mu\Theta h\left\{\frac{\eta}{3-2\eta h}(2a + |b|h) + ch\right\} < 1,$$

$$(3.7) \qquad \mu\bar{\Theta}\,(2a + \frac{2|b|h}{3-2\eta h} - ch^2) \leq 1,$$

where $\eta = \max(-p/\beta, -q/\gamma, 0)$. We put $-p/\beta = -\infty$ if $\beta = 0$, $-q/\gamma = -\infty$ if $\gamma = 0$. Note that here the boundary values U_0^m and U_J^m may depend, but need not depend isotonically on r, g, ϕ, ψ.

Let $u \in C^{4,2}(D)$ and let (3.5), (3.6), (3.7) be fulfilled. Then, using (2.13), one can prove convergence, which is of the order $O(h^2 + \tau)$, and stability of the scheme.

§ 4. Weakly coupled parabolic systems

On $[0,1]$, for $k = 1,2,\ldots, K$, we consider the differential equations

$$(4.1) \qquad \frac{\partial u_k}{\partial t} - a_k\frac{\partial^2 u_k}{\partial x^2} - b_k\frac{\partial u_k}{\partial x} - \sum_{k'=1}^{K} c_{k,k'}\,u_{k'} = r_k(x,t),\quad a_k > 0,$$

with initial conditions

$$(4.2) \qquad u_k(x, 0) = g_k(x)$$

and lateral boundary conditions

$$(4.3) \qquad -\frac{\partial u_k(0,t)}{\partial x} + \sum_{k'=1}^{K} p_{k,k'}\,u_{k'}(0,t) = \phi_k(t),$$

$$(4.4) \qquad \frac{\partial u_k(1,t)}{\partial x} + \sum_{k'=1}^{K} q_{k,k'}\,u_{k'}(1,t) = \psi_k(t)\,.$$

The coefficients a_k, b_k, c_k, $p_{k,k'}$, $q_{k,k'}$ are real constants. In [7] Douglas has mentioned the possibility of generalizing implicit difference schemes to systems (4.1).

We can generalize the method and the results of § 2 to this problem, with a certain modification. In (4.1) we again replace the x-derivatives by linear interpolation of central difference quotients between the t-levels $n\tau$ and $(n+1)\tau$, but replace the undifferentiated $u_{k'}$ by $U_{k',j}^n$ at the lower t-level. In (4.3) and (4.4) we replace the x-derivatives by central difference quotients and the u_k by $U_{k,j}$, the difference quotients and the $U_{k,j}$ linearly interpolated between the t-levels $n\tau$ and $(n+1)\tau$, whereas we replace the other $u_{k'}$, $k' \neq k$, by $U^n_{k',j}$ at the lower t-level ($j = 0$ and $j = J$). Then we eliminate the fictitious values $U^m_{k,j}$, $j = -1$ and $j = J+1$, from the scheme.

By this trick the difference scheme splits, for a given value of n, into K subsystems, one subsystem for each index k, each to be solved for the $U^{n+1}_{k,j}$, $j = 0,1,\ldots,J$, with tridiagonal matrices:

$$(4.5) \qquad (I - \mu\Theta M_k)U_k^{n+1} - ((1+\tau c_{k,k})I + \mu\bar{\Theta}M_k)U_k^n = S_k^n,$$

$$(4.6) \qquad U^o_{k,j} = g_k(jh),$$

where $U_k^m = (U^m_{k,o}, U^m_{k,1},\ldots, U^m_{k,J})'$,

$S_k^n = (s^n_{k,o}, s^n_{k,1},\ldots, s^n_{k,J})'$ with

$$s^n_{k,o} = \tau r^{n+\Theta}_{k,o} + \sum_{k'\neq k}\left\{\tau c_{k,k'} - \mu h(2a_k - hb_k)p_{k,k'}\right\} U^n_{k',o} + \mu h(2a_k - hb_k)\,\phi_k^{n+\Theta},$$

$$s^n_{k,j} = \tau r^{n+\Theta}_{k,j} + \tau \sum_{k'\neq k} c_{k,k'} U^n_{k',j} \qquad \text{for } 1 \leq j \leq J-1,$$

$$s^n_{k,J} = \tau r^{n+\Theta}_{k,J} + \sum_{k'\neq k}\left\{\tau c_{k,k'} - \mu h(2a_k + hb_k)q_{k,k'}\right\} U^n_{k',J} + \mu h(2a_k + hb_k)\,\psi_k^{n+\Theta},$$

$$M_k = \begin{pmatrix} -2a_k - (2a_k-hb_k)hp_{k,k}, & 2a_k & & & \\ a_k - \frac{h}{2}b_k, & -2a_k, & a_k + \frac{h}{2}b_k & & \\ & \cdots\cdots & \cdots\cdots & \cdots\cdots & \\ & & a_k - \frac{h}{2}b_k, & -2a_k, & a_k + \frac{h}{2}b_k \\ & & & 2a_k, & -2a_k - (2a_k+hb_k)hq_{k,k} \end{pmatrix}.$$

The scheme has order of accuracy $O(h^2 + \tau)$ for $1 \leq j \leq J-1$, $O(h + \tau)$ for $j = 0$ and $j = J$. (There is no gain in the order of accuracy if $\Theta = 1/2$). As monotonicity conditions we obtain

$$\left.\begin{array}{ll}(4.7) & c_{k,k'} \geq 0 \\ (4.8) & p_{k,k'} \leq 0\ , \quad q_{k,k'} \leq 0\end{array}\right\} \text{ for } k' \neq k\ ,$$

$$(4.9) \qquad h \max_k |b_k|/a_k \leq 2$$

$$(4.10) \qquad -\mu\theta h \min_k \min\left\{(2a_k - hb_k)p_{k,k},\ (2a_k + hb_k)q_{k,k}\right\} < 1,$$

$$(4.11) \qquad \mu \max_k\left\{\bar{\theta}(2a_k + h \max\{0, (2a_k - hb_k)p_{k,k}, (2a_k + hb_k)q_{k,k}\}) - h^2 c_{k,k}\right\} \leq 1.$$

Using (2.13) we obtain the

Theorem: Let (4.1), (4.2), (4.3), (4.4) have a solution $u \in C^{4,2}(D)$ and let the monotonicity conditions (4.7), (4.8), (4.9), (4.10), (4.11) be fulfilled. Then

$$|U^n_{k,j} - u_k(jh, n\tau)| \leq L(h^2 + \tau)$$

for $(k, j, n) \in \{1,2,\ldots,K\} \times \{0,1,\ldots,J\} \times \{0,1,\ldots,N\}$, with a constant L independent of h and τ.

Remark: It is not difficult to modify the scheme for the case that some or all of the 2K lateral boundary conditions are replaced by Dirichlet conditions. Then the respective $p_{k,k}$ and $q_{k,k}$ are to be replaced by 0 in (4.10) and (4.11).

§ 5 Weakly coupled parabolic systems under less severe restrictions

The restrictions (4.7) and (4.8) are rather severe. They correspond to the "quasi-monotonicity" assumptions of Walter [23] for the Nagumo-Westphal lemma. We cannot expect the difference scheme to be of monotonic type if they are not satisfied. The author nevertheless thinks the scheme to be convergent in the maximum norm, if (4.7) and (4.8) are not valid. It may be that then μ has to be restricted more severely for $0 \leq \theta < 1/2$. It should be possible to relax the restrictions for μ in the case $\theta \geq 1/2$ by an eigenvalue or an energy method.

At present we can only offer a fragmentary result obtained by a standard method. Consider the problem (4.1), (4.2), (4.3), (4.4) and the scheme (4.5), (4.6). Put $w^n_{k,j} = U^n_{k,j} - u_k(jh, n\tau)$ and take, in analogy to (1.7), $\varepsilon^n_{k,j}$ as local truncation errors. Assume

$$(5.1)\qquad p_{k,k'} = q_{k,k'} = 0 \text{ for } k' \neq k, \quad p_{k,k} \geq 0 , \quad q_{k,k} \geq 0 ,$$

$$(5.2)\qquad h \max_k |b_k|/a_k \leq 2 ,$$

$$(5.3)\qquad \mu \max_k \{2\bar{\theta}\, a_k - h^2 c_{k,k}\} \leq 1 .$$

We do not assume (4.7).

Then rearrange the difference equations with the terms $(1-\mu\Theta m_{k,j,j}) \times w_{k,j}^{n+1}$ to the left, all other terms to the right of the equality signs. The $m_{k,j,j}$ are the diagonal elements of M_k. Because now the coefficients of all $w_{k,j}^n$ and $w_{k,j}^{n+1}$ are non-negative and the coefficients of the $w_{k',j}^n$ for $k' \neq k$ are $O(\tau)$, we can pass over to absolute values of the $w_{k,j}^m$. Elementary estimates and rearrangements yield

$$(5.4)\qquad ||w^{n+1}|| \leq (1 + O(\tau))||\, w^n \,|| + \tau \,||\, \varepsilon^n \,||,$$

where $\quad ||w^m|| = \max_{k,j} |\, w_{k,j}^m \,| , \quad ||\, \varepsilon^n \,|| = \max_{k,j} |\, \varepsilon_{k,j}^n \,| .$

If $u \in C^{3,2}(D)$ we have $||\, \varepsilon^n \,|| = O(h + \tau)$ and obtain convergence of the order $O(h + \tau)$ in the maximum norm from (5.4).

By this simple method we unfortunately loose a factor h in accuracy. Even for $u \in C^{4,2}(D)$ we do not get a better result. It seems very unlikely that this loss of accuracy really occurs, and a better proof remains to be researched for. Likewise the application of the scheme (4.5), (4.6) to the general problem (4.1), (4.2), (4.3), (4.4) without the restrictions (5.1) should be investigated.

References

[1] R.S. ANDERSSEN: On the reliability of finite difference representations. Australian Computer Journal 1 (1968), No. 3.

[2] I. BABUŠKA, M. PRÁGER, E. VITÁSEK: Numerical processes in differential equations. J. Wiley & Sons, London 1966.

[3] G.W. BATTEN: Second order correct boundary conditions for the numerical solution of the mixed boundary problem for parabolic equations. Math. of Comp. 17 (1963), 405 - 413.

[4] C.M. CAMPBELL and P. KEAST: The stability of difference approximations to a self-adjoint parabolic equation under derivative boundary conditions. Math. of Comp. 22 (1968), 336 - 346.

[5] L. COLLATZ: Funktionalanalysis und numerische Mathematik. Springer Verlag, Berlin 1964.

[6] E.T. COPSON and P. KEAST: On a boundary value problem for the equation of heat. J. Inst. Maths Applics 2 (1966), 358 - 363.

[7] J. DOUGLAS: A survey of numerical methods for parabolic differential equations. Advances in Computers 2 (1961), 1 - 54.

[8] R. GORENFLO: Diskrete Diffusionsmodelle und monotone Differenzenschemata für parabolische Differentialgleichungen. Methoden und Verfahren der mathematischen Physik 1 (Oktober 1969), 143 - 162. Bibliographisches Institut, Mannheim.

[9] R. GORENFLO: Monotonic difference schemes for weakly coupled systems of parabolic differential equations. J.L. Morris (editor). Conference on the numerical solution of differential equations (Dundee/Scotland 1969). Lecture Notes in Mathematics 109, 160 - 167. Springer-Verlag, Berlin 1969.

[10] R. GORENFLO: Differenzenschemata monotoner Art für lineare parabolische Randwertaufgaben. To appear in ZAMM.

[11] R. GORENFLO: Differenzenschemata monotoner Art für schwach gekoppelte Systeme parabolischer Differentialgleichungen mit gemischten Randbedingungen. To appear in Computing.

[12] E. ISAACSON: Error estimates for parabolic equations. Comm. Pure Appl. Math. 14 (1961), 381 - 389.

[13] W. KOLAR: Über allgemeine, monotone Differenzenverfahren zur Lösung des ersten Randwertproblems bei parabolischen Differentialgleichungen. Doctoral Thesis, RWTH Aachen, West Germany, 1970. Report Jül. - 672 - MA Kernforschungsanlage Jülich GmbH.

[14] R. KRAWCZYK: Über Differenzenverfahren bei parabolischen Differentialgleichungen. Arch. Rat. Mech. Anal. 13 (1963), 81 - 121.

[15] M. LEES: A priori estimates for the solutions of difference approximationes to parabolic partial differential equations. Duke Math. J. 27 (1960), 297 - 311.

[16] A.R. MITCHELL: Computational methods in partial differential equations. John Wiley & Sons, 1969.

[17] M.R. OSBORNE: The numerical solution of the heat conduction equation subject to separated boundary conditions. Computer J. 12 (1969), 82 - 87.

[18] R.D. RICHTMYER and K.W. MORTON: Difference methods for initial value problems. Second edition.Interscience (Wiley), New York 1967.

[19] V.S. RJABENKI and A.F. FILIPPOW: Über die Stabilität von Differenzengleichungen (Translated from the Russian). Deutscher Verlag der Wissenschaften, Berlin 1960.

[20] M.E. ROSE: On the integration of non-linear parabolic equations by implicit difference methods. Quarterly Appl. Math. 14 (1956), 237 - 248.

[21] P.J. TAYLOR: The stability of the Du-Fort-Frankel method for the diffusion equation with boundary conditions involving space derivatives. Computer J. 13 (1970), 92 - 97.

[22] R.S. VARGA: Matrix iterative analysis. Prentice Hall, Englewood Cliffs (New Jersey) 1962.

[23] W. WALTER: Differential- und Integralungleichungen. Springer-Verlag, Berlin 1964.

[24] B. WENDROFF: Theoretical numerical analysis. Academic Press, New York 1966.

CELL DISCRETIZATION

J. Greenstadt

1. Introduction

The method to be outlined here for solving partial differential equations is a modification and improvement of one described in a previous paper [1]. In the twelve years since that publication the methods of splines [2] and of finite elements [3] have been developed and popularized. As we shall see, the present method of "cell discretization" is rather similar to the former two methods, in that the emphasis is on subdomains rather than on nodes. However, the way in which the representation of the solution is constructed is rather different. It resembles more a method proposed by Hersch [4] in which the word "cell" was actually used.

2. Cellwise Representations

For definiteness, we shall consider a simple domain D (Fig. 1) with a boundary B. We shall then subdivide D into a set of subdomains $\{D_1, D_2, \ldots, D_K\}$ as shown, for example, in Fig. 2. For later convenience, the rest of the space is denoted by D_0.

Between each two subdomains, or cells, we assume there is one interface. For example, between cells D_k and D_m, there is F_{km}. The cell D_k may have several contiguous neighbours, viz. $\{D_{m_1}, D_{m_2}, \ldots, D_{m_P}\}$. We shall adopt the convention that if m is the label of a contiguous neighbour of D_k, then $m = \dot{m}[k]$.

The boundary B breaks up into the "boundary interfaces" $\{F_{ok}\}$, where $k = k[0]$.

The partial differential equation to be solved will involve the coordinates $\{x_i\}$ and a dependent variable $\Psi(x)$. In order to discretize, we shall represent $\Psi(x)$ in D_k by a function $\psi_k(x, \theta_k)$, whose functional form is given. The set of parameters $\{\theta_{k1}, \theta_{k2}, \ldots, \theta_{kM_k}\}$ have, on the other hand, unknown values, and we shall try to determine them so as to make $\psi_k(x, \theta_k)$ a good approximation to $\Psi(x)$ in D_k.

3. Initial Variational Formulation

For convenience we will restrict our attention to the elliptic* self adjoint equation, i.e., the Euler equation resulting from a variational problem with a quadratic functional.

$$\sum_{ij} \frac{\partial}{\partial x_i} (B^{ij} \frac{\partial}{\partial x_j} \Psi) - C \Psi = E \qquad (3.1)$$

so that the functional has the form:

$$I_0 = \int_D F(\Psi) \, dD \qquad (3.2)$$

where

$$F(\Psi) \equiv \tfrac{1}{2} \sum_{ij} B^{ij} (\frac{\partial \Psi}{\partial x_i})(\frac{\partial \Psi}{\partial x_j}) + \tfrac{1}{2} C \Psi^2 + E \Psi \qquad (3.3)$$

When we pass to the discretization, and replace $\Psi(x)$ in D_k by $\psi_k(x, \theta_k)$, the functional becomes:

$$I_0 = \sum_{k=1}^{K} \int_{D_k} F(\psi_k) \, dD_k \qquad (3.4)$$

The most convenient form for the representation $\psi_k(x,\theta_k)$ (in D_k) is a linear sum of basis functions $\{\varphi_{k\mu}(x)\}$ $(\mu = 1,\ldots,M_k)$, preassigned in D_k: so that

$$\psi_k(x,\theta_k) = \sum_{\mu=1}^{M_k} \theta_{k\mu} \varphi_{k\mu}(x). \qquad (3.5)$$

When we substitute these representations into I_0, we obtain a quadratic function of the θ's, namely:

$$I_0 = \sum_k \{\tfrac{1}{2} \sum_{\mu\nu} S_{k\mu\nu} \theta_{k\mu} \theta_{k\nu} - \sum_\mu T_{k\mu} \theta_{k\mu}\} \qquad (3.6)$$

where

$$S_{k\mu\nu} \equiv \int_{D_k} \{ \sum_{ij} B_k^{ij} (\frac{\partial}{\partial x_i} \varphi_{k\mu})(\frac{\partial}{\partial x_j} \varphi_{k\nu}) \qquad (3.7)$$

$$+ C_k \varphi_{k\mu} \varphi_{k\nu}\} \, dD_k$$

and

$$T_{k\mu} \equiv \int_{D_k} E_k \varphi_{k\mu} \, dD_k \qquad (3.8)$$

*The cell method can also be applied, in principle, to the other equation types [5].

4. Interface Conditions

As it stands now, the functional in (3.6) is simply a sum of unconnected, independent functionals, one for each cell. It is therefore necessary to "couple" the representations to one another, so as to obtain a unified representation for the function Ψ in D. This is done by means of interface conditions (with boundary conditions as a special case).

In classical treatments of second-order P.D.E.'s, the solutions are generally assumed to be continuous. This requirement would, in the present context, be expressed as follows:

$$[\psi_k]_{x \in F_{km}} - [\psi_m]_{x \in F_{km}} = 0 \tag{4.1}$$

which means that at every point on the interface F_{km}, the representations in the two contiguous cells, D_k and D_m, must match.

Since there are only a finite number of "degrees of freedom", $\{\theta_{k1},\theta_{k2},\ldots,\theta_{kM_k}\}$ in D_k - as well as in D_m, it would not in general be possible to have a match at every point on F_{km}. Hence, this requirement must be weakened. One way of doing this is to require a match only at a finite set of points $\{x_\alpha\}$ on F_{km}; this is the method of collocation [6]. We shall adopt a form of collocation for our interface conditions.

However, instead of requiring that the difference $(\psi_k - \psi_m)$ vanish on a set of points, we shall require instead that a certain set of moments (over F_{km}) of this difference vanish. Hence, we define a set of weight functions $\{\eta^\alpha_{km}(x)\}$, defined on F_{km} (with $\alpha = 1,2,\ldots,A_{km}$), and impose the conditions:

$$\int_{F_{km}} [\psi_k(x, \theta_k) - \psi_m(x, \theta_m)]\ \eta^\alpha_{km}\ dF_{km} = 0 \tag{4.2}$$

so that these conditions might be termed moment collocation.

If we now replace the representations for ψ_k and ψ_m by the expressions given by (3.5), we obtain:

$$\sum_{\mu=1}^{M_k} U^{\mu\alpha}_{km}\ \theta_{k\mu} - \sum_{\nu=1}^{M_m} U^{\nu\alpha}_{mk}\ \theta_{m\nu} = 0 \tag{4.3}$$

where

$$U^{\mu\alpha}_{km} \equiv \int_{F_{km}} \varphi_{k\mu} \, \eta^{\alpha}_{km} \, dF_{km} \tag{4.4}$$

$$U^{\nu\alpha}_{mk} \equiv \int_{F_{km}} \varphi_{m\nu} \, \eta^{\alpha}_{km} \, dF_{km} \quad . \tag{4.5}$$

This type of interface condition may readily be generalized to include differences of normal derivatives, mixed expressions, etc. For example, if we apply moment collocation to the normal derivatives of the ψ's, we obtain the same expression as in (4.3), except that the U's are defined as:

$$U^{\mu\alpha}_{km} \equiv \int_{F_{km}} \frac{\partial\varphi}{\partial n}{}_{k\mu} \, \eta^{\alpha}_{km} \, dF_{km} \tag{4.6}$$

$$U^{\nu\alpha}_{mk} \equiv \int_{F_{km}} \frac{\partial\varphi}{\partial n}{}_{m\nu} \, \eta^{\alpha}_{km} \, dF_{km} \tag{4.7}$$

the normal derivatives being taken in a common direction perpendicular to F_{km}. In fact, a wide variety of admissible interface conditions will take the same form as (4.3), provided, of course, they are linear. An inhomogeneous term may also be included, to provide for boundary conditions, for example. The general form would then be:

$$\sum_{\mu} U^{\mu\alpha}_{km} \, \theta_{k\mu} - \sum_{\nu} U^{\nu\alpha}_{mk} \, \theta_{m\nu} = W^{\alpha}_{km} \quad . \tag{4.8}$$

To condense the notation, we define matrices such as θ, S,T,U, etc. whose elements are given by the preceding formulas. These matrices will, of course, have orders appropriate to the sets of elements in question. The functional I_0 can then be expressed by:

$$I_0 = \sum_{k} \{\tfrac{1}{2} \, \theta^T_k \, S_k \, \theta_k - \theta^T_k \, T_k\} \tag{4.9}$$

and the interface conditions become:

$$U^T_{km} \, \theta_k - U^T_{mk} \, \theta_m = W_{km} \quad . \tag{4.10}$$

In view of its definition, the set $\{W_{km}\}$ is obviously antisymmetric in its indices.

The form of the boundary conditions may be readily found. We simply impose the condition that $\psi_0(x,\theta)$, defined in D_0, vanish. This means that the parameter vector θ vanishes, so that, if we set m = 0 in (4.10), we obtain:

$$U_{ko}^T \theta_k = W_{ko} \quad . \qquad (4.11)$$

W_{ko} is the discretized form, on F_{ok}, of the inhomogeneous part of the boundary condition on that boundary segment.

5. Use of Lagrange Multipliers

The interface (and boundary) conditions may be incorporated into a composite variational functional by the use of Lagrange multipliers. The result is:

$$I = I_0 + \sum_{k<m} \lambda_{km}^T (U_{km}^T \theta_k - U_{mk}^T \theta_m - W_{km}) \quad . \qquad (5.1)$$

The $\{\lambda_{km}\}$ are vectors of appropriate order $(A_{km} \times 1)$, and the summation is restricted to $k < m$ so as to avoid the redundancies consequent to the antisymmetry of the interface conditions.

The treatment which follows may be carried through on this basis, but the results will be the same if we define, for $k > m$

$$\lambda_{km} = -\lambda_{mk} \qquad (5.2)$$

and sum over all values of k and m. However, we must maintain the antisymmetry of $\{\lambda_{km}\}$. We now have:

$$I = \sum_k \{\tfrac{1}{2} \theta_k^T S_k \theta_k - \theta_k^T T_k\} \qquad (5.3)$$

$$+ \tfrac{1}{2} \sum_{k,m} \lambda_{km}^T \{U_{km}^T \theta_k - U_{mk}^T \theta_m - W_{km}\} \quad .$$

We now find the necessary conditions for a stationary I by differentiating with respect to θ_k and λ_{km}. The results are, respectively:

$$S_k \theta_k - T_k + \sum_{m[k]} U_{km} \lambda_{km} = 0 \qquad (5.4)$$

$$U_{km}^T \theta_k - U_{mk}^T \theta_m = W_{km} \qquad (5.5)$$

which are the discrete equations of the problem. Note that the summation in (5.4) is over those values of m for which D_m is contiguous to D_k. This occurs because θ's for non-contiguous cells do not appear in the interface conditions involving D_k.

6. Elimination of Lagrange Multipliers

When feasible, it is desirable to eliminate the λ's from (5.4) in order that first, the resulting "reduced" system be comparable with the conventional nodal systems and that second, this system may be recast into a form suitable for solution by relaxation (which we shall refer to as equations of "template" form).

It is not in general possible to solve for θ_k in (5.4), because S_k may be singular (as in the case of Laplace's equation!). Hence, it is necessary to solve first for the λ's. There are various algebraic difficulties which may arise, but the simplest case is that in which the set of U's associated with each cell has a property of linear independence, i.e., if the matrices $\{U_{km[k]}\}$ are collected into one matrix U_k:

$$U_k = [U_{km_1}, U_{km_2}, \ldots, U_{km_p}] \tag{6.1}$$

then the columns of U_k are linearly independent. Since the number of columns in U_{km} is A_{km}, the order of U_k is $(M_k \times A_k)$, where $A_k = \sum_{m[k]} A_{km}$. Clearly, $A_k \leq M_k$.

Under these circumstances, we can find a matrix V_k, also of order $M_k \times A_k$, with the property:

$$V_k^T U_k = I \tag{6.2}$$

where I is the unit matrix of appropriate order $(A_k \times A_k)$. A simple formula for V would be, for example:

$$V_k = U_k (U_k^T U_k)^{-1} \tag{6.3}$$

V_k, in turn, may be partitioned analogously to U_k, in terms of groupings $\{V_{km}\}$, so that we have:

$$V_{kp}^T U_{km} = \delta_{pm} I \tag{6.4}$$

where I is again a unit matrix of appropriate order, $(A_{km} \times A_{km})$ and δ_{pm} is the Kronecker delta.

We may now premultiply eqn. (5.4) by V_{kp}^T, and use (6.4) to simplify the result:

$$V_{kp}^T(S_k \theta_k - T_k) + \sum_m V_{kp}^T U_{km} \lambda_{km} \tag{6.5}$$

$$= V_{kp}^T(S_k \theta_k - T_k) + \sum_m \delta_{pm} \lambda_{km}$$

$$= V_{kp}^T(S_k \theta_k - T_k) + \lambda_{kp} = 0$$

so that λ_{kp} is given in terms of θ_k. Similarly, λ_{pk} is given by:

$$\lambda_{pk} = - V_{pk}^T(S_p \theta_p - T_p) \tag{6.6}$$

and, because of the required antisymmetry of λ_{kp}, we have a consistency condition on the θ's:

$$\lambda_{km} + \lambda_{mk} = V_{km}^T(S_k \theta_k - T_k) + V_{mk}^T(S_m \theta_m - T_m) \tag{6.7}$$

which constitutes additional interface conditions <u>induced</u> by the variation.

Since the columns of U_k may be regarded as vectors in a space of dimension M_k, and since they are linearly independent and A_k in number, they span a subspace of that dimension. Hence, there are $(M_k - A_k)$ additional vectors, also linearly independent, which lie in the subspace complementary to that of the columns of U_k (and V_k). If we arrange these vectors into a matrix Z_k, of order $M_k x(M_k - A_k)$, we have:

$$Z_k^T U_{km} = 0 \quad . \tag{6.8}$$

Now, if we premultiply eqn. (5.4) by Z_k^T, we obtain:

$$Z_k^T(S_k \theta_k - T_k) = 0 \tag{6.9}$$

because of (6.8).

The result of substituting λ_{kp} into (5.4), from (6.5) would have been a linearly dependent set of equations equivalent to (6.9).

To summarize, we have the following sets of equations linking θ_k and $\theta_{m[k]}$:

$$Z_k^T(S_k \theta_k - T_k) = 0 \tag{6.10}$$

$$U_{km}^T \theta_k = U_{mk}^T \theta_m + W_{km} \tag{6.11}$$

$$V^T_{km}(S_k\,\theta_k - T_k) = - V^T_{mk}(S_m\,\theta_m - T_m) \qquad (6.12)$$

with the proviso that (6.12) does not apply at boundary segments, but only at interfaces.

This is the reduced system we were seeking.

7. Equations in Template Form

We may interpret (6.10), (6.11) and (6.12) as equations from which we may solve for θ_k in terms of the θ's of its neighbours. Of course, we must allow the index m to range over all the neighbours of D_k. These would then be A_k equations (6.11) and, for an interior cell, the same number of the form of (6.12). Since there are $(M_k - A_k)$ equations (6.10), there are a total of $(M_k - A_k) + A_k + A_k$, or $M_k + A_k$ equations for θ_k. But since θ_k itself consists of only M_k components it would seem as if there are too many equations in D_k. This has arisen because we have, as it were, "preempted" <u>all</u> the interface equations, in which D_k is involved, for use in the solution for θ_k. This would mean that for the neighbour of D_k, these equations were no longer available. A solution to this difficulty would be to assign equations (6.11) to D_k and (6.12) to D_m, but this would inevitably lead to unsymmetric equations of template form.

Another solution to this difficulty is to rearrange (6.11) and (6.12) so as to form a new system more suitable to an equitable association of the equations with cells. We may add (and subtract) a multiple (b_{km}) of (6.12) to (6.11) to obtain:

$$U^T_{km}\,\theta_k + b_{km}\,V^T_{km}(S_k\,\theta_k - T_k) = U^T_{mk}\,\theta_m - b_{km}\,V^T_{mk}(S_m\,\theta_m - T_m) + W_{km} \qquad (7.1)$$

$$U^T_{km}\,\theta_k - b_{km}\,V^T_{km}(S_k\,\theta_k - T_k) = U^T_{mk}\,\theta_m + b_{km}\,V^T_{mk}(S_m\,\theta_m - T_m) + W_{km} \qquad (7.2)$$

If we notice that (7.2) is the same as (7.1) when k and m are interchanged (bearing in mind that $W_{mk} = - W_{km}$), we see that, by assigning (7.1) to D_k and (7.2) to D_m, we have a symmetric allocation of the original interface equations.

(Obviously, the $(A_{km} \times A_{km})$ matrix b_{km} must be the same as b_{mk}).

In the case of boundary cells, we do not have eqn. (6.12) available for its boundary faces, but, on the other hand, we do not have to share eqn. (6.11) with a neighbour cell, so that we still have an unambiguous allocation. This case may be regarded as a special case of (7.1) in which b_{ko} vanishes.

The new set of equations for θ_k is now:

$$Z_k^T S_k \theta_k = Z_k^T T_k \tag{7.3}$$

$$X_{km}^T \theta_k = Y_{mk}^T \theta_m + W_{km} + \alpha_{km} \tag{7.4}$$

where

$$X_{km}^T \equiv U_{km}^T + b_{km} V_{km}^T S_k \tag{7.5}$$

$$Y_{mk}^T \equiv U_{mk}^T - b_{km} V_{mk}^T S_m \tag{7.6}$$

$$\alpha_{km} \equiv b_{km} (V_{km}^T T_k + V_{mk}^T T_m) \tag{7.7}$$

with m ranging over all "neighbour-values". A set of equations for θ_m, identical in form with the above, is obtained by interchanging k and m, and allowing k to range over the neighbour-values of m.

Equations (7.3) and (7.4) may be solved for θ_k in terms of quantities $\{\xi_{km}\}$ and ζ_k, which have properties analogous to those of $\{V_{km}\}$ and Z_k, viz.,

$$\xi_{kp}^T X_{km} = \delta_{pm} I \tag{7.8}$$

$$\zeta_k^T X_{km} = 0 \quad . \tag{7.9}$$

After defining:

$$B_k \equiv \zeta_k (Z_k^T S_k \zeta_k)^{-1} \tag{7.10}$$

$$C_{km} \equiv [I - B_k Z_k^T S_k] \xi_{km} \tag{7.11}$$

the solution for θ_k is:

$$\theta_k = B_k Z_k^T T_k + \sum_{m[k]} C_{km}(W_{km} + \alpha_{km}) + \sum_{m[k]} C_{km} Y_{mk}^T \theta_m \quad . \tag{7.12}$$

The convergence of the relaxation iteration based on this template system naturally will depend on the choice of $\{b_{km}\}$. It is evidently very difficult to estimate the spectral radius of the associated iteration matrix of this iterative process, as a function of $\{b_{km}\}$. However, on an intuitive basis, it would seem that $\{b_{km}\}$ should be chosen so as to make $\|Y_{mk}\|$ small compared to $\|X_{km}\|$. It is in fact relatively simple to find b_{km} so as to minimize the ratio:

$$p \equiv \frac{\|Y_{km}\|^2 + \|Y_{mk}\|^2}{\|X_{km}\|^2 + \|X_{mk}\|^2} \tag{7.13}$$

(where quadratic norms are meant). However, the formulas are somewhat long,and will not be reproduced here. A value for b_{km} given by the unit matrix (of appropriate order) would result in just the sums and differences of the U_k's and (V_kS_k)'s. We shall use this value in the simple example which follows.

8. A Simple Numerical Example

With regard to numerical results, a previous variant of this method has been programmed and cases run, and although it was found possible to force convergence, it was at the cost of a rather delicate adjustment of various parameters. These parameters have been eliminated in the present scheme - or at least collected in a compact manner in $\{b_{km}\}$ - but there are as yet no program and no numerical results.

We must therefore be satisfied with exhibiting an almost trivial example in which the results were computable by hand. It will, however, demonstrate two things; (1) that the cell approximation is capable of giving good results for Ψ, and (2) that a relaxation scheme constructed on the above lines converges in a satisfactory manner.

The (one-dimensional) problem we shall solve is as follows:

$$\frac{d^2\Psi}{dx^2} = 1 \tag{8.1}$$

$$\Psi(0) = \Psi(6) = 0 \tag{8.2}$$

whose exact solution is:

$$\Psi(x) = -3x + \tfrac{1}{2}x^2 \tag{8.3}$$

We shall divide the interval [0,6] into 3 cells as shown in Fig. 3. We introduce local (intracell) coordinates u as follows:

$$\text{In } D_1 : x = u + 1 \; ; \quad \Psi = -5/2 - 2u + \tfrac{1}{2}u^2 \tag{8.4}$$

$$\text{"} \; D_2 : x = u + 3 \; ; \quad \Psi = -9/2 + \tfrac{1}{2}u^2$$

$$\text{"} \; D_3 : x = u + 5 \; ; \quad \Psi = -5/2 + 2u + \tfrac{1}{2}u^2$$

and we shall try to approximate these intracell functions with quadratic representations:

$$\psi_k(u) = \theta_{k1} + \theta_{k2}\,u + \theta_{k3}\,u^2 \tag{8.5}$$

The functional corresponding to eqn. (8.1) is:

$$I_0 = \int_0^6 \{\tfrac{1}{2}(\frac{d\Psi}{dx})^2 + \Psi\}\,dx \tag{8.6}$$

which breaks up into:

$$I_0 = \sum_{k=1}^{3} \int_{-1}^{1} \{\tfrac{1}{2}(\psi_k')^2 + \psi_k\}\,du \quad . \tag{8.7}$$

The boundary conditions are:

$$\psi_1(-1) = 0 \quad ; \quad \psi_3(1) = 0 \tag{8.8}$$

and for interface conditions, we shall impose only continuity:

$$\psi_1(1) - \psi_2(-1) = 0 \; ; \quad \psi_2(1) - \psi_3(-1) = 0 \quad . \tag{8.9}$$

When expressed in terms of the intracell representations, (8.8) becomes:

$$\theta_{11} - \theta_{12} + \theta_{13} = 0 \; ; \; \theta_{31} + \theta_{32} + \theta_{33} = 0 \tag{8.10}$$

so that

$$U_{10}^T = (1,-1,1) \; ; \; U_{30}^T = (1,1,1) \quad . \tag{8.11}$$

The interface conditions are:

$$\theta_{11} + \theta_{12} + \theta_{13} - (\theta_{21} - \theta_{22} + \theta_{23}) = 0 \tag{8.12}$$

$$\theta_{21} + \theta_{22} + \theta_{23} - (\theta_{31} - \theta_{32} + \theta_{33}) = 0$$

so that:

$$U_{12}^T = (1,1,1) \; ; \; U_{21}^T = (1,-1,1) \tag{8.13}$$

$$U_{23}^T = (1,1,1) \; ; \; U_{32}^T = (1,-1,1) \quad .$$

The V's and Z's are found to be:

$$V_{10}^T = (\tfrac{1}{2},-\tfrac{1}{2},0) \;;\; V_{12}^T = (\tfrac{1}{2},\tfrac{1}{2},0) \tag{8.14}$$

$$Z_1^T = (-1,0,1)$$

with similar expressions for $(V_{21}^T, V_{23}^T, Z_2^T)$ and $(V_{32}^T, V_{30}^T, Z_3^T)$. Also:

$$S_k = \begin{bmatrix} 0 & & \\ & 2 & \\ & & 8/3 \end{bmatrix} \quad ; \quad T_k = \begin{bmatrix} -2 \\ 0 \\ 2/3 \end{bmatrix} \tag{8.15}$$

and, denoting $V_{km}^T\, S_k$ by H_k^T, we have:

$$H_{10}^T \equiv V_{10}^T\, S_1 \;=\; (0,-1,0) \tag{8.16}$$

$$H_{12}^T \equiv V_{12}^T\, S_1 \;=\; (0,1,0)\;, \text{ etc.} \quad .$$

The induced interface conditions contain:

$$H_{12}^T\, \theta_1 + H_{21}^T\, \theta_2 = -\theta_{12} + \theta_{22} \tag{8.17}$$

$$H_{23}^T\, \theta_2 + H_{32}^T\, \theta_3 = -\theta_{22} + \theta_{32}$$

and

$$V_{12}^T\, T_1 = -1 \quad ; \quad V_{21}^T\, T_2 = -1 \tag{8.18}$$

so that

$$-\theta_{12} + \theta_{22} = -2 \quad ; \quad -\theta_{22} + \theta_{32} = -2 \quad . \tag{8.19}$$

If we now let $b_{km} = I$, as suggested previously (with $b_{ko} = 0$, of course), we obtain:

$$X_{10}^T = U_{10} \quad ; \quad X_{12}^T = U_{12}^T + H_{12}^T = (1,2,1) \tag{8.20}$$

$$Y_{12}^T = U_{12}^T - H_{12}^T = (1,0,1)$$

$$X_{21}^T = (1,-2,1) \;;\; Y_{21}^T = (1,0,1)$$

$$X_{23}^T = (1,\ 2,1) \;;\; Y_{23}^T = (1,0,1)$$

$$X_{30}^T = U_{30} \quad .$$

Hence,

$$\begin{bmatrix} \xi_{10}^T \\ \xi_{12}^T \end{bmatrix} = \begin{bmatrix} \frac{2}{3} & -\frac{1}{3} & 0 \\ \frac{1}{3} & \frac{1}{3} & 0 \end{bmatrix} \quad ; \; \zeta_1^T = (-1,0,1) \tag{8.21}$$

$$\begin{bmatrix} \xi_{21}^T \\ \xi_{23}^T \end{bmatrix} = \begin{bmatrix} \frac{1}{2} & -\frac{1}{2} & 0 \\ \frac{1}{2} & \frac{1}{2} & 0 \end{bmatrix} ; \; \zeta_2^T = (-1,0,1)$$

$$\begin{bmatrix} \xi_{32}^T \\ \xi_{30}^T \end{bmatrix} = \begin{bmatrix} \frac{1}{3} & -\frac{1}{3} & 0 \\ \frac{2}{3} & \frac{1}{3} & 0 \end{bmatrix} ; \; \zeta_3^T = (-1,0,1)$$

and

$$\alpha_{12} = \alpha_{21} = \alpha_{23} = \alpha_{32} = -2 \quad . \tag{8.22}$$

Finally:

$$B_k^T = 3/8\,(-1,0,1) \tag{8.23}$$

$$B_k\, Z_k^T\, T_k = \begin{bmatrix} -\frac{1}{2} \\ 0 \\ \frac{1}{2} \end{bmatrix}$$

and

$$(C_{10},\, C_{12}) = \begin{bmatrix} \frac{2}{3} & \frac{1}{3} \\ -\frac{1}{3} & \frac{1}{3} \\ 0 & 0 \end{bmatrix} ; \quad (C_{21},\, C_{23}) = \begin{bmatrix} \frac{1}{2} & \frac{1}{2} \\ -\frac{1}{2} & \frac{1}{2} \\ 0 & 0 \end{bmatrix} \tag{8.24}$$

etc. .

The template system is:

$$\theta_1 = \begin{bmatrix} -7/6 \\ -2/3 \\ \frac{1}{2} \end{bmatrix} + \begin{bmatrix} \frac{1}{3} & 0 & \frac{1}{3} \\ \frac{1}{3} & 0 & \frac{1}{3} \\ 0 & 0 & 0 \end{bmatrix} \times \theta_2 \tag{8.25a}$$

$$\theta_2 = \begin{bmatrix} -5/2 \\ 0 \\ \frac{1}{2} \end{bmatrix} + \begin{bmatrix} \frac{1}{2} & 0 & \frac{1}{2} \\ -\frac{1}{2} & 0 & -\frac{1}{2} \\ 0 & 0 & 0 \end{bmatrix} \times \theta_1 + \begin{bmatrix} \frac{1}{2} & 0 & \frac{1}{2} \\ \frac{1}{2} & 0 & \frac{1}{2} \\ 0 & 0 & 0 \end{bmatrix} \times \theta_3 \tag{8.25b}$$

$$\theta_3 = \begin{bmatrix} -7/6 \\ 2/3 \\ \frac{1}{2} \end{bmatrix} + \begin{bmatrix} \frac{1}{3} & 0 & \frac{1}{3} \\ -\frac{1}{3} & 0 & -\frac{1}{3} \\ 0 & 0 & 0 \end{bmatrix} \times \theta_2 \quad . \tag{8.25c}$$

We start the iteration with the initial values $\theta_1 = \theta_2 = \theta_3 = 0$. The progress of the iterations is shown in Table 1 below:

Table 1

Iter.	θ_1^T			θ_2^T			θ_3^T		
0	0,	0,	0	0,	0,	0	0,	0,	0
1	-1.17,	-.67,	.5	-2.83,	.33,	.5	-1.94,	1.44,	.5
2	-1.94,	-1.44,	.5	-3.94,	0,	.5	-2.32,	1.82,	.5
3	-2.32,	-1.82,	.5	-4.32,	0,	.5	-2.44,	1.94,	.5
4	-2.44,	-1.94,	.5	-4.44,	0,	.5	-2.48,	1.98,	.5
5	-2.48,	-1.98,	.5	-4.48,	0,	.5	-2.50,	2.00,	.5
6	-2.50,	-2.00,	.5	-4.50,	0,	.5			

The convergence factor of the iteration is about 1/3. The final parameters $\{\theta_k\}$ match those of the exact solution.

We shall now indicate a connection between the template system and the standard Laplace operator. When we refer to (8.25b), we find that:

$$\theta_{21} = -5/2 + \tfrac{1}{2}(\theta_{11} + \theta_{13}) + \tfrac{1}{2}(\theta_{31} + \theta_{33}) \tag{8.26}$$

and, since $\theta_{13} = \theta_{33} = \frac{1}{2}$ at all times:

$$\theta_{21} = -2 + \tfrac{1}{2}(\theta_{11} + \theta_{31}) \ . \tag{8.27}$$

But $\theta_{k1} = \psi_k(0)$, so that, rearranging (8.27):

$$\frac{\psi_1(0) + \psi_3(0) - 2\,\psi_2(0)}{4} = 1 \tag{8.28}$$

which is just the second <u>divided</u> central difference

$$\delta_h^2 = \frac{1}{h^2}(E + E^{-1} - 2) \tag{8.29}$$

for h = 2, which is the case here. This result suggests that there is at least a family relationship between cell discretization and finite difference discretization.

9. Discussion

Cell discretization, like all other methods, has advantages and disadvantages. Some of the advantages are:

1. The lack of ambiguity of cell relationships, viz., "nearest neighbours".

2. The possibility of retaining integral identities exactly, e.g., conservation laws.

3. The possibility of representing fairly complicated behaviour of the solution with relatively few cells.

Some of the disadvantages are:

1. Considerable calculation to prepare the reduced system.

2. Considerable calculation in the relaxation process.

3. Lack of a relaxation convergence guarantee.

4. No estimates for accuracy of approximation.

In addition, there seems to be a substantial difficulty in eliminating the λ's in the degenerate case, i.e., when the columns of U_k are linearly dependent.

Figures

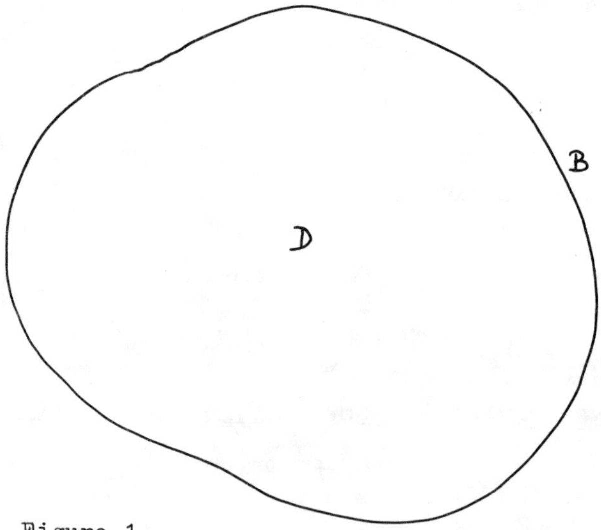

Figure 1

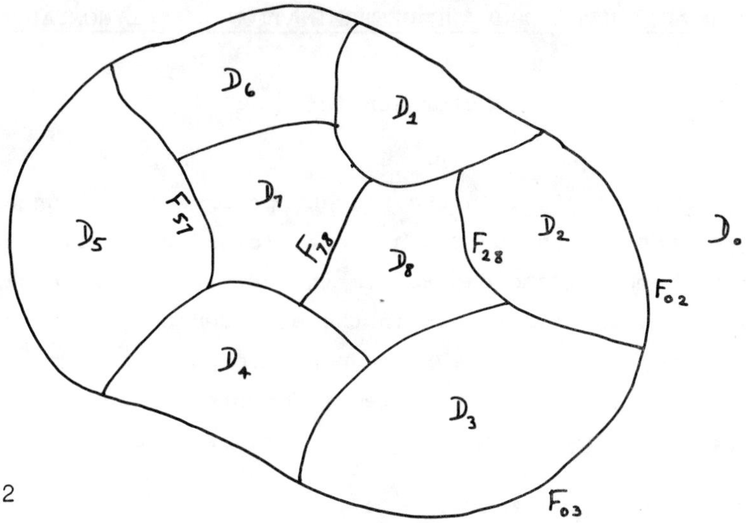

Figure 2

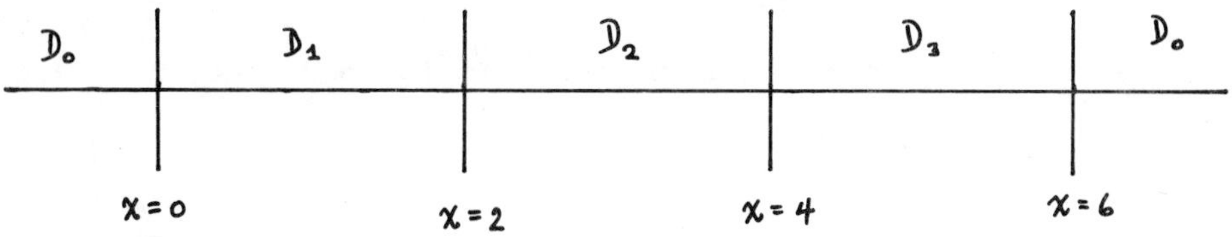

Figure 3

References

1. J. Greenstadt, "On the Reduction of Continuous Problems to Discrete Form", IBM Jour. of Res., Vol. 3, p. 355 (Oct. 1959).

2. G. Birkhoff, C. de Boor, B. Swartz and B. Wendroff, "Rayleigh-Ritz Approximation by Piecewise Cubic Polynomials", Jour. SIAM, Vol. 3, p. 188 (1966).

3. O. C. Zienkiewicz, The Finite Element Method, McGraw-Hill Ltd. London 1967.

4. J. Hersch, "Equations Differentielles et Fonctions de Cellules", C. R. Acad. Sci. Paris, Vol. 240, p. 1602 (1955).

5. J. Greenstadt, "Cell Discretization III, Treatment of Discrete Equations" IBM NY Sci. Centre Report, No. 320-2914 (Sept. 1967).

6. L. Collatz, The Numerical Treatment of Differential Equations, Springer-Verlag, 1959.

CIRCULAR ARITHMETIC AND THE DETERMINATION OF POLYNOMIAL ZEROS

Peter Henrici

1. Circular arithmetic. In the present paper we shall use lower case greek letters to denote real numbers, lower case latin letters to denote complex numbers, and capital latin letters to denote circular regions, i. e., closed point sets in the extended complex plane whose stereographic projection on the Riemann sphere is bounded by a circle. The letters i, j, k, m, n are reserved to denote nonnegative integers.

If Z is a circular region and a is a complex number, we define

$$a + Z := \{z + a : z \in Z\} ,$$

$$aZ := \{az : z \in Z\},$$

$$\frac{1}{Z} := \{ \frac{1}{z} : z \in Z\}.$$

All these sets are again circular regions. It follows that if ad - bc $\neq 0$, the set

$$\frac{aZ + b}{cZ + d}$$

is defined, and is a circular region. Moreover, if Z_1 and Z_2 are circular regions, then

$$Z_1 + Z_2 := \{z_1 + z_2 : z_1 \in Z_1, z_2 \in Z_2\}$$

is a circular region.

The above operations are easily parametrized if the circular regions involved are disks. If Z denotes the disk with center c and radius $\rho \geq 0$, we shall write

$$Z = [c; \rho], \qquad c = \text{mid } Z, \qquad \rho = \text{rad } Z.$$

Using this notation we have

$$a + Z = [a + c; \rho] ,$$

$$aZ = [ac; |a|\rho] ,$$

and if $0 \notin Z$,

$$\frac{1}{Z} = \left[\frac{\bar{c}}{c\bar{c} - \rho^2} ; \frac{\rho}{c\bar{c} - \rho^2}\right] .$$

(The bar denotes the complex conjugate.) If

$$Z_k = [c_k; \rho_k] \ , \ k = 1, 2, \dots , m,$$

then

$$\sum_{k=1}^{m} Z_k = \left[\sum_{k=1}^{m} c_k; \sum_{k=1}^{m} \rho_k\right] .$$

The above arithmetic for circular regions extends the so-called interval arithmetic [5, 6, 7] from the real line to the complex plane.

2. The problem. In this paper we wish to announce a solution to the following

PROBLEM: Let m and n be integers, $0 < m < n$, let z_0 be a complex number, and let $C_1, C_2, \dots , C_m; W_{m+1}, W_{m+2}, \dots , W_n$ be circular regions. Let the following be known about a polynomial p of degree n with zeros $w_1, w_2, \dots , w_n$:

(i) $w_k \in W_k$, $k = m + 1, \dots , n$;

(ii) the values of the functions

$$q_k(z) := \frac{1}{(k-1)!} \left(\frac{p'(z)}{p(z)}\right)^{(k-1)} , \ k = 1, \dots , m$$

at the point z_0 satisfy

$$q_k(z_0) \in C_k \ , \ k = 1, \dots , m.$$

What can be said about the location of the zeros $w_1, w_2, \dots , w_m$ of p?

The above question is motivated by certain algorithms for the simultaneous determination of all zeros of a polynomial which we have developed in earlier papers [2, 3]. Given any polynomial p of degree n, these algorithms produce sequences of arrays $\underline{W} = (W_1, W_2, \dots , W_n)$ of circular disks whose diameters tend to zero such that the zeros w_1, $\dots$, w_n of p at each stage satisfy $w_k \in W_k$, $k = 1, \dots , n$. Since these algorithms are costly in terms of arithmetical operations and since the convergence is at best linear, it is desirable to speed up convergence as soon as crude approximations to the zeros have been determined. For this it seems natural to use Newton's method, which only requires the computation of the Newton correction

$$-\frac{p(z)}{p'(z)} = -\frac{1}{q_1(z)} .$$

If allowance is made for an imprecise calculation of the Newton correction, say due to rounding errors, there results precisely the case $m = 1$ of the above problem. The case $m > 1$ similarly corresponds to the situation where m zeros are clustered close together and m derivatives are available to approximate them.

3. Solution of the problem in the case $m = 1$. Here the problem posed in §2 is solved completely by

THEOREM 1. *Let* z_0 *be a complex number, and let* C_1; $W_2, W_3, \ldots, W_n$ *be circular regions. Then the set of possible values of the zero* w_1 *of a polynomial* p *of degree* n *such that*

$$\frac{p'(z_0)}{p(z_0)} \in C_1$$

whose remaining zeros $w_2, \ldots, w_n$ *satisfy*

$$w_k \in W_k \;, \; k = 2, 3, \ldots, n,$$

is precisely the circular region

$$W_1 := z_0 - \frac{1}{C_1 - R_1} \,,$$

where

$$R_1 := \sum_{k=2}^{n} \frac{1}{z_0 - W_k} \,.$$

This result is a slight extension of Theorem 3a of a joint paper with I. Gargantini [1], and the proof runs along similar lines. By way of interpretation, we note that if the Newton correction is known exactly,

$$C_1 = \frac{p'(z_0)}{p(z_0)} \,,$$

and the formula defining W_1 may be written

$$W_1 = z_0 - \frac{p(z_0)}{p'(z_0) - p(z_0)R_1} \,.$$

If Newton's method were applied at the point z_0, there would result the new approximation $z_1 := z_0 - p(z_0)/p'(z_0)$. If z_0 is not contained in any of the disks $W_2, \ldots, W_n$, then R_1 is a disk. If, furthermore, z_0 is close to w_1, then $|p(z_0)|$ is small, and $p(z_0)R_1$ is a small disk close to the origin. Hence W_1 is a small disk near $z = z_1$, and the fact that $w_1 \in W_1$ may be looked at as a more precise version of Newton's method, obtained on the basis of information concerning the approximate location of the remaining zeros.

The following result due to Laguerre [4] is readily shown to be a corollary of Theorem 1:

THEOREM 2. *Let* p *be a polynomial of degree* n*, and let* z *be any complex number. Then any circular region whose boundary passes through the points* z *and* $z - n\,p(z)/p'(z)$ *contains at least one zero of* p.

4. *Newton's method with error bounds*. Here we apply Theorem 1 to the problem of refining a single zero of a polynomial. Let p be a polynomial of degree n, and let it be known that the disk $W_0 = [z_0; \varepsilon_0]$ contains precisely one zero, say w_1, of p, while no other zero is contained in the open disk $|z - z_0| < \rho$ where $\rho > \varepsilon_0$. Denoting by $U = \{z : |z - z_0| \geq \rho\}$ the set containing the remaining zeros $w_2, \ldots, w_n$, Theorem 1 suggests the following Newton-like algorithm for determining w_1 (the subscript in W_j now refers to the iteration step, and not to the zero contained in W_j):

For $j = 0, 1, 2, \ldots$, *let*

$$z_j = \text{mid}\, W_j ,$$

$$R_j = \frac{n-1}{z_j - U} ,$$

$$W_{j+1} = z_j - \frac{1}{q(z_j) - R_j} ,$$

where $q(z) = p'(z)/p(z)$.

By virtue of Theorem 1 w_1 is contained in all disks W_j thus constructed. Concerning convergence we have

THEOREM 3. *If* $6\varepsilon_0 < \eta$, *where* $(n-1)\eta = \rho$, *the numbers* $\varepsilon_j :=$ rad W_j *satisfy*

$$\frac{\varepsilon_{j+1}}{\eta} = 2\left(\frac{\varepsilon_j}{\eta}\right)^2 , \quad j = 0, 1, 2, \ldots ,$$

and hence converge to zero quadratically.

For the proof see [1].

5. *The simultaneous refinement of all zeros*. We may hope to do better than in Theorem 3 by iterating on all zeros simultaneously. Assume we have found an array $\underline{W}_0 = (W_{0,1}, W_{0,2}, \ldots, W_{0,n})$ of n non-overlapping disks such that the zeros w_k of the polynomial p of degree n satisfy $w_k \in W_k$, $k = 1, 2, \ldots, n$. Theorem 1 in this situation suggests the following algorithm for constructing a sequence of arrays $\underline{W}_j = (W_{j,1}, W_{j,2}, \ldots, W_{j,n})$ of disks covering the zeros:

<u>For each</u> $j = 0, 1, 2, \ldots$, <u>let for</u> $k = 1, 2, \ldots, n$

$$z_{j,k} = \text{mid } W_{j,k} ,$$

$$R_{j,k} = \sum_{\substack{m=1 \\ m \neq k}}^{n} \frac{1}{z_{j,m} - W_{j,k}} ,$$

$$W_{j+1,k} = z_{j,k} - \frac{1}{q(z_{j,k}) - R_{j,k}} ,$$

<u>where</u> $q(z) = p'(z)/p(z)$.

By Theorem 1, each zero w_k is contained in all disks $W_{j,k}$. To describe the convergence properties of the algorithm, we let

$$\varepsilon_j := \max_k \text{ rad } W_{j,k}$$

and define η by $(n - 1)\eta = \rho$ where

$$\rho := \min_{k \neq m} \{ |z_{0,k} - z| : z \in W_{0,m} \} .$$

We then have:

THEOREM 4. <u>If</u> $6\varepsilon_0 < \eta$, <u>the numbers</u> ε_j <u>satisfy</u>

$$\frac{\varepsilon_{j+1}}{\eta} \leqq 3 \left(\frac{\varepsilon_j}{\eta}\right)^3 , \quad j = 0, 1, 2, \ldots ,$$

<u>and thus converge to zero cubically</u>.

For the proof see again [1].

The above algorithm defines a "Gesamtschrittverfahren" inasmuch the new disks $W_{j+1,k}$ can be computed simultaneously. Convergence may improved even further if the algorithm is changed into an "Einzelschrittverfahren" by working with the circular regions

$$R'_{j,k} := \sum_{m=1}^{k-1} \frac{1}{z_{j,m} - W_{j+1,k}} + \sum_{m=k+1}^{n} \frac{1}{z_{j,m} - W_{j,k}}$$

in place of the regions $R_{j,k}$.

6. <u>A solution for the case</u> $m > 1$. In order to solve the problem stated in §2 in the general case, we have to define products of circular disks. If

$$Z_k = [c_k; \rho_k] \quad , k = 1, 2,$$

we put

$$Z_1 Z_2 := [c_1 c_2; |c_1|\rho_2 + |c_2|\rho_1 + \rho_1\rho_2] .$$

This product is commutative and associative. The circular disk Z_1Z_2 contains the set $\{z_1z_2 : z_1 \in Z_1, z_2 \in Z_2\}$, but is in general not contained in it. The distributive law holds in the form

$$Z_1(Z_2 + Z_3) \subset Z_1Z_2 + Z_1Z_3 .$$

Using the notation of §2, we now have the following result.

THEOREM 5. *Let* $z_0 \notin W_k$, $k = m+1, \ldots , n$. *For* $k = 1, 2, \ldots , m$, *let*

$$S_k := C_k + \sum_{j=m+1}^{n} \frac{1}{(W_j + z_0)^k} ,$$

and define the disks $B_0, B_1, \ldots , B_m$ *recursively by* $B_0 = [1; 0]$,

$$B_k := \frac{1}{k}(S_k + S_{k-1}B_1 + \ldots + S_1B_{k-1}) , \quad k = 1, \ldots , m.$$

Then the zeros $w_1, \ldots , w_m$ *of any polynomial* p *satisfying conditions* (i) *and* (ii) *of* §2 *are of the form*

$$w_k = z_0 + r_k , \quad k = 1, \ldots, m,$$

where $r_1, \ldots , r_m$ *are the zeros of a polynomial* $t(r) = 1 + b_1r + \ldots + b_mr^m$ *such that*

$$b_k \in B_k , \quad k = 1, \ldots , m.$$

The *proof* is very similar to the proof of Theorem 5b of [1].

For m = 1 Theorem 5 reduces precisely to Theorem 1. If rad $C_k = 0$, $k = 1, \ldots , m$, and rad $W_k = 0$, $k = m + 1, \ldots , n$, then rad $B_k = 0$, $k = 1, \ldots , m$, and Theorem 5 yields the *exact* zeros. In the general case the deviation of the numbers r_k from the zeros, say, of

$$t(r) = 1 + \sum_{k=1}^{m} (\text{mid } B_k)\, r^k$$

can be estimated in terms of max rad B_k , using results of Ostrowski [8].

References

[1] I. Gargantini and P. Henrici: Circular arithmetic and the determination of polynomial zeros. Submitted for publication.

[2] P. Henrici: Uniformly convergent algorithms for the simultaneous determination of all zeros of a polynomial. Studies in Numerical Analysis 2, 1 - 8 (1968).

[3] P. Henrici and I. Gargantini: Uniformly convergent algorithms for the simultaneous approximation of all zeros of a polynomial. Proc. Symp. on Constructive Aspects of the Fundamental Theorem of Algebra (B. Dejon and P. Henrici, eds.), Wiley-Interscience, London 1969, pp. 77 - 114.

[4] E. Laguerre: Sur la résolution des équations numériques. Nouvelles Annales des Mathématiques, sér. 2, 17 (1878).

[5] R. E. Moore: Interval Analysis. Prentice Hall, Englewood Cliffs 1966.

[6] K. Nickel: Ueber die Notwendigkeit einer Fehlerschrankenarithmetik bei Rechenautomaten. Numer. Math. 9, 69 - 79 (1966).

[7] J. Rokne and P. Lancaster: Complex Interval Arithmetic. Comm. Assoc. Comp. Mach. 14, 111 - 112 (1971).

[8] A. M. Ostrowski: Solution of Equations and Systems of Equations, 2nd ed. Academic Press, New York 1966.

COMPUTATION OF THE MOMENTS OF SOLUTIONS OF CERTAIN RANDOM TWO POINT BOUNDARY VALUE PROBLEMS

R. Huss and R. Kalaba

SUMMARY

Assume that the linear two-point boundary value problem

$$\ddot{x} + [p(t) + \lambda\, q(t)]\, x = -g(t), \qquad 0 \le t \le 1,$$

$$x(0) = 0, \qquad x(1) = c$$

possesses a unique solution for all λ in the interval $0 \le \lambda \le \Lambda$. Consider λ to be a random variable with probability density function $f(\lambda)$, $0 \le \lambda \le \Lambda$. A method for determining the moments

$$E[x^n(t,\lambda)] = \int_0^\Lambda x^n(t,\lambda)\, f(\lambda)\, d\lambda\,, \qquad n = 1, 2, \cdots,$$

is presented. Numerical experiments show the computational feasibility of the new approach.

1. INTRODUCTION

In the analysis of various physical phenomena, one is faced with solving two-point boundary value problems having stochastic aspects. Equations of this nature arise, for example, in the analysis of wave propagation through inhomogeneous media. In this paper we shall present a numerical technique for obtaining the statistical moments of the solutions of a class of such random differential equations through the application of invariant imbedding.

There exist numerous numerical schemes for the solution of initial value problems; see Bekey [2] and Berezin [4]. Many such schemes possess the requisite stability properties to be useful from the practical computational standpoint. It is therefore of potential benefit to transform systems of functional equations into Cauchy systems. Among the problems that have been treated in this way are two-point boundary value problems for ordinary differential equations (Bellman and Kalaba [3] and Kagiwada and Kalaba [13]), integral equations (Kagiwada and Kalaba [12] and Casti and Kalaba [6]), variational problems (Casti, et al. [7] and Kagiwada, et al. [8], and potential problems (Kalaba and Ruspini [15] and Buell, et al. [5]). However, computational schemes for the treatment of random differential equations are not so readily available (Adomian [1]). We shall show how one such class of equations, however, can be handled, and we shall also present numerical results.

In this study we regard a parameter in a two-point boundary value problem as a new independent variable, and then transform the original two-point boundary

---

The authors were supported by the National Institutes of Health under Grants Nos. GM-16437-02 and GM-01724-05.

problem into a Cauchy system. In particular, a Cauchy system for the Green's function (Courant and Hilbert [8]) is obtained. The validity of this technique has been demonstrated previously (Huss, *et al.* [9, 10]). In this paper, the method is extended to cover a stochastic two-point boundary value problem. A numerical scheme is developed for obtaining not only the statistical moments of the solution of the equation, but also the moments of the Green's function.

An alternative Cauchy system for a Green's function is given in Kagiwada and Kalaba [11].

2. A TWO-POINT BOUNDARY VALUE PROBLEM

Consider the linear two-point boundary value problem

(2.1) $\ddot{x} + [p(t) + \lambda q(t)]x = -g(t), \qquad 0 \leq t \leq 1,$

(2.2) $x(0) = 0,$

(2.3) $x(1) = c.$

Preliminarily λ is viewed as a constant; later it will be considered a random variable. We assume that a unique solution exists for all λ in the interval $(0, \Lambda)$.

Introduce the auxiliary functions u and w that are the solutions of the linear two-point boundary value problems

(2.4) $\ddot{u} + [p(t) + \lambda q(t)]\,u = 0, \qquad 0 \leq t \leq 1,$

(2.5) $u(0) = 0,$

(2.6) $u(1) = 1;$

(2.7) $\ddot{w} + [p(t) + \lambda q(t)]\,w = -g(t), \qquad 0 \leq t \leq 1,$

(2.8) $w(0) = 0,$

(2.9) $w(1) = 0.$

It is clear that the desired function x can be written in terms of the functions u and w in the form

(2.10) $x = w + cu, \qquad 0 \leq t \leq 1.$

In the remaining sections we shall be concerned only with the function w. The determination of the function u closely parallels these procedures, as has previously been shown (Huss [9, 10]).

3. DERIVATION OF THE CAUCHY SYSTEM

We shall first consider a special case of Eq. (2.7), where we let $p(t) \equiv 0$. In a later section, the general case will be treated. The system for the function w is now written as

(3.1) $\ddot{w} + \lambda q(t)\,w = -g(t), \qquad 0 \leq t \leq 1,$

(3.2) $w(0) = 0, \qquad 0 \leq \lambda \leq \Lambda,$

(3.3) $w(1) = 0,$

where λ is a parameter.

In this discussion, w will be regarded as a function of λ as well as t, and we shall therefore write, when needed,

$$w = w(t, \lambda), \qquad 0 \le t \le 1, \quad 0 \le \lambda \le \Lambda . \tag{3.4}$$

Differentiate both sides of Eq. (3.1) with respect to λ to obtain

$$(w_\lambda)'' + q(t)w + \lambda q(t) w_\lambda = 0. \tag{3.5}$$

Here we regard w_λ as a new function of t and λ,

$$w_\lambda = w_\lambda(t, \lambda), \qquad 0 \le t \le 1, \quad 0 \le \lambda \le \Lambda .$$

The boundary conditions are

$$w_\lambda(0) = w_\lambda(0, \lambda) = 0 \tag{3.6}$$

and

$$w_\lambda(1) = w_\lambda(1, \lambda) = 0. \tag{3.7}$$

Next introduce the Green's function G, in terms of which the solution of Eqs. (3.1)-(3.3) is

$$w(t, \lambda) = \int_0^1 G(t, y, \lambda)\, g(y)\, dy, \qquad 0 \le t \le 1, \quad 0 \le \lambda \le \Lambda . \tag{3.8}$$

In terms of this Green's function, the solution of Eqs. (3.5)-(3.7) is

$$w_\lambda(t, \lambda) = \int_0^1 G(t, y, \lambda)\, q(y)\, w(y, \lambda)\, dy , \qquad 0 \le t \le 1, \quad 0 \le \lambda \le \Lambda . \tag{3.9}$$

Eq. (3.9) is the first basic differential equation of the Cauchy system. It is an equation for the function w. The initial condition at $\lambda = 0$ is

$$w(t, 0) = \int_0^1 G(t, y, 0)\, g(y)\, dy , \tag{3.10}$$

while the Green's function at $\lambda = 0$ is

$$G(t, y, 0) = \begin{cases} y(1-t), & 0 \le y \le t , \\ t(1-y), & t \le y \le 1. \end{cases} \tag{3.11}$$

Next, we shall derive a Cauchy system for the Green's function G. We have already seen that

$$w_\lambda(t, \lambda) = \int_0^1 G(t, y, \lambda)\, q(y)\, w(y, \lambda)\, dy . \tag{3.12}$$

Eq. (3.8) is an expression for $w(y, \lambda)$. This is substituted in Eq. (3.12) to yield

(3.13) $$w_\lambda(t, \lambda) = \int_0^1 G(t, y, \lambda)\, q(y) \int_0^1 G(y, y', \lambda)\, g(y')\, dy'\, dy.$$

This equation shows the explicit dependence of the function w_λ on the forcing function g.

A second expression for w_λ can be obtained by differentiating both sides of Eq. (3.8) with respect to λ. This operation yields

(3.14) $$w_\lambda(t, \lambda) = \int_0^1 G_\lambda(t, y, \lambda)\, g(y)\, dy\, .$$

Eqs. (3.13) and (3.14) are expressions for w_λ and must therefore be equal. It therefore follows, since the function g is arbitrary, that

(3.15) $$G_\lambda(t, y, \lambda) = \int_0^1 G(t, y', \lambda)\, q(y')\, G(y', y, \lambda)\, dy'\, .$$

This is the second basic differential equation of the Cauchy system, and is interpreted as a differential equation for G. The initial condition on the function G at $\lambda = 0$ is given in Eq. (3.11).

The necessary differential equations and initial conditions for the functions w and G have now been determined. In the following section we shall state the Cauchy system and discuss the statistical aspects of the problem.

4. DETERMINATION OF STATISTICAL MOMENTS

The functions w and G are now defined to be solutions of the differential equations

(4.1) $$w_\lambda(t, \lambda) = \int_0^1 G(t, y', \lambda)\, q(y')\, w(y', \lambda)\, dy'\, ,$$

(4.2) $$G_\lambda(t, y, \lambda) = \int_0^1 G(t, y', \lambda)\, q(y')\, G(y', y, \lambda)\, dy'\, ,$$

$$0 \le t \le 1,$$

$$0 \le \lambda \le \Lambda\, .$$

The initial conditions at $\lambda = 0$ are

(4.3) $$w(t, 0) = \int_0^1 G(t, y', 0)\, g(y')\, dy',$$

(4.4) $$G(t, y, 0) = \begin{cases} y(1-t), & 0 \le y \le t, \\ t(1-y), & t \le y \le 1\, . \end{cases}$$

We assume that this Cauchy system has a unique solution for $0 \le \lambda \le \Lambda$, that the function w has a continuous second derivative with respect to t, and that the function G is continuous in t but has a jump discontinuity in $\dot{G}$ at $t = y$. The Cauchy system of Eqs. (4.1)-(4.4) is a solution of the boundary value problem of Eqs. (3.1)-(3.3), and G is indeed the Green's function (Kagiwada and Kalaba [13] and Kagiwada, Kalaba and Thomas [14]).

We now consider the statistical aspects of the problem. The parameter λ is now regarded as a random variable with a known density function. For each value of λ there exists a specific solution of the boundary value problem of Eqs. (3.1)-(3.3). We can express the average solution, therefore, in the form

$$(4.5) \qquad E[w(t, \lambda)] = \int_0^{\Lambda} w(t, \lambda)\, f(\lambda)\, d\lambda\,, \qquad 0 \le t \le 1.$$

Here $f(\lambda)$ is the assumed known density function of λ. Higher statistical moments can similarly be expressed as

$$(4.6) \qquad E[w^n(t, \lambda)] = \int_0^{\Lambda} w^n(t, \lambda)\, f(\lambda)\, d\lambda\,, \qquad n = 2, 3, \cdots.$$

In general, the solution $w(t, \lambda)$ of the boundary value problem cannot be found analytically. Thus, Eqs. (4.5) and (4.6) are not ordinarily directly usable. However, since the Cauchy system of Eqs. (4.1)-(4.4) uses λ as the variable of integration, the moments are obtainable directly by adjoining to this Cauchy system additional equations for the statistical moments.

Introduce the new functions m and v defined by

$$(4.7) \qquad m(t, \lambda) = \int_0^{\lambda} w(t, r)\, f(r)\, dr,$$

$$(4.8) \qquad v(t, \lambda) = \int_0^{\lambda} [w(t, r)]^2\, f(r)\, dr,$$

$$0 \le t \le 1,$$

$$0 \le \lambda \le \Lambda\,.$$

When λ attains the value Λ, m is the mean value and v the mean squared value of $w(t)$ over $[0, \Lambda]$. If we differentiate Eqs. (4.7) and (4.8) with respect to λ, we obtain the desired equations,

$$(4.9) \qquad m_\lambda(t, \lambda) = w(t, \lambda)\, f(\lambda)\,,$$

$$(4.10) \qquad v_\lambda(t, \lambda) = [w(t, \lambda)]^2\, f(\lambda)\,,$$

with initial conditions

$$(4.11) \qquad m(t, 0) = 0\,,$$

$$(4.12) \qquad v(t, 0) = 0\,.$$

Eqs. (4.9) and (4.10) are adjoined to the Cauchy system of Eqs. (4.1)-(4.4). When the independent variable λ in the numerical integration procedure reaches Λ, $m(t, \Lambda)$ will be the mean value of $w(t)$, and $v(t, \Lambda)$ will be its mean square. The variance, $s^2(t)$, is found through use of the standard formula

$$(4.13) \qquad s^2(t) = v(t, \Lambda) - [m(t, \Lambda)]^2\,.$$

In a similar fashion one can find the average Green's function using the system

$$(4.14) \qquad Y_\lambda(t, y, \lambda) = G(t, y, \lambda)\, f(\lambda)\,,$$

$$(4.15) \qquad Y(t, y, 0) = 0.$$

For $\lambda = \Lambda$, $Y(t, y, \Lambda)$ will be the average Green's function. Higher moments can also be found in this way.

We consider now the more general case of Eqs. (2.7)-(2.9). We rewrite them here for convenience,

$$\ddot{w} + [p(t) + \lambda q(t)] w = -g(t) , \quad 0 \leq t \leq 1, \quad 0 \leq \lambda \leq \Lambda , \tag{4.16}$$

$$w(0) = 0 , \tag{4.17}$$

$$w(1) = 0. \tag{4.18}$$

Proceeding as before, differentiate Eq. (4.16) with respect to λ to obtain

$$(w_\lambda)'' + [p(t) + \lambda q(t)] w_\lambda + q(t)w = 0 , \quad 0 \leq t \leq 1, \quad 0 \leq \lambda \leq \Lambda , \tag{4.19}$$

$$w_\lambda(0, \lambda) = 0 , \tag{4.20}$$

$$w_\lambda(1, \lambda) = 0 . \tag{4.21}$$

As in the previous case, w_λ again satisfies the same homogeneous equation and boundary conditions as w, hence, has the same Green's function. The Cauchy system for this case is precisely the same as that for the previous case, except for the initial conditions. For $\lambda = 0$, the function $w(t, 0)$ and the Green's function $G(t, y, 0)$ are associated with the problem

$$\ddot{w} + p(t)w = -g(t), \quad 0 \leq t \leq 1, \tag{4.22}$$

$$w(0) = 0 , \tag{4.23}$$

$$w(1) = 0 . \tag{4.24}$$

If we introduce a new parameter μ into Eq. (4.22) we have

$$\ddot{w} + \mu p(t)w = -g(t), \quad 0 \leq t \leq 1, \quad \mu = 1 . \tag{4.25}$$

But this is precisely the problem of Eqs. (3.1)-(3.3) for which a Cauchy system has been derived, with $p(t)$ replacing $q(t)$.

To solve the boundary value problem of Eqs. (4.14)-(4.16), then, we first establish a Cauchy system for Eqs. (4.22)-(4.24) and allow μ to vary from 0 to 1. Using the values of $w(t)$ and $G(t, y)$ resulting from this operation as initial conditions, we establish a second Cauchy system for Eqs. (3.1)-(3.3), and adjoin to it Eqs. (4.9)-(4.12) for m and v, and integrate this system numerically from 0 to Λ.

5. COMPUTATIONAL TECHNIQUES

One numerical technique that lends itself readily to the numerical evaluation of the function w and the Green's function G is the well-known method of lines (Berezin [4]). In this method, Eqs. (4.1)-(4.2) are integrated over $[0, \Lambda]$ along lines of constant values of t. The Cauchy system for w contains as many differential equations as there are values of t. At each step in the numerical integration, the right-hand sides of these equations are evaluated by an appropriate quadrature formula using the values of w at each of the lines of constant t. The type and order of the quadrature formula dictate the locations of the lines of constant t.

Several numerical examples will now be presented. In these examples, Simpson's rule of various interval sizes is used for the evaluation of the definite integrals, and a fourth order Adams-Moulton procedure is used to integrate the differential equations. All numerical values refer to results using an IBM 360-44 digital computer run in single precision.

6. SOME NUMERICAL RESULTS

The equation selected to test the numerical technique is Eq. (3.1), with $q(t) = +1$ and $g(t) = +1$. The system is

$$(6.1) \qquad \ddot{w} + \lambda w = -1 , \qquad 0 \le t \le 1 , \quad 0 \le \lambda \le \Lambda < \pi^2,$$

$$(6.2) \qquad w(0) = 0 ,$$

$$(6.3) \qquad w(1) = 0 .$$

The solution for $w(t, \lambda)$ is easily found to be

$$(6.4) \qquad w(t, \lambda) = \frac{\sin(t\sqrt{\lambda}) - \sin(\sqrt{\lambda}) + \sin\left[(1-t)\sqrt{\lambda}\right]}{\lambda \sin(\sqrt{\lambda})} .$$

In order to find the true (analytic) average function, we must evaluate

$$(6.5) \qquad m(t, \Lambda) = \int_0^{\Lambda} w(t, \lambda)\, f(\lambda)\, d\lambda .$$

A density function $f(\lambda)$ may be chosen such that Eq. (6.5) can be readily evaluated in closed form. This density function is

$$(6.6) \qquad f(\lambda) = A(\Lambda)\sqrt{\lambda}\, \sin\sqrt{\lambda} ,$$

where $A(\Lambda)$ is determined from the condition

$$(6.7) \qquad \int_0^{\Lambda} A(\Lambda)\sqrt{\lambda}\, \sin\sqrt{\lambda}\, d\lambda = 1 .$$

The average function is evaluated numerically using several quadrature orders and step sizes. In general, for reasonably well-behaved functions, it is found that the solutions are accurate to three significant figures at $\Lambda = 5.0$ using twenty interval quadrature and a step size of 0.05 in λ.

The second moment is evaluated numerically.

TABLE 1.

Analytic and Numerical Values of the Average Function $m(t, \Lambda)$ for $\Lambda = 5$.

t	Analytic Value	Numerical Values* N = 20, $\Delta\lambda$ = 0.05	Numerical Values* N = 10, $\Delta\lambda$ = 0.05
0.0	0.0	0.0	0.0
0.1	0.064605	0.064697	0.067030
0.2	0.117161	0.117330	0.117848
0.3	0.155971	0.156199	0.159459
0.4	0.179774	0.180039	0.180850
0.5	0.187796	0.188073	0.191663
0.6	0.179774	0.180039	0.180850
0.7	0.155971	0.156199	0.159459
0.8	0.117161	0.117330	0.117848
0.9	0.064605	0.064697	0.067030
1.0	0.0	0.0	0.0

* N = Number of intervals in Simpson's rule,

$\Delta\lambda$ = Step size in λ.

It is of interest also to illustrate the folly of using the average value of λ over $[0, \Lambda]$ and presuming that the solution of Eq. (6.1) using this value, will be the average function. The average value of λ for $\Lambda = 5$ is found, using the density function of Eq. (6.6), to be 3.01848. The function $w(t)$, using this value of λ, is obtained directly by substitution in Eq. (6.4). Table 2 shows these results compared with the analytical values of the average. For other equations or density functions, the discrepancy could, of course, be greater or less than this.

7. DISCUSSION

We have indicated a method for determining the statistical moments of solutions of certain two-point boundary value problems. We have also demonstrated, with examples, that the method will produce the true values of this function, to a precision determined by the order of quadrature. We have also indicated a method for evaluating the average Green's function.

It is not intended that the particular integration schemes (i.e., Simpson's rule, Adams-Moulton) used in this paper are an essential part of the method. They are selected because of their ease of implementation. In view of the fact that the first derivative of the Green's function exhibits a discontinuity, it is no doubt the case that other numerical methods will yield more precise results in fewer steps. The selection of the method should be based on the equation to be solved. Using the methods of this paper one may expect two figures of accuracy using twenty interval Simpson's rule integration and a step size of 0.05 in a fourth order Adams-Moulton integration scheme.

TABLE 2.

Comparison of True Average Solution of $\ddot{w} + \lambda w = -1$ with Soution Using Average Value ($\bar{\lambda}$) of λ, for $\Lambda = 5$.

$\bar{\lambda} = 3.01848$

t	True Solution $\bar{w}(t, \lambda)$	$w(t, \bar{\lambda})$
0.0	0.0	0.0
0.1	0.064605	0.06271
0.2	0.117161	0.11357
0.3	0.155971	0.15102
0.4	0.179774	0.17396
0.5	0.187796	0.18168
0.6	0.179774	0.17396
0.7	0.155971	0.15102
0.8	0.117161	0.11357
0.9	0.064605	0.06271
1.0	0.0	0.0

REFERENCES

1. Adomian, G., J. Math. Phys. 11, 1069-1084 (1970).

2. Bekey, G.A. and W. Karplus, Hybrid Computation (John Wiley & Sons, New York, 1968).

3. Bellman, R. and R. Kalaba, Proc. Nat. Acad. Sci. USA 47, 336-338 (1961).

4. Berezin, I.S. and N.P. Zhidkhov, Computing Methods (Addison-Wesley Publishing Company, Palo Alto, 1965).

5. Buell, J., R. Kalaba and E. Ruspini, Int. J. Engin. Sci. 7, 1167-1172 (1969).

6. Casti, J. and R. Kalaba, Information Sciences 2, 51-67 (1970).

7. Casti, J., R. Kalaba and B. Vereeke, J. Opt. Th. Applic. 3, 81-88 (1969).

8. Courant, R. and D. Hilbert, Methods of Mathematical Physics (Interscience Publishers, New York, 1953).

9. Huss, R. and R. Kalaba, "Invariant Imbedding and the Numerical Determination of Green's Functions", J. Opt. Th. Applic., in press.

10. Huss, R., H. Kagiwada and R. Kalaba, "A Cauchy System for the Green's Function and the Solution of Two-Point Boundary Value Problems", J. Franklin Institute, in press.

11. Kagiwada, H. and R. Kalaba, J. Opt. Th. Applic. 1, 33-39 (1967).

12. Kagiwada, H. and R. Kalaba, J. Math. Anal. Applic. 23, 540-550 (1968).

13. Kagiwada, H. and R. Kalaba, J. Opt. Th. Applic. 2, 378-385 (1968).

14. Kagiwada, H., R. Kalaba and Y. Thomas, J. Opt. Th. Applic. 5, 11-21 (1970).

15. Kalaba, R. and E. Ruspini, Int. J. Engin. Sci. 7, 1091-1101 (1969).

STABILITE ET PRECISION DES SCHEMAS DSN POUR L'EQUATION DE TRANSPORT EN GEOMETRIE SPHERIQUE

P. Lascaux et P.A. Raviart

Nous étudions les schémas DSN de Carlson pour l'équation de transport en géométrie sphérique. Nous supposerons que le terme de source (le second membre) est connu, et nous voulons établir une majoration de l'erreur entre la solution exacte et la solution du schéma aux différences finies. Une telle majoration ne peut être obtenue qu'en deux temps.

Premièrement, il faut établir la stabilité du schéma, ce qui signifie établir une majoration de la norme de la solution discrète, en fonction de la norme du terme de source. Nous avons envisagé deux normes différentes aux paragraphes 3 et 4.

Deuxièmement, on applique cette majoration non plus à la solution discrète, mais à l'erreur, c'est-à-dire à la différence entre la solution exacte et la solution discrète. Le terme de source est alors remplacé par l'erreur de troncature (différence entre les dérivées et les quotients différentiels) et nous devons donc étudier l'erreur de troncature. Le problème est complexe à cause de la singularité des coefficients de l'équation en géométrie sphérique.

Nous montrons au paragraphe 5 que le choix des coefficients intervenant dans le schéma DSN conduit à une certaine opposition entre stabilité et précision, mais on peut néanmoins conclure sur l'intérêt de prendre le schéma le plus précis au voisinage de la singularité r=0, bien qu'il soit un peu moins stable.

INTRODUCTION

L'équation de transport des neutrons est l'équation de conservation que vérifie la densité φ des neutrons /1/. φ dépend des coordonnées $\vec{x}$ de la position et de la vitesse $\vec{v}$ des neutrons. Nous nous intéresserons au cas où il existe une symétrie sphérique et où les neutrons ont même module de vitesse. Dans ce cas là φ ne dépend plus que de la distance au centre r et du cosinus μ de l'angle que font la vitesse et le rayon vecteur.

L'équation s'écrit :

(1) $$A\varphi = \mu \frac{\partial \varphi}{\partial r} + \frac{1-\mu^2}{r} \frac{\partial \varphi}{\partial \mu} + \sigma \varphi = S(\varphi) \qquad \text{pour } -1 \leq \mu \leq 1, \quad 0 \leq r \leq R$$

σ est une fonction de r et μ telle $\sigma(r,\mu) \geq \sigma_0 > 0$

S représente les sources et en général dépend de φ.

Il faut ajouter une condition aux limites, par exemple :

(2) $$\varphi(R,\mu) = 0 \qquad \text{pour } \mu < 0$$

qui signifie qu'aucun neutron ne pénètre dans la sphère de rayon R.

On peut écrire (1) sous la forme :

$$\frac{d\varphi}{ds} + \sigma \varphi = S$$

où $\frac{d}{ds}$ représente la dérivée le long des courbes caractéristiques $r^2(1-\mu^2) = C^{te}$ ce qui met en évidence le rôle de (2) comme condition initiale de cette équation différentielle (figure 1)[1].

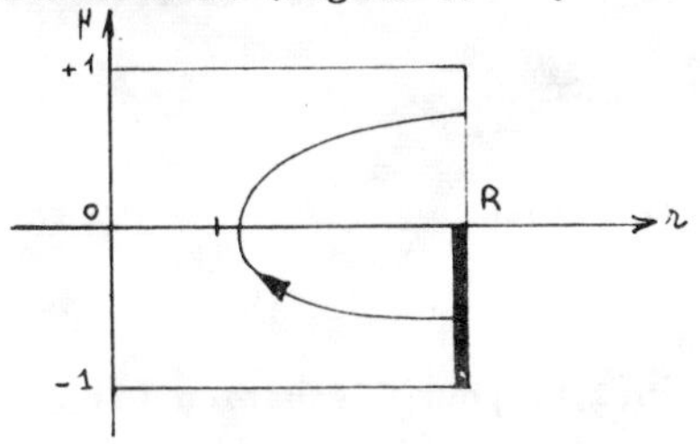

Figure 1

Nous allons nous intéresser ici à la discrétisation de l'équation (1) et plus spécialement à celle du premier membre, en supposant que le second membre est connu (soit parce qu'il est donné, soit qu'il résulte de l'itération précédente au cours d'un calcul itératif, par exemple, $A\varphi^{\nu+1} = S(\varphi^{\nu})$).

En fait, on a intérêt , pour assurer de façon discrète certaines propriétés physiques de conservation, à discrétiser l'équation (1) sous la "forme conservative" suivante ($r^2 dr$ étant l'élément de volume en géométrie sphérique) (Cf. /1/, p. 177):

(3) $$\mu \frac{\partial}{\partial r}(r^2\varphi) + r\frac{\partial}{\partial \mu}\left((1-\mu^2)\varphi\right) + r^2\sigma\varphi = r^2 S$$

Nous étudierons la stabilité et la précision des schémas introduits par Carlson et connus sous le nom de schémas D.S.N. Pour cela nous utiliserons deux égalités de l'énergie dans L^2 avec poids que l'on obtient en multipliant (3) soit par 2φ, soit par $\frac{2\varphi}{r^2(1-\mu^2)}$ et en intégrant sur le domaine. Ces deux égalités sont :

(4) $$R^2\int_{-1}^{+1}\mu\,\varphi^2(R,\mu)\,d\mu + 2\iint \sigma r^2\varphi^2\,dr\,d\mu = 2\iint r^2 S\varphi\,dr\,d\mu$$

(5) $$\int_{-1}^{+1}\frac{\mu}{1-\mu^2}\,\varphi^2(R,\mu)\,d\mu + 2\iint\frac{\sigma\varphi^2}{1-\mu^2}\,dr\,d\mu = 2\iint\frac{S\varphi}{1-\mu^2}\,dr\,d\mu \quad .$$

(1) - On n'utilisera pas cette propriété, car on veut écrire un schéma dans lequel on puisse faire apparaître les quantités discrètes représentant les différents moments en μ de $\varphi(r,\mu)$ étant donné que ce sont surtout les premiers moments qui ont une importance physique (Cf. /1/, p. 176)

Les intégrales simples de (4) et (5) sont positives grâce à la condition (2).

Le plan de l'étude est le suivant :

§1 - Description des schémas D.S.N. pour l'équation (3).

§2 - Principes de démonstration de la stabilité et de la précision pour le schéma DSN appliqué à une équation modèle plus simple $\frac{\partial\varphi}{\partial t} + a\frac{\partial\varphi}{\partial x} + \sigma\varphi = 0$.

§3 - Stabilité du schéma pour l'équation du transport en utilisant la norme $\iint r^2 \varphi^2 dr\, d\mu$.

§4 - Stabilité du schéma centré en utilisant la norme $\iint \frac{\varphi^2}{1-\mu^2} dr\, d\mu$.

§5 - Quelques indications sur la précision.

§1 - DESCRIPTION DES SCHEMAS DSN POUR L'EQUATION DU TRANSPORT

On introduit des points de discrétisation en r et μ, avec des pas en général non uniformes :

$$0 = r_0 < r_1 < \dots < r_I = R$$

$$-1 = \mu_{-J} < \mu_{-(J-1)} < \dots < \mu_J = 1 \quad .$$

De plus, on introduit aussi des points intermédiaires $r_{i+\frac{1}{2}}$ et $\mu_{j+\frac{1}{2}}$ tels que :

$$r_i < r_{i+\frac{1}{2}} < r_{i+1}$$

$$\mu_j < \mu_{j+\frac{1}{2}} < \mu_{j+1} \quad .$$

Dans les calculs effectifs, on choisit toujours :

$$\mu_j = -\mu_{-j} \quad \text{et} \quad \mu_{j+\frac{1}{2}} = -\mu_{-(j+1/2)} \quad .$$

Les inconnues sont les densités φ évaluées aux points suivants :

$\varphi_{i,j+\frac{1}{2}}$ qui approche $\varphi(r_i, \mu_{j+\frac{1}{2}})$ que nous noterons pour simplifier, φ_i
$\varphi_{i+\frac{1}{2},j}$ qui approche $\varphi(r_{i+\frac{1}{2}}, \mu_j)$ que nous noterons pour simplifier, φ_j
$\varphi_{i+\frac{1}{2},j+\frac{1}{2}}$ qui approche $\varphi(r_{i+\frac{1}{2}}, \mu_{j+\frac{1}{2}})$ que nous noterons pour simplifier, $\overline{\varphi}$

A l'intérieur de la cellule $[r_i, r_{i+1}] \times [\mu_j, \mu_{j+1}]$ (Figure 2)

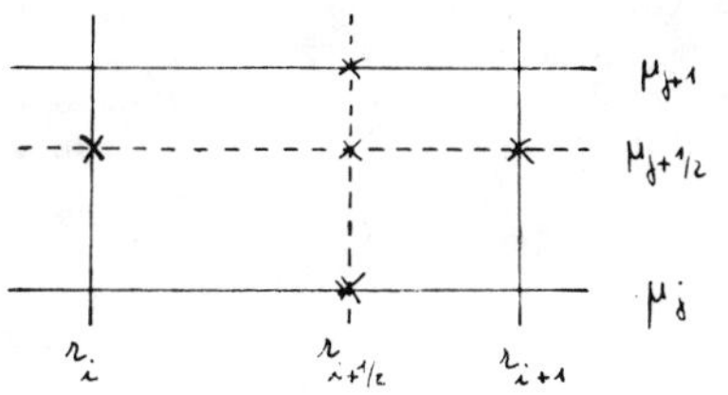

Figure 2

on remplace l'équation (3) par :

$$\mu_{j+\frac{1}{2}} \frac{r_{i+1}^2 \varphi_{i+1} - r_i^2 \varphi_i}{r_{i+1} - r_i} + \frac{(r_{i+1} + r_i)}{2} \frac{\beta_{j+1}\varphi_{j+1} - \beta_j \varphi_j}{\mu_{j+1} - \mu_j} + \bar{r}^2_{i+\frac{1}{2}} \sigma_{i+\frac{1}{2}} \bar{\varphi} = \bar{r}^2_{i+\frac{1}{2}} \bar{S} \qquad (6)$$

où i) $\bar{r}^2_{i+\frac{1}{2}} = \frac{1}{3} \cdot \frac{r_{i+1}^3 - r_i^3}{r_{i+1} - r_i}$

ii) les β_j vérifient $\beta_{j+1} - \beta_j = -2\mu_{j+\frac{1}{2}} (\mu_{j+1} - \mu_j)$, $\beta_{\pm J} = 0$

de telle sorte que β_j est une approximation de $1 - \mu_j^2$. Si les μ_j et $\mu_{j+\frac{1}{2}}$ sont symétriques autour de 0, alors $\beta_j = \beta_{-j}$.

Par cellule, il y a trois inconnues (puisque les 4 densités sur le bord sont chacune communes à deux cellules adjacentes). En plus de (6) il faut donc ajouter deux équations que nous appellerons équations complémentaires :

$$\begin{cases} \bar{\varphi} = \theta_{i+\frac{1}{2}} \varphi_{i+1} + (1 - \theta_{i+\frac{1}{2}}) \varphi_i & 0 \leq \theta_{i+\frac{1}{2}} \leq 1 \\ \bar{\bar{\varphi}} = \alpha_{j+\frac{1}{2}} \varphi_{j+1} + (1 - \alpha_{j+\frac{1}{2}}) \varphi_j & 0 \leq \alpha_{j+\frac{1}{2}} \leq 1 \end{cases} \qquad (7)$$

Dans ces relations, a priori, $\theta_{i+\frac{1}{2}}$ (resp. $\alpha_{j+\frac{1}{2}}$) peut dépendre de j (resp. i).

Numériquement, le calcul des φ vérifiant (6) et (7) se fait explicitement dans le sens suivant (qui correspond au sens de la caractéristique) :

A) - Si $\mu_{j+\frac{1}{2}} < 0$, connaissant φ_{i+1} et φ_j , on élimine, grâce à (7), φ_i et φ_{j+1} de (6).

On tire alors de (6) la valeur de $\bar{\varphi}$, qui, reportée dans (7), permet d'obtenir φ_i et φ_{j+1} (à condition que $\alpha_{j+1/2} \neq 0$ et $\theta_{i+1/2} \neq 1$). On procède ainsi de proche en proche en partant de φ_I donné, nul d'après (2), et de φ_{-J} . Pour éliminer φ_{-J} on utilise l'une des deux possibilités suivantes :

1°) On prend $\alpha_{-(J-1/2)} = 1$ auquel cas φ_{-J} n'intervient pas dans (7), ni dans (6), puisque $\beta_{-J} = 0$.

2°) On discrétise l'équation (3) pour $\mu = -1$, c'est-à-dire :

(8) $$-\frac{d}{dr}\left(r^2\varphi\right) + 2r\varphi + \sigma r^2 \varphi = r^2 S .$$

B) - Si $\mu_{j+1/2} > 0$, connaissant φ_i et φ_j , on élimine, grâce à (7), φ_{i+1} et φ_{j+1} de (6).

On tire alors de (6) la valeur de $\overline{\varphi}$, qui, reportée dans (7), permet d'obtenir φ_{i+1} et φ_{j+1} (à condition que $\alpha_{j+1/2} \neq 0$ et $\theta_{j+1/2} \neq 0$). On procède ainsi de proche en proche, en partant de φ_j pour $j=0$ calculé en A) au cours de l'intégration dans le domaine $\mu < 0$ et de φ_i pour $i=0$ que l'on élimine de l'une des deux façons suivantes :

1°) En prenant $\theta_{1/2} = 1$.

2°) En imposant $\varphi_{0,j+1/2} = \varphi_{0,-(j+1/2)}$ déjà calculé (en effet physiquement la solution est telle que $\varphi(0,\mu) = \varphi(0,-\mu)$).

§2 - STABILITE ET PRECISION DANS LE CAS D'UNE EQUATION MODELE

Nous verrons aux paragraphes suivants que les majorations de stabilité et l'étude de la précision sont assez complexes à cause de la singularité des coefficients de l'équation du transport. Aussi nous préférons, dans un premier temps, exposer les principes de démonstration de la stabilité et de la précision des schémas DSN appliqués à l'équation modèle ci-dessous (qui schématise l'équation de transport d'évolution en géométrie plane).

Considérons l'équation modèle :

$$(9)\qquad \begin{cases} \dfrac{\partial \varphi}{\partial t} + a\,\dfrac{\partial \varphi}{\partial x} + \sigma \varphi = S & 0 < x < 1, \quad 0 < t < T \\ \varphi(x,0) = \varphi^0(x) & \\ \varphi(0,t) = 0 & \end{cases}$$

où a et σ sont des constantes positives.

On introduit les points de discrétisation $t^n = n\Delta t$ et $x_i = i\Delta x$, ainsi que $t^{n+1/2}$ et $x_{i+1/2}$, et on remplace (9) par :

$$(10)\qquad \frac{1}{\Delta t}\left(\varphi_{i+1/2}^{n+1} - \varphi_{i+1/2}^{n}\right) + \frac{a}{\Delta x}\left(\varphi_{i+1}^{n+1/2} - \varphi_{i}^{n+1/2}\right) + \sigma \varphi_{i+1/2}^{n+1/2} = S_{i+1/2}^{n+1/2}$$

$$(11)\qquad \begin{cases} \bar{\varphi} = \varphi_{i+1/2}^{n+1/2} & = \theta^{n+1/2}\, \varphi_{i+1/2}^{n+1} + (1-\theta^{n+1/2})\, \varphi_{i+1/2}^{n} \\ & = \alpha_{i+1/2}\, \varphi_{i+1}^{n+1/2} + (1-\alpha_{i+1/2})\, \varphi_{i}^{n+1/2} \end{cases}$$

.

Posons :

$$|\bar{\varphi}|_h = \left(\Delta x \, \Delta t \sum_{i,n} \left| \varphi_{i+1/2}^{n+1/2} \right|^2 \right)^{1/2} .$$

On a le :

THEOREME 1

Sous les conditions $\theta^{n+1/2} \geq \frac{1}{2}$ et $\alpha_{i+1/2} \geq \frac{1}{2}$, la solution du schéma (10,11) vérifie, quels que soient Δx et Δt :

$$\Delta x \sum_{i=0}^{I-1} \left| \varphi_{i+1/2}^N \right|^2 + \left| \bar{\varphi} \right|_h^2 \leq c^{te} \left\{ \Delta x \sum_{i=0}^{I-1} \left| \varphi_{i+1/2}^0 \right|^2 + \left| \bar{S} \right|_h^2 \right\} .$$

Démonstration - On multiplie (10) par $2\bar{\varphi} \, \Delta x \, \Delta t$ et on se sert de l'identité :

$$(12) \quad 2(\varphi^{n+1} - \varphi^n)\bar{\varphi} = 2(\varphi^{n+1} - \varphi^n)(\theta\varphi^{n+1} + (1-\theta)\varphi^n) = |\varphi^{n+1}|^2 - |\varphi^n|^2 + (2\theta - 1)\,|\varphi^{n+1} - \varphi^n|^2$$

ainsi que de l'identité analogue par rapport à i. On obtient :

$$(13) \quad \Delta x \left\{ |\varphi^{n+1}|^2 - |\varphi^n|^2 + (2\theta^{n+1/2} - 1)|\varphi^{n+1} - \varphi^n|^2 \right\} + a\,\Delta t \left\{ |\varphi_{i+1}|^2 - |\varphi_i|^2 + (2\alpha_{i+1/2} - 1)|\varphi_{i+1} - \varphi_i|^2 \right\} + 2\sigma \, \Delta x \, \Delta t \, \bar{\varphi}^2 = 2\Delta x \, \Delta t \, \bar{S}\bar{\varphi} .$$

Supposons que $2\theta^{n+1/2} - 1 \geq 0$ et $2\alpha_{i+1/2} - 1 \geq 0$, alors en sommant (13) en n et i, on obtient l'inégalité :

$$(14) \quad \Delta x \sum_{i=0}^{I-1} \left| \varphi_{i+\frac{1}{2}}^N \right|^2 + a\,\Delta t \sum_{n=0}^{N-1} \left| \varphi_I^n \right|^2 + 2\sigma \, \Delta x \, \Delta t \sum_{i,n} \left| \varphi_{i+1/2}^{n+1/2} \right|^2 \leq \Delta x \sum_{i=0}^{I-1} \left| \varphi_i^0 \right|^2 + 2\Delta x \, \Delta t \sum_{i,n} \varphi_{i+1/2}^{n+1/2} S_{i+1/2}^{n+1/2} \leq \Delta x \sum_{i=0}^{I-1} \left| \varphi_i^0 \right|^2 + \sigma \, \Delta x \, \Delta t \sum_{i,n} \left| \bar{\varphi} \right|^2 + \frac{1}{\sigma} \Delta x \, \Delta t \sum_{i,n} \left| \bar{S} \right|^2 \quad \blacksquare$$

MAJORATION DE L'ERREUR

On peut envisager d'obtenir une majoration de l'erreur en posant $\varepsilon = \varphi - \varphi_{ex}$, où φ_{ex} est la solution exacte de l'équation (9). On vérifie que ε est la solution d'un schéma identique à (10,11) dans lequel au second membre de (10) et (11) on a des termes supplémentaires représentant l'erreur de troncature.

Si $\theta^{n+1/2} = \alpha_{i+1/2} = 1/2$ -ce que l'on appelle le schéma diamand- et si $\varphi_{ex} \in C^3$, les termes d'erreur de troncature sont de l'ordre de $\Delta x^2 + \Delta t^2$.

Par la même technique de majoration que celle utilisée dans la démonstration du théorème 1, on montre que :

$$\left(\Delta x \sum_{i=0}^{I-1} \left| \varepsilon_{i+\frac{1}{2}}^{N} \right|^2 \right)^{1/2} \leq C^{te} (\Delta x^2 + \Delta t^2) .$$

REMARQUE 1

Si on avait choisi $\theta^{n+\frac{1}{2}} = \alpha_{i+1/2} = 1$ -ce que l'on appelle schéma en escalier- on aurait obtenu un schéma d'ordre 1, de type positif (c'est-à-dire que $\varphi^0 \geq 0$ et $S \geq 0$ entraînent que les φ^n, φ_i, et $\bar{\varphi}$ sont tous positifs) donc stable pour la norme du maximum, $|\varphi| = \max_i |\varphi_i|$.

REMARQUE 2

Les conditions de stabilité $\theta \geq \frac{1}{2}$, $\alpha \geq \frac{1}{2}$ indiquent qu'au cours du calcul dans une cellule, le $\bar{\varphi}$ des relations complémentaires (11) est une combinaison linéaire des φ inconnus (φ^{n+1} et φ_{i+1}) avec un poids supérieur ou égal à celui des φ connus (φ^n et φ_i) (Figure 3).

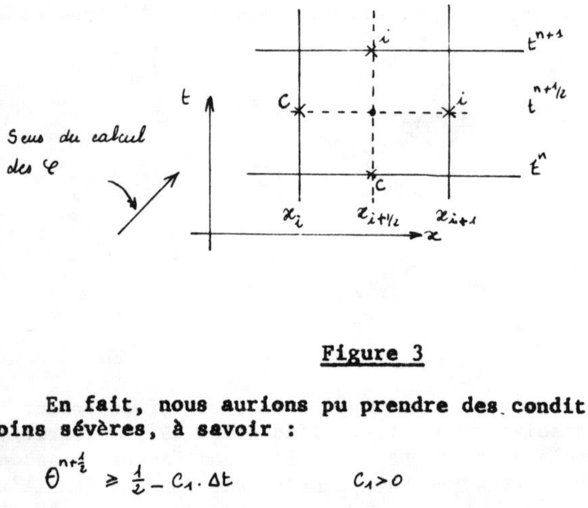

Figure 3

En fait, nous aurions pu prendre des conditions de stabilité un peu moins sévères, à savoir :

$$\theta^{n+\frac{1}{2}} \geq \frac{1}{2} - C_1 . \Delta t \qquad C_1 > 0$$

$$\alpha_{i+\frac{1}{2}} \geq \frac{1}{2} - C_2 . \Delta x \qquad C_2 > 0 .$$

Dans ce dernier cas, on obtient encore une majoration de stabilité du type :

$$\Delta x \sum_{i=0}^{I-1} |\varphi_{i+\frac{1}{2}}^{N}|^2 \leq C^{te} \cdot \Delta x \sum_{i=0}^{I-1} |\varphi_{i+\frac{1}{2}}^{0}|^2 \quad \text{(si par exemple } S = 0 \text{)}$$

mais avec une constante plus grande qui dépend de c_1 et c_2.

§3 - PREMIERE MAJORATION DE STABILITE DES SCHEMAS DSN POUR L'EQUATION DE TRANSPORT

Revenons au schéma DSN donné par les équations (6) et (7).

Définissons une première norme L^2 discrète par :

(15) $$|\bar{\varphi}|_1 = \left[\sum_{i,j} \frac{1}{3}(r_{i+1}^3 - r_i^3)(\mu_{j+1} - \mu_j)\, \varphi^2_{i+\frac{1}{2},j+\frac{1}{2}} \right]^{1/2}$$

qui est l'équivalent de : $\left(\iint r^2 \varphi^2(r,\mu)\, dr\, d\mu \right)^{1/2}$.

THEOREME 2

Sous les conditions :

(16) $$\begin{cases} r_i \theta_{i+\frac{1}{2}} \geq r_{i+1}(1-\theta_{i+1/2}) & \text{pour } \mu_{j+\frac{1}{2}} > 0 \\ r_i \theta_{i+\frac{1}{2}} \leq r_{i+1}(1-\theta_{i+\frac{1}{2}}) & \text{pour } \mu_{j+\frac{1}{2}} < 0 \\ \sqrt{\beta_j}\, \alpha_{j+\frac{1}{2}} \geq \sqrt{\beta_{j+1}}\,(1-\alpha_{j+\frac{1}{2}}) & . \end{cases}$$

La solution du schéma DSN (6,7) vérifie :

(17) $$|\bar{\varphi}|_1 \leq \frac{1}{\sigma_0} |\bar{S}|_1$$

quelles que soient les abscisses r_i et μ_j (donc quels que soient les pas de discrétisation).

Démonstration - Nous allons nous servir des deux identités suivantes (dans lesquelles on a supprimé l'indice de θ et α pour alléger l'écriture) :

(18) $$2(r_{i+1}^2\varphi_{i+1} - r_i^2\varphi_i)\bar{\varphi} = 2(r_{i+1}^2\varphi_{i+1} - r_i^2\varphi_i)(\theta\varphi_{i+1} + (1-\theta)\varphi_i)$$
$$= r_{i+1}^2\varphi_{i+1}^2 - r_i^2\varphi_i^2 + \underline{(r_{i+1}^2 - r_i^2)\bar{\varphi}^2} + [r_i^2\theta^2 - r_{i+1}^2(1-\theta)^2](\varphi_{i+1} - \varphi_i)^2$$

$$(19)\quad 2(\beta_{j+1}\varphi_{j+1}-\beta_j\varphi_j)\bar{\varphi} = 2(\beta_{j+1}\varphi_{j+1}-\beta_j\varphi_j)(\alpha\varphi_{j+1}+(1-\alpha)\varphi_j)$$

$$= \beta_{j+1}\varphi_{j+1}^2-\beta_j\varphi_j^2+\underline{(\beta_{j+1}-\beta_j)\bar{\varphi}^2}+\left[\beta_j\alpha^2-\beta_{j+1}(1-\alpha)^2\right](\varphi_{j+1}-\varphi_j)^2.$$

Multiplions (6) par $2(r_{i+1}-r_i)(\mu_{j+1}-\mu_j)\bar{\varphi}$ **, on obtient à l'aide de (18) et (19)**

$$\mu_{j+\frac{1}{2}}(\mu_{j+1}-\mu_j)\left[r_{i+1}^2\varphi_{i+1}^2-r_i^2\varphi_i^2+(r_i^2\theta^2-r_{i+1}^2(1-\theta)^2)(\varphi_{i+1}-\varphi_i)^2\right]$$

$$(20)\quad +\frac{(r_{i+1}+r_i)}{2}(r_{i+1}-r_i)\left[\beta_{j+1}\varphi_{j+1}^2-\beta_j\varphi_j^2+(\beta_j\alpha^2-\beta_{j+1}(1-\alpha)^2)(\varphi_{j+1}-\varphi_j)^2\right]$$

$$+\sigma_{i+\frac{1}{2}}\cdot\frac{2}{3}(r_{i+1}^3-r_i^3)(\mu_{j+1}-\mu_j)\bar{\varphi}^2 = \frac{2}{3}(r_{i+1}^3-r_i^3)(\mu_{j+1}-\mu_j)\bar{\varphi}\bar{S}$$

car la somme des termes provenant des termes soulignés dans (18) et (19) est nulle compte tenu de 6)ii).

Supposons que les conditions (16) sont satisfaites et sommons l'équation (20) en i **et** j **, il vient :**

$$(21)\quad R^2\sum_j(\mu_{j+1}-\mu_j)\mu_{j+1/2}\varphi_{I,j+\frac{1}{2}}^2+\frac{2}{3}\sum_{i,j}\sigma_{i+1/2}(r_{i+1}^3-r_i^3)\bar{\varphi}^2 \leq \frac{2}{3}\sum_{i,j}(r_{i+1}^3-r_i^3)(\mu_{j+1}-\mu_j)\bar{\varphi}\bar{S}.$$

Le premier terme de (21) est positif grâce à la condition aux limites, ce qui entraîne :

$$\sigma_0|\bar{\varphi}|_1^2 \leq (\bar{\varphi},\bar{S})_1 \leq |\bar{\varphi}|_1|\bar{S}|_1.$$

Donc : $|\bar{\varphi}|_1 \leq \frac{1}{\sigma_0}|\bar{S}|_1$. ∎

<u>REMARQUE 3</u>

Compte tenu du sens de résolution des équations (6,7), défini au paragraphe 1, les conditions (16) expriment qu'au cours du calcul dans chaque cellule, $\bar{\varphi}$ **est une combinaison linéaire des** φ **connus et inconnus, le poids des** φ **inconnus étant supérieur ou égal à une certaine borne inférieure. On retrouve exactement ce qui a été dit à la remarque 2.**

REMARQUE 4

Si l'on veut que $\theta_{i+\frac{1}{2}}$ soit indépendant de j , d'après (16), il faut qu'il vérifie :

$$r_i \theta_{i+\frac{1}{2}} = r_{i+1} (1 - \theta_{i+\frac{1}{2}}) \text{ .}$$

Il apparaît donc une difficulté supplémentaire due au fait que l'on ne peut pas prendre $\theta_{1/2} = 1$ pour $j < 0$ (comme on l'a noté en A au paragraphe 1), sinon les équations (6) et (7) sont incompatibles dans la maille $(0 , r_1)$. On est alors obligé de changer la relation complémentaire dans cette maille, en posant :

(7 bis) $$\overline{\varphi}_{\frac{1}{2}, j+\frac{1}{2}} = \varphi_{1, j+\frac{1}{2}} + \Psi_{j+\frac{1}{2}}$$

et on impose alors la relation de pseudo-symétrie :

$$\Psi_{j+\frac{1}{2}} = \Psi_{-(j+\frac{1}{2})}$$

étant donné que $r_1 \Psi_{j+\frac{1}{2}}$ est alors une approximation de $\lim_{r \to 0} r \varphi(r, \mu_{j-1/2})$.

On voit également que les conditions (16) imposent de prendre $\alpha_{-(J-\frac{1}{2})} = 1$ et dans ce cas $\varphi_{i+\frac{1}{2}, -J}$ n'intervient pas dans le schéma.

§4 - UNE DEUXIEME MAJORATION DE STABILITE POUR LE SCHEMA DSN

Les conditions de stabilité (16) nous empêchent donc de prendre des relations complémentaires centrées ($\alpha = \theta = \frac{1}{2}$), ce qui serait souhaitable pour accroître la précision. Nous allons chercher à obtenir une majoration de stabilité pour une norme L^2 avec un poids différent, qui nous permettra de prendre $\alpha = \theta = \frac{1}{2}$.

Introduisons une seconde norme L^2 discrète par :

$$(22) \quad |\bar{\varphi}|_2 = \left(\sum_{i,j} (r_{i+1} - r_i)(\mu_{j+1} - \mu_j) \cdot \frac{\bar{r}^2_{i+1/2}}{(1-\theta_{i+1/2}) r^2_{i+1} + \theta_{i+\frac{1}{2}} \cdot r^2_i} \cdot \frac{\varphi^2_{i+\frac{1}{2}, j+\frac{1}{2}}}{(1-\alpha_{j+\frac{1}{2}})\beta_{j+1} + \alpha_{j+\frac{1}{2}} \beta_j} \right)^{1/2}$$

qui est l'équivalent discret de : $\left(\iint \frac{\varphi^2}{1-\mu^2} \, dr \, d\mu \right)^{1/2}$.

THEOREME 3

Sous les conditions :

$$(23) \quad \begin{cases} \theta_{i+\frac{1}{2}} \geq \frac{1}{2} & \text{si } \mu_{j+\frac{1}{2}} > 0 \\ \theta_{i+\frac{1}{2}} \leq \frac{1}{2} & \text{si } \mu_{j+\frac{1}{2}} < 0 \\ \alpha_{j+\frac{1}{2}} \geq \frac{1}{2} & \end{cases}$$

la solution du schéma DSN (6,7) vérifie :

$$(24) \quad |\bar{\varphi}|^2_2 \leq C^{te} \left\{ |\bar{S}|^2_2 + \sum_i \frac{1}{2} \frac{r^2_{i+1} - r^2_i}{(1-\theta) r^2_{i+1} + \theta r^2_i} \varphi^2_{i+\frac{1}{2}, -J} \right\}$$

quelles que soient les abscisses r_i et μ_j (donc quels que soient les pas de discrétisation).

<u>Démonstration</u> - Nous allons nous servir des identités suivantes :

(25) $$2(r_{i+1}^2\varphi_{i+1}-r_i^2\varphi_i)\,\bar\varphi = 2(r_{i+1}^2\varphi_{i+1}-r_i^2\varphi_i)(\theta\varphi_{i+1}+(1-\theta)\varphi_i)$$
$$= \underline{2(r_{i+1}^2-r_i^2)\bar\varphi^2} + \left[(1-\theta)r_{i+1}^2+\theta r_i^2\right]\left[\varphi_{i+1}^2-\varphi_i^2+(2\theta-1)(\varphi_{i+1}-\varphi_i)^2\right],$$

(26) $$2(\beta_{j+1}\varphi_{j+1}-\beta_j\varphi_j)\,\bar\varphi = 2(\beta_{j+1}\varphi_{j+1}-\beta_j\varphi_j)(\alpha\varphi_{j+1}+(1-\alpha)\varphi_j)$$
$$= \underline{2(\beta_{j+1}-\beta_j)\bar\varphi^2} + \left[(1-\alpha)\beta_{j+1}+\alpha\beta_j\right]\left[\varphi_{j+1}^2-\varphi_j^2+(2\alpha-1)(\varphi_{j+1}-\varphi_j)^2\right].$$

Posons pour simplifier :

$$r_\theta^2 = (1-\theta_{i+1/2})r_{i+1}^2+\theta_{i+1/2}r_i^2 \quad \text{et} \quad \beta_\alpha = (1-\alpha_{j+1/2})\beta_{j+1}+\alpha_{j+1/2}\beta_j \;.$$

Multiplions (6) par $2(r_{i+1}-r_i)(\mu_{j+1}-\mu_j)\bar\varphi$, on obtient à l'aide de (25) et (26) :

(27) $$\mu_{j+\frac12}(\mu_{j+1}-\mu_j)r_\theta^2\left[\varphi_{i+1}^2-\varphi_i^2+(2\theta-1)(\varphi_{i+1}-\varphi_i)^2\right]+\frac{r_{i+1}^2-r_i^2}{2}\beta_\alpha\left[\varphi_{j+1}^2-\varphi_j^2+(2\alpha-1)(\varphi_{j+1}-\varphi_j)^2\right]$$
$$+\ \sigma_{i+\frac12}\cdot\frac{2}{3}(r_{i+1}^3-r_i^3)(\mu_{j+1}-\mu_j)\,\bar\varphi^2 = \frac{2}{3}(r_{i+1}^3-r_i^3)(\mu_{j+1}-\mu_j)\,\bar\varphi\,\bar S$$

car la somme des termes provenant des termes soulignés dans (25) et (26) est nulle.

Supposons que les conditions (23) sont satisfaites, divisons (27) par $r_\theta^2\,\beta_\alpha$ et sommons cette égalité sur les i et les j, il vient :

(28) $$\sum_j\frac{(\mu_{j+1}-\mu_j)\mu_{j+1/2}}{\beta_\alpha}\varphi_{I,j+\frac12}^2-\sum_j\frac{(\mu_{j+1}-\mu_j)\mu_{j+1/2}}{\beta_\alpha}\varphi_{0,j+1/2}^2+\sum_i\frac{(r_{i+1}^2-r_i^2)}{2r_\theta^2}\varphi_{i+\frac12,J}^2$$
$$-\sum_i\frac{(r_{i+1}^2-r_i^2)}{2r_\theta^2}\varphi_{i+\frac12,-J}^2+\frac{2}{3}\sigma_0\sum_{i,j}\frac{(r_{i+1}^3-r_i^3)}{r_\theta^2\cdot\beta_\alpha}(\mu_{j+1}-\mu_j)\bar\varphi^2\le\frac{2}{3}\sum_{i,j}\frac{(r_{i+1}^3-r_i^3)}{r_\theta^2\cdot\beta_\alpha}(\mu_{j+1}-\mu_j)\,\bar\varphi\cdot\bar S\;.$$

Le premier terme du membre de gauche est positif grâce à la condition aux limites.

Le deuxième terme du membre de gauche est nul si on a posé :

$$\varphi_{0,j+\frac12} = \varphi_{0,-(j+\frac12)} \quad \text{et} \quad \alpha_{j+\frac12} = 1-\alpha_{-(j+\frac12)}$$

car alors compte tenu de la symétrie des μ_j on a $(\beta_\alpha)_{j+\frac12} = (\beta_\alpha)_{-(j+\frac12)}$.

On obtient donc l'inégalité :

$$(29) \qquad 2\sigma_0 |\bar{\varphi}|_2^2 \leq 2(\bar{\varphi}, \bar{S})_2 + \sum_{i=0}^{I-1} \frac{(r_{i+1}^2 - r_i^2)}{2r_\theta^2} \varphi_{i+\frac{1}{2},-J}^2$$

d'où l'on déduit (24) par application de l'inégalité de Cauchy Schwarz à $(\bar{\varphi}, \bar{S})_2$ ■.

REMARQUE 5

Le terme supplémentaire dans le membre de droite de (24) provient du fait que maintenant on ne peut plus choisir $\alpha_{-(J-\frac{1}{2})} = 1$ (sinon on aurait $(\beta\omega)_{-(J-\frac{1}{2})} = 0$) et que par conséquent on se sert nécessairement des valeurs de départ $\varphi_{i+\frac{1}{2},-J}$, qui sont obtenues par la discrétisation de (8).

REMARQUE 6

Si $S(r, \pm 1) \neq 0$, alors $|\bar{S}|_2$ n'est pas borné lorsque les pas de discrétisation tendent vers zéro, à cause de la discrétisation de $\frac{1}{1-\mu^2}$ dans l'expression de $|\bar{S}|_2$. En réalité il faudrait appliquer la majoration précédente directement à l'erreur $\varepsilon = \varphi - \varphi_{ex}$, auquel cas, le second membre $\bar{S}$ serait remplacé par l'erreur de troncature. Nous sommes donc amener à étudier comment se comporte l'erreur de troncature.

§5 - QUELQUES CONSIDERATIONS SUR L'ERREUR DE TRONCATURE

Les résultats de ce paragraphe font suite à un article de LATHROP et REED /2/. Ils ne constituent pas une approche définitive de la question et soulèvent de nouveaux problèmes sur la manière dont on peut les utiliser d'après la remarque 6.

Les difficultés résidant évidemment au voisinage de $r = 0$, nous devons convenir d'étudier l'erreur de troncature des différents schémas (c'est-à-dire pour des choix différents des α et θ) sur la même forme. Nous choisirons la forme discrétisée similaire à l'équation (1).

Nous pouvons écrire l'équation (6) sous la forme (32) en nous servant des deux identités suivantes :

(30) $$r_{i+1}^2 \varphi_{i+1} - r_i^2 \varphi_i = (r_{i+1}^2 - r_i^2)(\theta \varphi_{i+1} + (1-\theta)\varphi_i) + [(1-\theta) r_{i+1}^2 + \theta r_i^2](\varphi_{i+1} - \varphi_i)$$

(31) $$\beta_{j+1}\varphi_{j+1} - \beta_j \varphi_j = (\beta_{j+1} - \beta_j)(\alpha \varphi_{j+1} + (1-\alpha)\varphi_j) + [(1-\alpha)\beta_{j+1} + \alpha \beta_j](\varphi_{j+1} - \varphi_j)$$

(32) $$\frac{\mu_{j+1/2}}{(1-\alpha)\beta_{j+1} + \alpha\beta_j} \frac{\varphi_{i+1} - \varphi_i}{r_{i+1} - r_i} + \frac{\frac{1}{2}(r_{i+1} + r_i)}{(1-\theta) r_{i+1}^2 + \theta r_i^2} \cdot \frac{\varphi_{j+1} - \varphi_j}{\mu_{j+1} - \mu_j} + \frac{\bar{r}_{i+1/2}^2}{r_\theta^2 \cdot \beta_\alpha} \sigma_{i+\frac{1}{2}} \bar{\varphi} = \frac{\bar{r}_{i+1/2}^2}{r_\theta^2 \cdot \beta_\alpha} \bar{S}$$

où $r_\theta^2 = (1-\theta) r_{i+1}^2 + \theta r_i^2$ et $\beta_\alpha = (1-\alpha)\beta_{j+1} + \alpha\beta_j$.

Réécrivons également les relations complémentaires (7) :

(7) $$\bar{\varphi} = \theta\varphi_{i+1} + (1-\theta)\varphi_i = \alpha \varphi_{j+1} + (1-\alpha)\varphi_j \quad .$$

A priori, il est tentant de choisir $\alpha = \theta = \frac{1}{2}$ et d'évaluer l'erreur de troncature en faisant un développement de Taylor au voisinage du point $\frac{r_{i+1}+r_i}{2}$, $\mu_{j+1/2}$. Mais cela conduit à une erreur importante au voisinage de $r=0$. Evaluons en effet l'erreur de troncature sur le terme $\frac{1}{r}\frac{d\varphi}{d\mu}$ (qui reste borné lorsque r tend vers zéro, comme cela peut se voir sur l'équation (1)):

(33) $$\frac{\frac{1}{2}(r_{i+1}+r_i)}{\frac{1}{2}(r_{i+1}^2+r_i^2)} - \frac{1}{\frac{r_{i+1}+r_i}{2}} = \frac{(r_{i+1}+r_i)^2 - 2(r_{i+1}^2+r_i^2)}{(r_{i+1}^2+r_i^2)(r_{i+1}+r_i)} = -\frac{(r_{i+1}-r_i)^2}{(r_{i+1}+r_i)(r_{i+1}^2+r_i^2)} .$$

Donc :

(34) $$\frac{r_{i+1}+r_i}{r_{i+1}^2+r_i^2}\,\frac{\varphi_{j+1}-\varphi_j}{\mu_{j+1}-\mu_j} - \frac{1}{\frac{r_{i+1}+r_i}{2}}\,\frac{\varphi_{j+1}-\varphi_j}{\mu_{j+1}-\mu_j} = -\frac{(r_{i+1}-r_i)^2}{r_{i+1}^2+r_i^2}\cdot\frac{1}{r_{i+1}+r_i}\cdot\frac{\varphi_{j+1}-\varphi_j}{\mu_{j+1}-\mu_j} .$$

Mais $\frac{1}{r_{i+1}+r_i}\,\frac{\varphi_{j+1}-\varphi_j}{\mu_{j+1}-\mu_j}$ reste borné lorsque $i \to 0$ en tant que discrétisation de $\frac{1}{r}\frac{d\varphi}{d\mu}$.

On voit donc que l'erreur de troncature est bien d'ordre 2 pour $r_i \geq a > 0$, mais qu'elle devient d'ordre zéro lorsque $r_i \longrightarrow 0$.

On peut donc décider d'évaluer l'erreur de troncature en faisant le développement de Taylor autour du point :

(35) $$r_{i+\frac{1}{2}} = \theta\, r_{i+1} + (1-\theta)\, r_i$$

mais en choisissant $\theta \neq \frac{1}{2}$ et par exemple de telle sorte que :

(36) $$\frac{\frac{1}{2}(r_{i+1}+r_i)}{(1-\theta)r_{i+1}^2+\theta r_i^2} = \frac{1}{r_{i+1/2}} .$$

Cela est possible, une fois les r_i fixés, et conduit à ;

(37) $$\theta_{i+\frac{1}{2}} = \frac{1}{2} + \frac{1}{6}\cdot\frac{r_{i+1}-r_i}{r_{i+1}+r_i} .$$

Il en résulte d'ailleurs -Oh miracle !- que :

(38) $$\bar{r}_{i+\frac{1}{2}}^2 = \frac{1}{3}\,\frac{r_{i+1}^3-r_i^3}{r_{i+1}-r_i} = \bar{r}_\theta^2 .$$

Dans ce dernier cas - θ donné par (37)- on ne commet aucune erreur sur le terme en $\frac{1}{r}$ dans l'approximation de $\frac{1}{r}\frac{d\varphi}{d\mu}$, ni sur le coefficient 1 devant $\sigma\varphi$ grâce à (38). Mais il faut évaluer l'erreur sur le premier terme :

$$\frac{\varphi_{i+1}-\varphi_i}{r_{i+1}-r_i}-\frac{d}{dr}\varphi_{ex}(r_{i+1/2}) .$$

Si la solution exacte φ_{ex} est suffisamment régulière, on a :

$$\frac{\varphi_{i+1}-\varphi_i}{r_{i+1}-r_i} = \frac{d\varphi}{dr}\left(\frac{r_{i+1}-r_i}{2}\right)+O(r_{i+1}-r_i)^2$$

$$= \frac{d\varphi}{dr}(r_{i+\frac{1}{2}})+O\left(\frac{r_{i+1}+r_i}{2}-r_{i+\frac{1}{2}}\right)+O(r_{i+1}-r_i)^2 .$$

Or :

(39) $$\frac{1}{2}(r_{i+1}+r_i)-r_{i+\frac{1}{2}} = \frac{1}{6}\frac{(r_{i+1}-r_i)^2}{r_{i+1}+r_i} .$$

Donc l'erreur sur la dérivée $\frac{d\varphi}{dr}$ est toujours d'ordre 2 pour $r_i \geq a > 0$ mais reste d'ordre 1 lorsque $r_i \longrightarrow 0$.

En conclusion, on peut dire que l'erreur de troncature en r est au voisinage de $r=0$, d'ordre zéro pour $\theta=\frac{1}{2}$, et d'ordre 1 pour θ donné par (37). Comme dans les deux cas elle est d'ordre 2 pour $r_i \geq a$, il en résulte que l'on a intérêt à choisir θ donné par (37).

En ce qui concerne l'erreur de troncature en μ , on peut procéder de la même façon en posant :

(40) $$\mu_{j+\frac{1}{2}} = \alpha\,\mu_{j+1}+(1-\alpha)\mu_j$$

et chercher si on peut déterminer $\alpha_{j+\frac{1}{2}}$ pour que :

(41) $$(1-\alpha_{j+\frac{1}{2}})\beta_{j+1}+\alpha_{j+\frac{1}{2}}\,\beta_j = 1-\mu^2_{j+\frac{1}{2}}$$

en n'oubliant pas que nous avions imposer en (6)ii) :

(42) $$\beta_{j+1} - \beta_j = -2\,\mu_{j+\frac{1}{2}}\,(\mu_{j+1} - \mu_j) \quad .$$

LATHROP et REED /2/ont montré que l'on pouvait effectivement choisir les $\alpha_{j+\frac{1}{2}}$, une fois fixés les μ_j de telle sorte que (40), (41) et (42) soient vérifiées simultanément. Dans ce cas on ne commet évidemment aucune erreur sur la discrétisation du terme $\frac{1}{1-\mu^2}$, et celle que l'on commet sur la discrétisation de $\frac{d\varphi}{d\mu}$ est :

(43) $$\frac{\varphi_{j+1}-\varphi_j}{\mu_{j+1}-\mu_j} - \frac{d\varphi}{d\mu}\,(\mu_{j+\frac{1}{2}}) = O\left(\tfrac{1}{2}(\mu_{j+1}+\mu_j) - \mu_{j+\frac{1}{2}}\right) .$$

On peut montrer le lemme suivant (on se limite au cas $\mu > 0$ puisque de toute façon on choisit toujours $\mu_j = -\mu_{-j}$ et $\beta_j = \beta_{-j}$) :

LEMME

Supposons que $\mu_{j+1}-\mu_j \leq \mu_j - \mu_{j-1}$ pour $J-1 \geq j \geq 1$ ($\mu_J = 1$, $\mu_0 = 0$) alors on a :

(44) $$0 < \mu_{j+\frac{1}{2}} - \tfrac{1}{2}\,(\mu_{j+1} + \mu_j) \; .$$

De plus si $\mu_{j+1}-\mu_j = \Delta\mu = \frac{1}{J}$ alors $\mu_{j+\frac{1}{2}} - \frac{1}{2}(\mu_{j+1}+\mu_j) < \frac{C\,\Delta\mu^2}{\sqrt{\mu_{j+1}}}$.

Démonstration - Voir annexe A. ■

REMARQUE 7

Il résulte de (44) que pour $j > 0$ on a $\alpha_{j+\frac{1}{2}} > \frac{1}{2}$. Au contraire pour $j < 0$ on aura $\alpha_{j+\frac{1}{2}} < \frac{1}{2}$, ce qui pose un problème nouveau pour la stabilité du schéma puisqu'au théorème 3 nous avions fait l'hypothèse $\alpha_{j+\frac{1}{2}} > \frac{1}{2}$.

REMARQUE 8

LATHROP et REED ont montré /2/ que l'on avait la relation suivante :

(45) $$\frac{1}{3} \sum_{j=0}^{J-1} (\mu_{j+1} - \mu_j)\,\mu_{j+\frac{1}{2}}^2 = 1 \quad .$$

On dit que le schéma vérifie "l'approximation de la diffusion", ce qui signifie que si l'on suppose que φ_{ex} est un polynome du premier degré en μ , alors on ne commet aucune erreur de discrétisation en μ (Cf. /1/ p.181).

Résumons finalement quel peut être un choix optimal de α et θ .

Ayant fixé les r_i on pose :

(46)
$$\begin{cases} \theta_{i+\frac{1}{2}} = \frac{1}{2} + \frac{1}{6} \dfrac{r_{i+1}-r_i}{r_{i+1}+r_i} \\ r_{i+\frac{1}{2}} = \theta_{i+\frac{1}{2}}\, r_{i+1} + (1-\theta_{i+\frac{1}{2}})\, r_i \, . \end{cases}$$

Les μ_j ayant été fixés, on détermine les $\alpha_{j+\frac{1}{2}}$ et β_j , pour qu'ils vérifient :

(47)
$$\begin{cases} \beta_{j+1} - \beta_j = -2\mu_{j+\frac{1}{2}}(\mu_{j+1}-\mu_j) \qquad , \quad \beta_J = 0 \\ \mu_{j+\frac{1}{2}} = \alpha_{j+\frac{1}{2}}\mu_{j+1} + (1-\alpha_{j+\frac{1}{2}})\mu_j \\ 1-\mu^2_{j+\frac{1}{2}} = (1-\alpha_{j+\frac{1}{2}})\beta_{j+1} + \alpha_{j+\frac{1}{2}}\beta_j \, . \end{cases}$$

On verra dans l'annexe A, la manière d'effectuer ce calcul.

Alors l'équation discrétisée sous forme conservative peut en fait s'écrire :

(48)
$$\left(\frac{\mu}{1-\mu^2}\right)_{j+\frac{1}{2}} \cdot \frac{\varphi_{i+1}-\varphi_i}{r_{i+1}-r_i} + \frac{1}{r_{i+\frac{1}{2}}} \cdot \frac{\varphi_{j+1}-\varphi_j}{\mu_{j+1}-\mu_j} + \sigma_{i+1/2} \cdot \frac{\overline{\varphi}}{1-\mu^2_{j+\frac{1}{2}}} = \frac{\overline{S}}{1-\mu^2_{j+\frac{1}{2}}}$$

$$\overline{\varphi} = \theta_{i+1/2}\varphi_{i+1} + (1-\theta_{i+1/2})\varphi_i = \alpha_{j+\frac{1}{2}}\varphi_{j+1} + (1-\alpha_{j+1/2})\varphi_j \, .$$

Dans le demi domaine $\mu > 0$, les hypothèses du théorème 3 sont satisfaites ($\alpha_{j+1/2} > \frac{1}{2}$, $\theta_{j+\frac{1}{2}} > \frac{1}{2}$). Malheureusement ces hypothèses ne sont pas satisfaites dans le demi domaine $\mu < 0$ bien que l'on y ait :

$$(1-\theta_{i+\frac{1}{2}}) \geq \frac{1}{2} - O(\Delta r) \quad \text{sauf au voisinage de } r = 0$$

$$\alpha_{j+\frac{1}{2}} \geq \frac{1}{2} - O(\Delta\mu) \quad \text{sauf au voisinage de } \mu = 0 \, .$$

Il y a donc une difficulté non résolue pour savoir s'il est encore possible d'obtenir une majoration de stabilité pour une certaine norme L^2 . En fait si on regarde ce qui se passe sur l'équation (8) pour $\mu = -1$, on peut voir que pour

$$\theta = \frac{1}{2} + \frac{1}{6} \frac{r_{i+1}-r_i}{r_{i+1}+r_i}$$

on peut obtenir la stabilité de $r^\beta \varphi$ avec $\beta = \frac{1}{3}$ (Cf. Annexe B).

CONCLUSION

Les essais numériques faits par LATHROP et REED /2/ semblent prouver que le schéma (48) est encore stable pour les choix de α et θ donnés par (46) et (47). Par ailleurs, la discussion de l'erreur de troncature semble recouvrir un problème réel. En effet, on a mis en évidence (par des essais numériques faits avec divers choix de Δr et de $\Delta \mu$) que l'on a bien une erreur d'ordre 2 en Δr pour $r \geq a > 0$ mais qu'au voisinage de $r = 0$ l'erreur est d'ordre strictement inférieur et comprise entre 1 et 2, lorsque θ est donné par (46), alors qu'elle est d'ordre 1 si on choisit $\theta = \frac{1}{2}$.

Il est probable qu'il se passe le phénomène suivant : En passant de $\theta = \frac{1}{2}$ à $\theta = \frac{1}{2} + \frac{1}{6}\frac{r_{i+1}-r_i}{r_{i+1}+r_i}$, l'erreur de troncature au centre passe de $O(1)$ à $O(\Delta r)$ mais l'erreur sur la solution passe de $O(\Delta r)$ à $O(\Delta r^{1+\varepsilon})$ seulement avec $0 < \varepsilon < 1$, parce que dans le deuxième cas, le schéma n'est pas stable dans L^2 pour φ ,mais seulement pour $r^\beta \varphi$, avec $0 < \beta < 1$.

ANNEXE A

On suppose que les μ_j sont donnés tels que :

$$0 = \mu_0 < \mu_1 < \dots < \mu_J = 1 \quad .$$

On cherche alors s'il est possible de déterminer des β_j et $\alpha_{j+1/2}$ tels que l'on ait simultanément les relations suivantes :

(1) $$\beta_{j+1} - \beta_j = -2\,\mu_{j+\frac{1}{2}}(\mu_{j+1} - \mu_j) \quad , \quad \beta_J = 0$$

(2) $$\mu_{j+\frac{1}{2}} = \alpha_{j+\frac{1}{2}}\,\mu_{j+1} + (1 - \alpha_{j+\frac{1}{2}})\,\mu_j$$ avec $0 < \alpha_{j+\frac{1}{2}} < 1$

(3) $$1 - \mu^2_{j+\frac{1}{2}} = (1 - \alpha_{j+\frac{1}{2}})\beta_{j+1} + \alpha_{j+\frac{1}{2}}\,\beta_j \quad .$$

On procède de proche en proche en partant de $j = J-1$ et en faisant décroître l'indice j . Nous avons les relations suivantes :

$$\begin{aligned} 1 - \mu^2_{j+\frac{1}{2}} &= \beta_{j+1} - \alpha_{j+\frac{1}{2}}(\beta_{j+1} - \beta_j) \\ &= \beta_{j+1} + 2\,\alpha_{j+\frac{1}{2}}\,\mu_{j+\frac{1}{2}}(\mu_{j+1} - \mu_j) \\ &= \beta_{j+1} + 2\mu_{j+\frac{1}{2}}\left[\mu_{j+\frac{1}{2}} - \mu_j\right] \quad . \end{aligned}$$

Donc :

(4) $$3\,\mu^2_{j+\frac{1}{2}} - 2\mu_j\,\mu_{j+\frac{1}{2}} + \beta_{j+1} - 1 = 0$$

et aussi en utilisant (1)

(5) $$3\,\mu^2_{j+\frac{1}{2}} - 2\,\mu_{j+1}\,\mu_{j+\frac{1}{2}} + \beta_j - 1 = 0 \quad .$$

De (4), on tire :

(6) $$\mu_{j+\frac{1}{2}} = \frac{1}{3}\left[\mu_j + \sqrt{\mu_j^2 + 3\,(1 - \beta_{j+1})}\right]$$

qui permet d'obtenir $\mu_{j+\frac{1}{2}}$ en fonction de β_{j+1} , et (1) donne β_j .

Par ailleurs nous savons qu'il est important pour la stabilité de situer $\alpha_{J+\frac{1}{2}}$ par rapport à $\frac{1}{2}$.

D'après (6), on a, en posant $\mu_{J-1} = a$:

$$\mu_{J-\frac{1}{2}} = \frac{1}{3}\left(a + \sqrt{3 + a^2}\right)$$

$$2\mu_{J-\frac{1}{2}} - (\mu_J + \mu_{J-1}) = \frac{2}{3}\left(a + \sqrt{3+a^2}\right) - (1+a) .$$

Vérifions que cette quantité est positive. Cela est équivalent à :

$$a + \sqrt{3 + a^2} > \frac{3}{2}(1+a)$$

Donc : $(3 + a^2) > \left(\frac{3+a}{2}\right)^2$.

Donc : $4(3+a^2) - (9 + 6a + a^2) > 0$.

Donc : $3 - 6a + 3a^2 > 0$ Donc : $a \neq 1$.

Il en résulte que, quels que soient μ_{J-1} , on a :

$$\mu_{J-\frac{1}{2}} > \frac{1}{2}\left(\mu_J + \mu_{J-1}\right) \quad \text{soit :} \quad \alpha_{J-\frac{1}{2}} > \frac{1}{2} .$$

Par ailleurs, si on prend $J = 2$ avec $\mu_1 = \frac{1}{4}$, on obtient :

$$\mu_{\frac{3}{2}} = \frac{1}{3}\left[\frac{1}{4} + \sqrt{\frac{1}{16} + 3}\right] = \frac{2}{3}$$

$$\beta_1 = 2\mu_{\frac{3}{2}}(\mu_2 - \mu_1) = 2 . \frac{2}{3} . \left(1 - \frac{1}{4}\right) = 1$$

$$\mu_{\frac{1}{2}} = 0$$

donc :

$$\alpha_{\frac{1}{2}} = 0 .$$

On voit donc qu'il faut faire une hypothèse supplémentaire sur les μ_j pour avoir l'inégalité voulue $\alpha_{j+\frac{1}{2}} > \frac{1}{2}$. Cette hypothèse est que la longueur des intervalles $(\mu_{j+1} - \mu_j)$ va en croissant lorsque j varie de $J-1$ à 0. Cette hypothèse est vérifiée dans les deux seuls cas utilisés en pratique, à savoir :

1°) Les μ_j sont équirépartis donc : $\mu_{j+1} - \mu_j = \Delta\mu = \frac{1}{J}$.

2°) Les μ_j sont les racines du polynome de Legendre.

Nous allons donc montrer le :

<u>LEMME 1</u>

Supposons que $(\mu_{j+1} - \mu_j) \leq (\mu_j - \mu_{j-1})$ pour $J-1 \geq j \geq 1$ alors on peut déterminer des β_j et $\alpha_{j+\frac{1}{2}}$ vérifiant (1), (2), (3). De plus, on a :

$$1 > \alpha_{j+\frac{1}{2}} > \frac{1}{2}$$

$$0 < \beta_j < 1 \qquad \text{pour} \quad J-1 \geq j \geq 1 \qquad \text{et} \quad \beta_0 > 1 .$$

<u>**Démonstration**</u>

Nous allons montrer par récurrence que :

(7) $$4(1 - \mu_j^2 - \beta_j) + (\mu_{j+1} - \mu_j)^2 > 0$$

et que cette relation entraîne :

$$\mu_{j-\frac{1}{2}} > \frac{1}{2}(\mu_j + \mu_{j-1})$$

donc : $$\alpha_{j-\frac{1}{2}} > \frac{1}{2}$$

Supposons que (7) soit vraie pour $j \geq j_0$, alors puisque $\mu_{j_0} - \mu_{j_0-1} \geq \mu_{j_0+1} - \mu_{j_0}$ on a :

$$4(1 - \mu_{j_0}^2 - \beta_{j_0}) + (\mu_{j_0} - \mu_{j_0-1})^2 > 0$$

soit :

$$12(1 - \beta_{j_0}) + 3\mu_{j_0-1}^2 - 6\mu_{j_0}\mu_{j_0-1} - 9\mu_{j_0}^2 > 0$$

$$\mu_{j_0-1}^2 + 3(1 - \beta_{j_0}) > \left(\frac{3\mu_{j_0} + \mu_{j_0-1}}{2}\right)^2$$

donc :

$$\mu_{j_0-\frac{1}{2}}=\frac{1}{3}\left(\mu_{j_0-1}+\sqrt{\mu_{j_0-1}^2+3(1-\beta_{j_0})}\right)>\frac{1}{3}\left(\mu_{j_0-1}+\frac{3\mu_{j_0}+\mu_{j_0-1}}{2}\right)=\frac{1}{2}\left(\mu_{j_0}+\mu_{j_0-1}\right).$$

Par ailleurs, d'après (5), on a :

$$\mu_{j_0-\frac{1}{2}}=\frac{1}{3}\left[\mu_{j_0}+\sqrt{\mu_{j_0}^2+3(1-\beta_{j_0-1})}\right].$$

Ecrivons que $2\mu_{j_0-\frac{1}{2}} > \mu_{j_0}+\mu_{j_0-1}$ **; il vient :**

$$\sqrt{\mu_{j_0}^2+3(1-\beta_{j_0-1})}>\frac{\mu_{j_0}+3\mu_{j_0-1}}{2}.$$

Soit :

$$4\left(\mu_{j_0}^2+3(1-\beta_{j_0-1})\right)>\mu_{j_0}^2+6\mu_{j_0}\mu_{j_0-1}+9\mu_{j_0-1}^2.$$

Donc :

$$4\left(1-\beta_{j_0-1}-\mu_{j_0-1}^2\right)+\left(\mu_{j_0}-\mu_{j_0-1}\right)^2>0$$

et la relation de récurrence (7) se trouve vérifiée pour $j=j_0-1$ **.**

Par ailleurs nous avons vu précédemment que $2\mu_{J-1}>\mu_J+\mu_{J-1}$ **, donc, en utilisant (5) comme ci-dessus, on en déduit que la relation de récurrence est vraie pour** $j=J-1$ **.**

On a d'après (1) : $\beta_j=\beta_{j+1}+2\mu_{j+\frac{1}{2}}\left(\mu_{j+1}-\mu_j\right)$

donc d'après ce que nous venons de démontrer :

$$\beta_j>\beta_{j+1}+\mu_{j+1}^2-\mu_j^2>\beta_{j+1}$$

soit :

$$1-\beta_j-\mu_j^2<1-\beta_{j+1}-\mu_{j+1}^2 \ \cdots\ <1-\beta_J-\mu_J^2=0 .$$

Donc pour tout j **:** $\beta_j>1-\mu_j^2$.

et d'après l'hypothèse de récurrence qui est vérifiée comme nous l'avons montré ci-dessus :

$$1-\mu_j^2<\beta_j<1-\mu_j^2+\left(\frac{\mu_{j+1}-\mu_j}{2}\right)^2$$

Mais $\mu_2-\mu_1 < \mu_1-\mu_0 = \mu_1$, donc :

$$\beta_j \leqslant \beta_1 < 1 - \frac{3\mu_1^2}{H} \qquad \text{pour } j \geqslant 1$$

et :

$$1 = 1-\mu_0^2 < \beta_0 < 1-\mu_0^2 + \left(\frac{\mu_1-\mu_0}{2}\right)^2 = 1 + \frac{\mu_1^2}{4} .$$

De plus, on a :

$$1-\beta_{j-1} < 1-\beta_j < \mu_j^2 .$$

Donc :

$$\sqrt{\mu_j^2 + 3(1-\beta_{j-1})} < 2\mu_j$$

et :

$$\mu_{j-\frac{1}{2}} = \frac{1}{3}\left[\mu_j + \sqrt{\mu_j^2 + 3(1-\beta_{j-1})}\right] < \mu_j \qquad \text{soit : } \alpha_{j-\frac{1}{2}} < 1 \ ■$$

Nous allons maintenant démontrer un deuxième lemme dans le cas où les μ_j sont équirépartis, qui nous permettra d'évaluer l'erreur de troncature en μ , c'est-à-dire d'évaluer $\mu_{j+\frac{1}{2}} - \frac{1}{2}(\mu_{j+1}+\mu_j)$.

LEMME 2

Supposons que $\mu_{j+1}-\mu_j = \Delta\mu = \frac{1}{J}$ pour $j = 0, \ldots, J-1$
Alors :

$$0 < \mu_{j+\frac{1}{2}} - \frac{1}{2}(\mu_{j+1}+\mu_j) < \frac{C.\ \Delta\mu^2}{\sqrt{\mu_{j+1}}} .$$

Démonstration - Posons : $E_j = 1-\mu_j^2 - \beta_j + \frac{1}{4}\Delta\mu^2$. Nous avons vu au lemme 1 que :

(8) $$0 < E_j < \frac{1}{4}\Delta\mu^2 .$$

Calculons $\mu_{j-\frac{1}{2}} - \frac{1}{2}(\mu_j + \mu_{j-1})$ en fonction de E_j. On a :

$$3(1-\beta_j) + \mu_{j-1}^2 = 3E_j + \left(\frac{3\mu_j + \mu_{j-1}}{2}\right)^2$$

$$\mu_{j-\frac{1}{2}} = \frac{1}{3}\left(\mu_{j-1} + \sqrt{\mu_{j-1}^2 + 3(1-\beta_j)}\right) = \frac{1}{3}\mu_{j-1} + \frac{3\mu_j + \mu_{j-1}}{6}\left\{1 + \frac{3E_j}{\left(\frac{3\mu_j+\mu_{j-1}}{2}\right)^2}\right\}^{1/2}$$

$$= \frac{1}{2}(\mu_j + \mu_{j-1}) + \frac{3\mu_j + \mu_{j-1}}{6}\left[\left(1 + \frac{12E_j}{(3\mu_j + \mu_{j-1})^2}\right)^{1/2} - 1\right].$$

Donc :

(9) $$\mu_{j-\frac{1}{2}} - \frac{1}{2}(\mu_j + \mu_{j-1}) = \frac{3\mu_j + \mu_{j-1}}{6}\left[\left(1 + \frac{12E_j}{(3\mu_j + \mu_{j-1})^2}\right)^{1/2} - 1\right].$$

Par ailleurs :

$$E_j = E_{j+1} + \mu_{j+1}^2 + \beta_{j+1} - \mu_j^2 - \beta_j .$$

Mais :

$$\beta_{j+1} - \beta_j = -2\mu_{j+1/2}(\mu_{j+1} - \mu_j) = -(\mu_{j+1}^2 - \mu_j^2) + (\mu_{j+1} + \mu_j - 2\mu_{j+1/2})(\mu_{j+1} - \mu_j).$$

Donc :

$$E_j = E_{j+1} + (\mu_{j+1} - \mu_j)(\mu_{j+1} + \mu_j - 2\mu_{j+1/2}).$$

Soit :

(10) $$E_j = E_{j+1} + (\mu_{j+1} - \mu_j)\frac{3\mu_{j+1} + \mu_j}{3}\left[1 - \left(1 + \frac{12E_j}{(3\mu_{j+1} + \mu_j)^2}\right)^{1/2}\right].$$

Posons $\varepsilon = \dfrac{12E_j}{(3\mu_{j+1} + \mu_j)^2} > 0$ et servons nous de l'inégalité :

$$(1+\varepsilon)^{1/2} > 1 + \frac{\varepsilon}{2} - \frac{\varepsilon^2}{8} \quad \text{pour } \varepsilon > 0$$

il vient :

$$E_j \leq E_{j+1} - \frac{2E_{j+1}(\mu_{j+1} - \mu_j)}{3\mu_{j+1} + \mu_j} + 6\frac{E_{j+1}^2(\mu_{j+1} - \mu_j)}{(3\mu_{j+1} + \mu_j)^3}$$

(11) $$E_j \leq \frac{\mu_{j+1} + 3\mu_j}{3\mu_{j+1} + \mu_j}E_{j+1} + 6\frac{E_{j+1}^2(\mu_{j+1} - \mu_j)}{(3\mu_{j+1} + \mu_j)^3}.$$

Posons $\varepsilon_j = \dfrac{E_j}{3\mu_j + \mu_{j-1}}$ **et divisons (11) par** $3\mu_j + \mu_{j-1}$ **, on obtient :**

$$\varepsilon_j \leq \frac{\mu_{j+1} + 3\mu_j}{3\mu_j + \mu_{j-1}} \cdot \varepsilon_{j+1} \left(1 + \frac{6(\mu_{j+1} - \mu_j) E_{j+1}}{(3\mu_{j+1} + \mu_j)^2 (\mu_{j+1} + 3\mu_j)} \right).$$

Mais $E_{j+1} < \frac{1}{4} \Delta\mu^2$ **et** $\mu_j = j\Delta\mu$ **, il en résulte que :**

$$\varepsilon_j \leq \frac{4j+1}{4j-1} \; \varepsilon_{j+1} \left[1 + \frac{3}{2(4j+3)^2(4j+1)} \right] \quad \textbf{pour } j \geq 1 .$$

Supposons que η_j **vérifie** $\eta_j = \dfrac{4j+1}{4j-1} \eta_{j+1}$ **et** $\eta_J = \varepsilon_J$ **et posons** $\varphi_j = \dfrac{\varepsilon_j}{\eta_j}$ **.**
Alors :

$$\begin{cases} \varphi_j \leq \varphi_{j+1} \left(1 + \dfrac{3}{2(4j+3)^2(4j+1)} \right) & j \geq 1 . \\ \varphi_J = 1 \end{cases}$$

On obtient :

$$\varphi_j \leq \prod_{\ell=j}^{J-1} \left[1 + \frac{3}{2(4\ell+3)^2(4\ell+1)} \right] \leq C^{te}$$

car on a un produit convergent.

Il en résulte que $\varepsilon_j \leq C^{te} \eta_j$ **donc nous devons évaluer** η_j**.**
Or :

$$\eta_j = \prod_{\ell=j}^{J-1} \cdot \frac{4\ell+1}{4\ell-1} \; \varepsilon_J \leq \frac{1}{4J^2} \cdot \prod_{\ell=j}^{J-1} \left(\frac{4\ell+1}{4\ell-1} \right) .$$

Posons $S_j = \prod_{\ell=j}^{J-1} \left(\dfrac{4\ell+1}{4\ell-1} \right)$ **. On a :**

(12) $$\operatorname{Log} S_j = \sum_{\ell=j}^{J-1} \operatorname{Log} \frac{4\ell+1}{4\ell-1} \leq \int_{j-1}^{J-1} \operatorname{Log} \frac{4x+1}{4x-1} \, dx = \left[x \operatorname{Log} \frac{4x+1}{4x-1} + \frac{1}{4} \operatorname{Log} (16x^2 - 1) \right]_{j-1}^{J-1} .$$

Donc :

$$\begin{aligned} \operatorname{Log} S_2 &\leq (J-1) \operatorname{Log} \frac{4J-3}{4J-5} + \frac{1}{4} \operatorname{Log} \left(16(J-1)^2 - 1 \right) - \operatorname{Log} \frac{5}{3} - \frac{1}{4} \operatorname{Log} 15 \\ &\sim \frac{1}{2} - \operatorname{Log} \frac{5}{3} + \frac{1}{2} \operatorname{Log} (J-1) . \end{aligned}$$

Par ailleurs :

$$\operatorname{Log} S_1 = \operatorname{Log} \frac{5}{3} + \operatorname{Log} S_2 \sim \frac{1}{2} + \frac{1}{2} \operatorname{Log} (J-1) .$$

Donc : $S_1 \sim c\sqrt{J}$.

On tire en fait de (12) que :

$$\text{Log } S_j = (J-1) \text{ Log } \frac{4J-3}{4J-5} - (j-1) \text{ Log } \frac{4j-3}{4j-5} + \frac{1}{4} \text{ Log } \frac{(J-1)^2 - \frac{1}{16}}{(j-1)^2 - \frac{1}{16}}$$

$$\sim c + \frac{1}{2} \text{ Log } \frac{J}{j} \sim c + \frac{1}{2} \text{ Log } \frac{1}{\mu_j} .$$

Donc :

$$S_j \sim \frac{c}{\sqrt{\mu_j}} .$$

Il en résulte que :

$$\varepsilon_j \sim \frac{c \Delta \mu^2}{\sqrt{\mu_j}}$$

soit :

$$2 \mu_{j+\frac{1}{2}} - (\mu_{j+1} + \mu_j) \sim \frac{2 E_{j+1}}{3\mu_{j+1} + \mu_j} = 2 \varepsilon_{j+1} \sim \frac{c \Delta \mu^2}{\sqrt{\mu_{j+1}}} . \blacksquare$$

ANNEXE B

Nous voulons montrer ici comment on peut obtenir la stabilité de $r^{4/3}\varphi$ pour la discrétisation de l'équation du transport en $\mu = -1$, par une technique susceptible de s'appliquer à l'équation (48) dans le demi domaine $\mu < 0$. Le résultat n'est pas intéressant en lui-même, car par une autre technique, plus simple, on peut montrer directement que φ est stable.

Soit donc l'équation :

(1) $$-\frac{d}{dr}(r^2\varphi) + 2r\varphi + r^2\sigma\varphi = r^2 S \quad , \quad \varphi(R) = 0$$

discrétisée en :

(2) $$-\frac{r_{i+1}^2\varphi_{i+1} - r_i^2\varphi_i}{r_{i+1}-r_i} + (r_{i+1}+r_i)\bar{\varphi} + \bar{r}_{i+1/2}^2\,\sigma_{i+1/2}\,\bar{\varphi} = \bar{r}_{i+1/2}^2\,\bar{S}$$

où : $\bar{\varphi} = \theta\varphi_{i+1} + (1-\theta)\varphi_i$ avec : $\theta = \frac{1}{2} + \frac{1}{6}\,\frac{r_{i+1}-r_i}{r_{i+1}+r_i} > \frac{1}{2}$

que l'on peut réécrire, sous une forme analogue à (48), d'après (30) :

(3) $$-\frac{\varphi_{i+1}-\varphi_i}{r_{i+1}-r_i} + \sigma_{i+1/2}\bar{\varphi} = \bar{S} \quad .$$

Nous avons l'identité :

(4) $$(\varphi_{i+1}-\varphi_i)(\theta\varphi_{i+1}+(1-\theta)\varphi_i) = \theta\varphi_{i+1}^2 - (1-\theta)\varphi_i^2 + (1-2\theta)\varphi_{i+1}\varphi_i$$

et la majoration :

(5) $$|(2\theta-1)\varphi_i\varphi_{i+1}| \leq (\theta-\tfrac{1}{2})\left[\frac{1-\theta}{\theta}\varphi_i^2 + \frac{\theta}{1-\theta}\varphi_{i+1}^2\right]$$

ce qui entraîne :

(6) $$-(\varphi_{i+1}-\varphi_i)\bar{\varphi} \geq \frac{1-\theta}{2\theta}\varphi_i^2 - \frac{\theta}{2(1-\theta)}\varphi_{i+1}^2 \quad .$$

Multiplions (3) par $(r_{i+1}-r_i)\bar{\varphi}$ et utilisons (6) ; il vient :

(7) $$\frac{1-\theta}{2\theta}\varphi_i^2 - \frac{\theta}{2(1-\theta)}\varphi_{i+1}^2 + \sigma_{i+1/2}(r_{i+1}-r_i)\bar{\varphi}^2 \leq (r_{i+1}-r_i)\bar{S}\bar{\varphi}$$

Par ailleurs, on montre que :

(8) $$\frac{1-\theta}{\theta} > \left(\frac{r_i}{r_{i+1}}\right)^{1/3} .$$

En effet, en posant $x = \frac{r_i}{r_{i+1}}$, cela revient à vérifier que :

(9) $$\frac{1-\theta}{\theta} = \frac{1+2x}{2+x} > x^{1/2} \quad \text{pour} \quad 0 \leq x < 1 .$$

En multipliant (7) par $\frac{2\theta}{1-\theta}$, on obtient une inégalité du type :

(10) $$r_i^{2/3}\varphi_i^2 - r_{i+1}^{2/3}\varphi_{i+1}^2 \leq (r_{i+1}-r_i)\, r_{i+1}^{2/3}\,\bar{\varphi}\,\bar{S} .$$

Soit, en sommant de i à I :

(11) $$r_i^{2/3}\varphi_i^2 \leq \sum_{j=i}^{(I-1)} (r_{j+1}-r_j)\, r_{j+1}^{2/3}\, \varphi_{j+1/2}\, S_{j+1/2}$$

qui est l'équivalent discret de :

(12) $$r^{2/3}\varphi^2(r) \leq \int_r^R t^{2/3}\varphi(t)\,S(t)\,dt .$$

Quelques manipulations supplémentaires donnent alors :

(13) $$r_i^{1/3}|\varphi_i| \leq \sum_{j=i}^{I-1} (r_{j+1}-r_j)\, r_{j+1}^{1/3}\, |S_{j+1/2}|$$

ce qui prouve que :

(14) $$\max_i\, r_i^{1/3}|\varphi_i| \leq c^{te} .$$

REMARQUE

Pour obtenir la majoration précédente, nous ne nous sommes pas servis du terme en facteur de σ, contrairement à ce qui est fait dans la démonstration des théorèmes 2 et 3.

REFERENCES

/1/ - K. LATHROP et B. CARLSON - "Numerical Solution of the Boltzmann Transport Equation" - J. Comput. Physics, 2 (1967), pp. 173-197.

/2/ - W.H. REED et K.D. LATHROP - "Truncation Error Analysis of Finite Difference Approximations to the Transport Equation" - Nuclear Sc. Enfin, 41 (1970), pp. 237-248.

SOME APPLICATIONS OF THE NUMERICAL SOLUTION OF INTEGRAL EQUATIONS TO BOUNDARY VALUE PROBLEMS

Ben Noble

1. Introduction. The motivation for the development of the theory of integral equations around 1900 came largely from the fact that the Dirichlet and Neumann problems could be formulated as Fredholm integral equations. This enabled mathematicians to obtain insight into existence-uniqueness problems associated with elliptic partial differential equations. Nystrom realized around 1930 that numerical solution of the integral equations provided a practical method for solving boundary value problems but, as in many other contexts, effective exploitation of this idea had to await the advent of the digital computer.

We remind the reader of some of the background connected with solution of boundary value problems by integral equations. Consider

$$\lambda f(s) + \int_S k(s,t)f(t)d\sigma = g(s), \quad s \in S, \tag{1}$$

where, for the one-dimensional equation, corresponding to a two-dimensional boundary value problem, s is a single variable in a range $a \leq s \leq b$, and, for the two-dimensional equation, s represents a pair of numbers (x,y) that are typically parametric coordinates on a surface S in three-dimensional space. The integral equation is first or second kind, depending on whether $\lambda = 0$ or $\lambda \neq 0$, respectively.

A given boundary value problem can be formulated in terms of either first or second kind Fredholm equations. To solve the Dirichlet problem, for example, we can think in physical terms of representing the potential by a single or double layer on the boundary. This leads to first or second kind equations, respectively. (These can of course be obtained by Green's function methods.)

Most numerical treatments of boundary value problems use the formulation in terms of second-kind equations since first-kind equations tend to be more difficult to deal with, both theoretically and numerically. The reason for this can be seen, from one point of view, by considering, in operator notation,

$$\lambda f + Kf = g, \tag{2}$$

where K is a compact hermitian (or normal) operator. (Equation (1) is a special case of this if k(s,t) is symmetric and continuous in s,t.) There then exists a complete set of eigenfunctions ϕ_i corresponding to eigenvalues λ_i, $K\phi_i = \lambda_i\phi_i$. We assume that

$$g = \sum g_i \phi_i \quad .$$

It is then easy to see that, if a solution of (2) exists, it is given by

$$f = \sum \frac{g_i}{\lambda + \lambda_i} \phi_i \quad . \tag{3}$$

The point now is that if K is compact, λ_j tends to zero as j tends to infinity. If $\lambda \neq 0$, the rate of convergence of the series for f is the same as the rate of convergence of the series for g. If $\lambda = 0$, the rate for f is <u>worse</u> than that for g because of the factor $(1/\lambda_j)$ that then multiplies the terms in (3).

The key question in deciding whether a given first-kind integral equation will be troublesome to solve numerically is to decide how fast the eigenvalues λ_j go to zero as j tends to infinity. Some insight can be obtained from the case of the one-dimensional difference kernel, for which k(s,t) = s - t. The asymptotic behavior of λ_j can be studied by Fourier methods, and it turns out that the <u>smoother</u> the kernel, the <u>faster</u> the λ_j tend to zero as j increases. (This is related to the fact that if f(x) is periodic and k-times continuously differentiable, the jth Fourier coefficient decreases at least as fast as j^{-k-1}.) When solving first-kind equations it therefore helps to have a nice singular kernel - the smoother the kernel, the more troublesome will be the numerical solution. (This is reflected in ill-conditioning in the sets of linear equations to which the equations are reduced.) Fortunately the kernels that appear when boundary value problems are solved by integral equations of the first kind have kernels that are infinite when s = t. The corresponding eigenvalues λ_j tend to zero as j increases only as a small inverse power of j. (This can be checked by looking at cases where the original boundary value problem can be solved exactly by separation of variables.) It turns out that the difficulty in solving boundary value problems by numerical solution of integral equations of the first kind lies in technical difficulties associated with evaluation of integrals rather than ill-conditioning associated with first-kind equations.

The advantage of the integral equation approach to the numerical solution of boundary value problems, as opposed to finite-difference methods, is that the dimension of the problem is reduced by one, because the problem is formulated in terms of an unknown function defined on the boundary of the region. It is debatable whether there is any real advantage in using integral equations, as opposed to finite differences, in a bounded region. The integral equation approach is more difficult to automate (though this may be simply because little work has been done on developing standard computer routines). Also the integral equation method is somewhat more sophisticated (though this may be a function of familiarity).

The integral equation method has a clear advantage over finite differences when the region is of infinite extent, and the integral equation for the boundary value problem involves a function defined over a region of finite extent. A problem involving an infinite region is then replaced by the problem of finding a function over a finite region of dimension one less than the original.

The natural context in which to exploit this advantage is to formulate boundary value problems in terms of integral equations of the <u>second</u> kind, since the theory of these equations has been settled definitively, and their numerical solution has been studied extensively (see Atkinson [1], where references will be found to earlier work of Anselone and others). Thus the McDonnell-Douglas aircraft company has a large computer program for calculating the potential flow round bodies of arbitrary shape (see Hess and Smith [3]). Of several related references we mention only Lynn and Timlake [11].

Several programs have been developed in recent years for computing the sound field produced by a radiating body of arbitrary shape (see Schenck [17]). Here again there are several related references, of which we mention Kussmaul [7]), and Kussmaul and Werner [8]. One interesting feature of the radiation (or diffraction) problem is that although the physical problem has a unique solution, the integral equations run into uniqueness trouble at eigenvalues associated with a boundary value problem for the interior region enclosed by the bounding surface. These remarks are included simply to remind the reader that the numerical solution of Fredholm integral equations of the second kind is a well-established technique for solving boundary value problems.

The objective of this paper is to describe two situations in which it is convenient to solve boundary value problems numerically by formulating them in terms of integral equations of the first kind.

We first discuss the computation of the capacitance of a rectangular solid in free space. The basic idea goes back at least as far as Maxwell, 1879 [13]. A later reference is Hildebrand, 1941 [4]. Surprisingly, similar ideas were not exploited in elasticity until quite recently by Jaswon and his students, [5].

We next discuss an approach to the numerical solution of mixed boundary value problems via integral equations involving Abel-type integrals. This was first exploited by one of my students, P. Linz (see [9] which is one-half of his thesis). D. A. Spence has recently obtained extensive numerical results on the elastic punch with slip, using a similar basic idea. It is particularly appropriate to talk about this in Scotland where much work has been done on mixed boundary value problems by I. N. Sneddon and colleagues.

2. The charged rectangular lamina in free space. We first establish the integral equation governing the electrostatic charge distribution on a flat conducting rectangular lamina in free space. Suppose that the total

(i.e., two sided) charge density at a point (ξ,η) on the lamina, or plate of zero thickness, lying in $z = 0$, $-a \leq x \leq a$, $-b \leq y \leq b$, is $f(\xi,\eta)$. By superposition, the potential produced at any point (x,y,z) by this charge is

$$\phi = \int_{-a}^{a}\int_{-b}^{b} \frac{f(\xi,\eta)d\eta d\xi}{[(x-\xi)^2 + (y-\eta)^2 + z^2]^{1/2}} .$$

Since the lamina is conducting, the charge will distribute itself so that the potential on the plate is a constant. An integral equation for the unknown function $f(\xi,\eta)$ is obtained by letting (x,y,z) tend to any point $(x,y,0)$ on the plate. If the plate is assumed to be at unit potential, this gives

$$\int_{-a}^{a}\int_{-b}^{b} \frac{f(\xi,\eta)d\eta d\xi}{[(x-\xi)^2 + (y-\eta)^2]^{1/2}} = 1, \quad \begin{cases} -a \leq x \leq a \\ -b \leq x \leq b \end{cases} . \tag{4}$$

This is an integral equation of the first kind.

To solve this equation numerically we approximate $f(x,y)$ by piecewise constant functions. We need consider only $0 \leq x \leq a$, $0 \leq y \leq b$, since $f(x,y)$ is clearly symmetric about $x = 0$, and $y = 0$. Introduce constants a_r, b_s (the choice of which will be discussed later), such that:

$$0 = a_0 < a_1 < \ldots < a_m = a, \quad 0 = b_0 < b_1 < \ldots < b_n = b.$$

We take, as an approximation for $f(x,y)$, the piecewise constant function:

$$f(x,y) \approx \sum_{r=1}^{m} \sum_{s=1}^{n} c_{rs} f_{rs}(x,y), \tag{5}$$

where

$$f_{rs}(x,y) = \begin{cases} 1, & a_{r-1} \leq x \leq a_r,\ b_{s-1} \leq y \leq b_s, \\ 0, & \text{elsewhere} \end{cases}$$

The unknown constants c_{rs} are determined by the condition that the expression resulting from the substitution of (5) into the left-hand side of (4) should equal unity for the mn values of x,y corresponding to the centers of the rectangles defining the $f_{rs}(x,y)$. This gives

$$\sum_{r=1}^{m}\sum_{s=1}^{n} A_{ij,rs}c_{rs} = 1, \quad i=1,\dots,m;\ j=1,\dots,n,$$

where

$$A_{ij,rs} = J_{rs}(x_i,y_j) + J_{rs}(-x_i,y_j) + J_{rs}(x_i,-y_j) + J_{rs}(-x_i,-y_j),$$

$$J_{rs}(x,y) = \int_{a_{r-1}}^{a_r}\int_{b_{s-1}}^{b_s} \frac{d\eta d\xi}{[(x-\xi)^2 + (x-\eta)^2]^{1/2}} ,$$

$$x_i = \tfrac{1}{2}(a_{i-1}+a_i) , \quad y_j = \tfrac{1}{2}(b_{j-1}+b_j) .$$

An explicit expression for $J_{rs}(x,y)$ can be obtained by specialization of (7) below.

We now discuss the optimum choice of the subdivisions a_r, b_s. On physical grounds, it is known that the charge distribution, i.e., the unknown function $f(x,y)$, tends to infinity as $x \to a$ (for fixed y) and as $y \to b$ (for fixed x). This indicates that the widths of the subrectangles should be reduced near $x = a$ and $y = b$. Following a method that I used in [14], in connection with a variational principle for the same problem, we choose

$$\frac{a_r}{a} = \frac{(m-r+1)^k + \cdots + (m-1)^k + m^k}{1^k + 2^k + \cdots + (m-1)^k + m^k} , \quad r = 1,2,\dots,m. \tag{6}$$

Here m is the number of subdivisions, and k is a constant that is also at our disposal. The larger k, the narrower the rectangles near $x = a$.

There is of course no guarantee that the optimum subdivisions found by the variational method in [14] will also be the best for the colloctation method described above, so we proceed empirically, by finding the rate of convergence of the results as m,n increase, for fixed k in (6). A suitable quantity to use in judging the most rapid rate of convergence is the total charge

$$C_{mn} = 4\sum_{r=1}^{m}\sum_{s=1}^{n} c_{rs}(a_r-a_{r-1})(b_s-b_{s-1}) .$$

When m=n it is found that the value of C_{mm} varies symmetrically with m in a way that is consistent with the hypothesis that, for fixed k and varying m, the difference between the true capacity and the estimated capacity varies as pm^{-q}, where p and q are suitable constants. This means that Aitken's δ^2-extrapolation can be used to obtain an improved estimate of the capacity from estimates for m = 2,4,8. Rapid convergence is found for k = 3,4, and these value of m and k yield the results given in Table 1 for the capacity, in e.s.u., of rectangular laminas with a shorter side equal to 2 cms. (a=1 cm.) The internal consistency of the results for various m and k indicate that these should be accurate to about 1 in 5000. The capacity of a circular lamina of radius 1 cm. is $2/\pi$ e.s.u. To convert results in e.s.u. to μμF, divide by 0.9. For comparison we give results derived in [14] using the variational method. The agreement is gratifying.

Table 1 : Capacity in e.s.u. of a rectangular lamina of shorter side 2 cms.

b:a	1:1	2:1	3:1	4:1
Present method	0.7337	1.0640	1.6189	2.5698
Variational [14]	0.734	1.065	1.619	2.570

3. The capacity of a rectangular solid in free space. The method used for the lamina in the last section can be extended in an obvious way to deal with the rectangular solid in free space. If the solid lies in $-a \le x \le a$, $-b \le y \le b$, $-c \le z \le c$, we now have three sets of subdivisions a_r, b_s, c_t. The coefficients of the simultaneous linear equations can be expressed in terms of the integral

$$\begin{aligned} I(\alpha,\beta;x,y,z) &= \int_{-\alpha}^{\alpha}\int_{-\beta}^{\beta} \frac{d\xi\, d\eta}{[(x-\xi)^2+(y-\eta)^2+z^2]^{1/2}} \\ &= H(x+\alpha,\ y+\beta,\ z) - H(x-\alpha,\ y+\beta,\ z) \\ &\quad - H(x+\alpha,\ y-\beta,\ z) + H(x-\alpha,\ y-\beta,\ z) \end{aligned} \qquad (7)$$

where

$$H(p,q,z) = p \log \{q + (p^2+q^2+z^2)^{1/2}\} + q \log \{p + (p^2+q^2+z^2)^{1/2}\}$$

$$- z \tan^{-1} \frac{pq}{z(p^2+q^2+z^2)^{1/2}} ,$$

and this result holds for all x,y,z.

Some numerical results for the cube are given in Table 2. This gives estimates of the capacity for various m and k. It is found that the most rapid convergence is given by k = 2 or 3 instead of k = 3 or 4 for the lamina. This is understandable since, as we go towards the edges and corners of a cube, the charge density tends to zero less rapidly than for a lamina.

From Table 2, the estimated capacity of a cube of side 2 cms. in free space is 1.32136 e.s.u. The most accurate previously published estimate is 1.322 given by Greenspan [2].

The column marked "Maxwell method" in Table 2 is obtained by dividing the side of the cube into equal squares. The potential at the midpoint of any square is the sum of contributions from all other squares, and from the square itself. The contributions from the other squares are calculated as if the charges on the other squares are concentrated at their centers. The contribution from the square on itself is calculated as if the charge were equally distributed over the square. This can be regarded as a crude method for solving the integral equation. It has been called the "method of subareas" by T. J. Higgins (see [17]). Maxwell applied the method only to the lamina, and actually he added an ingenious twist. He realized that "fudge-factors" should be introduced to allow for the concentration of charge near the edges and corners.

Table 2 : Estimates of the capacity of a cube of side 2 cms. obtained by various methods.

m	Maxwell method	Collocation k = 0	2	3
1	1.2658	1.2730	1.2730	1.2730
2	1.3011	1.3008	1.3138	1.3146
3	1.3112	1.3090	1.3190	1.3192
4	1.3152	1.3128	1.32034	1.32042
6	1.3185	1.3163	1.32106	1.32106
8	1.3198	1.3179	1.32123	1.32123

Table 3: Estimates of the capacity of a rectangular solid in free space with sides 2a, 2b, 2c, largest side 2a = 2 cms. The case c = 0 is a lamina.

	c/a = 1	1/2	1/4	1/8	0
b/a = 1	1.321	1.082	0.939	0.853	0.734
$\frac{1}{2}$		0.860	0.727	0.646	0.532
$\frac{1}{4}$			0.598	0.520	0.405
$\frac{1}{8}$				0.441	0.321

Estimates of the capacities of rectangular solids for various ratios of $a : b : c$ are given in Table 3. These were derived from results for $m = 1,2,4$ and $k = 2,3$. The results for $m = 4$, $k = 2,3$, agreed with each other and with the δ^2-extrapolated values to within one digit in the last figure quoted in the table. The results quoted are the mean of the δ^2-extrapolated values for $k = 2,3$, rounded to three decimals.

4. The charge distribution on an annulus. The remainder of this paper is concerned with mixed boundary value problems. There is an extensive literature, most of which is concerned with somewhat restricted classes of axially symmetric problems. For potential theory the literature has been well summarized by Sneddon [19], and for static elasticity see Sneddon and Lowengrub [20]. The point that we are going to make is that, particularly for numerical purposes, it is often convenient to deal directly with integral equations that involve Abel-type integrals. The basic idea was first exploited in a thesis by a student of mine, P. Linz [9]. The particular device I will use to derive the equations (which is neater than the original method) was suggested to me by D. A. Spence.

Consider an axially symmetric potential ϕ in cylindrical coordinates (r,z). Suppose that an annulus lies in $z = 0$, $\alpha \leq r \leq 1$. We need consider only the upper half-space:

$$\frac{1}{r}\frac{\partial}{\partial r}\, r\,\frac{\partial \phi}{\partial r} + \frac{\partial^2 \phi}{\partial z^2} = 0\,, \quad z \geq 0,\ 0 \leq r < \infty$$

with the following boundary conditions on $z = 0$:

$$\begin{aligned} \phi &= 1, && \alpha \leq r \leq 1 \\ \frac{\partial \phi}{\partial z} &= 0, && 0 \leq r < \alpha \quad 1 < r < \infty \end{aligned} \qquad (8)$$

Also $\phi \to 0$ as $r \to \infty$. The potential in the upper half-space has the well-known representation

$$\phi = \int_0^\infty tA(t)e^{-tz} J_0(rt)dt.$$

We first proceed quite generally. Suppose that

$$\phi = f(r), \quad \frac{\partial\phi}{\partial z} = g(r), \; z = 0, \quad 0 \leq r < \infty. \tag{9}$$

Then

$$f(r) = \int_0^\infty tA(t)J_0(rt)dt, \tag{10}$$

$$g(r) = -\int_0^\infty t^2A(t)J_0(rt)dt. \tag{11}$$

If we invert each of these expressions and eliminate A(t) we find

$$\int_0^\infty r\, f(r)\, J_0(rt)dt = -\frac{1}{t}\int_0^\infty r\, g(r)J_0(rt)dt. \tag{12}$$

Now comes the device mentioned earlier. We take the Fourier sine transform of both sides with respect to t, and use the standard integrals:

$$\int_0^\infty \sin xt\, J_0(rt)dt = \begin{cases} 0 & , \quad r > x, \\ \dfrac{1}{(x^2-r^2)^{1/2}} & , \quad r < x. \end{cases} \tag{13}$$

$$\int_0^\infty \frac{\sin xt}{t} J_0(rt)dt = \begin{cases} \text{arc}\sin(x/r), & r > x, \\ \pi/2, & r < x. \end{cases} \tag{14}$$

On differentiating the resulting equations with respect to x we find

$$\frac{d}{dx}\int_0^x \frac{rf(r)}{(x^2-r^2)^{1/2}}\, dr = -\int_x^\infty \frac{rg(r)}{(r^2-x^2)^{1/2}}\, dr \quad . \tag{15}$$

We note in passing that problems involving parallel disks or a disk between parallel planes give rise to the following generalization of (10):

$$f(r) = \int_0^\infty t\{1 + H(t)\}A(t)J_0(rt)dt, \tag{16}$$

where $H(t) \to 0$ as $t \to \infty$. Then (12) is replaced by

$$\int_0^\infty r\, f(r) J_0(rt)dt = -\frac{1}{t}\{1 + H(t)\} \int_0^\infty r\, g(r) J_0(rt)dt. \tag{17}$$

The Fourier sine transform of this equation leads to:

$$\frac{d}{dx}\int_0^x \frac{rf(r)}{(x^2-r^2)^{1/2}}\, dr = -\int_x^\infty \frac{rg(r)}{(r^2-x^2)^{1/2}}\, dr - \int_0^\infty rg(r)h(r,x)dr, \tag{18}$$

where

$$h(r,x) = \int_0^\infty H(t)\cos xt\, J_0(rt)dt. \tag{19}$$

The utility of the present approach depends to a large extent on whether we can handle this integral. The case of charged parallel disks is considered in Linz [10].

A second note-in-passing concerns two-dimensional problems corresponding to axially symmetric problems. These involve trignometric instead of Bessel functions. Thus consider the following where, to avoid divergencies, the first equation has been differentiated with respect to x:

$$f'(x) = -\int_0^\infty t\{1 + H(t)\}\, A(t)\, \sin xt\, dt,$$

$$g(x) = -\int_0^\infty t\, A(t)\, \cos rt\, dt.$$

Inverting and eliminating A(t) we obtain (compare (17):

$$\int_0^\infty f'(x)\sin xt\, dx = \{1 + H(t)\}\int_0^\infty g(x)\, \cos xt\, dx\ .$$

If we multiply both sides by $J_1(rt)$, integrate with respect to t, and evaluate the resulting integrals in t, we find

$$\int_0^r \frac{xf'(x)}{(r^2-x^2)^{1/2}}\,dx = \int_0^\infty g(x)dx - \int_r^\infty \frac{xg(x)}{(x^2-r^2)^{1/2}}\,dx \tag{20}$$

$$+ \; r\int_0^\infty g(x)\; h(r,x)dx,$$

where

$$h(r,x) = \int_0^\infty H(t)\;\cos xt\; J_1(rt)dt. \tag{21}$$

These bear an interesting resemblance to (18), (19). Again the main practical problem lies in handling the integral (21).

Coming back to the problem of the annulus, if we insert the boundary conditions (8) in (15), we find

$$f(0) + x\int_0^x \frac{f'(r)}{(x^2-r^2)^{1/2}}\,dr = -\int_\alpha^1 \frac{rg(r)}{(r^2-x^2)^{1/2}}\,dr, \quad 0 \leq x \leq \alpha,$$

$$f(0) + x\int_0^\alpha \frac{f'(r)}{(x^2-r^2)^{1/2}}\,dr = -\int_x^1 \frac{rg(r)}{(r^2-x^2)^{1/2}}\,dr, \quad \alpha \leq x \leq 1.$$

These are coupled integral equations of an unusual type for the unknown potential $f(r)$ in $0 \leq r \leq \alpha$ and the unknown charge distribution $g(r)$ in $\alpha \leq r \leq 1$.

The numerical solution of these equations presents no great difficulty. We can subdivide the range $0 \leq r \leq 1$, assume appropriate representations for the unknown functions in the subintervals, and integrate explicitly over the kernel singularities. Physically it is clear that $f'(r)$ and $g(r)$ will have singularities near $r = \alpha - 0$ and $r = \alpha + 0,\ 1 - 0$, respectively. To obtain accurate results these must be taken care of, either by choosing small subdivisions near $r = \alpha, 1$ (compare the earlier part of this paper), or by building the singularities into the assumed representations for the unknown functions.

In practice it is found that the numerical solution of the integral equations is stable and presents no fundamental difficulty. Some numerical results for the annulus are given in Linz [9], and for the parallel plate condenser (which involves equations related to (18)) in Linz [10].

5. Diffraction by a disk and annulus. Consider, instead of Laplace's equation, the steady-state wave equation

$$\frac{1}{r}\frac{\phi}{\delta r} r \frac{\delta\phi}{\delta r} + \frac{\delta^2\phi}{\delta z^2} + k^2\phi = 0, \qquad z \geq 0,\ 0 \leq r < \infty,$$

where ϕ satisfies the radiation condition at infinity. The representation for ϕ in the upper half-space is;

$$\phi = \int_0^\infty t\, A(t) e^{-\alpha z}\, J_0(rt)dt,$$

where

$$\alpha = (t^2-k^2)^{1/2}, \quad t \geq k; \qquad -i(k^2-t^2)^{1/2}, \qquad t \leq k.$$

Proceeding as before, using the notation [9], we find

$$\int_0^\infty f\, f(r)\, J_0(rt)dr = -\frac{1}{\alpha}\int_0^\infty r\, s(r)\, J_0(rt)dr. \tag{22}$$

Two courses of action are now possible. We can obtain a "static approximation" by writing equation (21) in the form (16) with

$$H(t) = \frac{t}{(t^2-k^2)^{1/2}} - 1.$$

Contour integration can be used to reduce the resulting integral (19) to an integral from 0 to k that is convenient for numerical work.

However we can do better than the static approximation. Instead of taking the sine transform of (22) we multiply through by

$$\frac{t}{(t^2-k^2)^{1/2}} \quad \sin\,(t^2-k^2)^{1/2} \;,$$

integrate with respect to t from 0 to ∞, differentiate with respect to x, and use the results:

$$\int_0^\infty \frac{t}{(t^2-k^2)^{1/2}} \quad \sin x(t^2-k^2)^{1/2}\, J_0(rt)dt = \begin{cases} \dfrac{\cosh k(x^2-r^2)^{1/2}}{(x^2-r^2)^{1/2}} \;, & (r < x) \\ 0 \;, & (r > x) \end{cases}$$

$$\frac{d}{dx}\int_0^\infty \frac{t}{t^2-k^2} \quad \sin x(t^2-k^2)^{1/2}\, J_0(rt)dt = A + \frac{i \sin k(r^2-x^2)^{1/2}}{(r^2-x^2)^{1/2}} \;,$$

where

$$A = \begin{cases} 0 \;, & r < x \;, \\ \dfrac{\cos k(r^2-x^2)^{1/2}}{(r^2-x^2)^{1/2}} \;, & r > x \;. \end{cases}$$

The final result is:

$$\frac{d}{dx}\int_0^x r\, f(r)\, \frac{\cosh\ k(x^2-r^2)^{1/2}}{(x^2-r^2)^{1/2}}\, dr = -\int_x^\infty rg(r)\, \frac{\cos\ k(r^2-x^2)^{1/2}}{(r^2-x^2)^{1/2}}\, dr$$

$$-i\int_0^\infty rg(r)\, \frac{\sin k\,(r^2-x^2)^{1/2}}{(r^2-x^2)^{1/2}}\, dr.$$

This reduces to (15) when $k = 0$, as it must do.

We can obtain integral equations for the diffraction of waves normally incident on a soft annulus by setting $g(r) = 0$, $0 \leq r < \alpha$, $1 \leq r < \infty$, and $f(r) = 1$, $\alpha \leq r \leq 1$. These are convenient for numerical solution. If $\alpha = 0$, i.e., we are dealing with a disk, we can recover a well-known integral equation due to D. S. Jones [6].

6. Indentation with friction. A typical axially-symmetric contact problem concerns the indentation of an elastic half-space $z \geq 0$ by a rigid body, symmetric about the z-axis, and exerting a prescribed force P normal to the surface. When there is no friction between the body and the half-space, this type of problem can be solved by methods surveyed in Sneddon and Lowengrub [20]. We quote formulae involving Abel-type integral equations analogous to those derived for the potential and wave equations in the last two sections.

Using cylindrical coordinates (r,z), consider the elastic half-space $z \geq 0$, and denote the surface values of the normal and shear stresses by $\sigma(r)$, $\tau(r)$, and the normal and radial surface displacements by $w(r)$, $u(r)$. A typical indentation problem is such that the normal and shear stresses are zero on $z = 0$ for $r > 1$, and then the required equations are found to be:

$$\int_x^1 \frac{s\sigma(s)ds}{(s^2-x^2)^{1/2}} - \gamma\left\{\int_0^1 \tau(s)ds - x\int_0^x \frac{\tau(s)ds}{(x^2-s^2)^{1/2}}\right\} = -\frac{\mu}{1-\nu}\frac{d}{dx}\int_0^x \frac{sw(s)ds}{(x^2-s^2)^{1/2}} .$$

$$\gamma\int_0^x \frac{s\sigma(s)ds}{(x^2-s^2)^{1/2}} - x\int_x^1 \frac{\tau(s)ds}{(s^2-x^2)^{1/2}} = \frac{\mu}{1-\nu}\frac{d}{dx}\, x\int_0^x \frac{u(s)ds}{(x^2-s^2)^{1/2}} .$$

In these equations, ν is Poisson's ratio, $\gamma = (1-2\nu)/2(1-\nu)$, and $\mu = E/2(1+\nu)$, where E is Young's modulus.

The utility of these for numerical work seems to have been first exploited by Linz [9]. A derivation of the equations by the method used in the last two sections is given in [16],where related equations are also developed including equations for the two-dimensional case. Linz used the equations to compute solutions for the punch with adhesion. D. A. Spence has recently obtained some important results using this type of equation, where the coefficient of friction is finite, so that there is adhesion for part of the region of contact, and slip for the remainder. It would be a straightforward matter to compute similar results for an annular indentor.

7. Concluding remarks. We have drawn attention to two situations where the numerical solution of integral equations is particularly useful in solving boundary-value problems, one involving an integral equation of the first kind with unbounded kernel, the other involving Abel-type integrals. Both techniques would seem to deserve much more exploitation than they have received in the literature so far.

The calculation of the capacity of a rectangular solid in free space presented little difficulty because both the geometry and the equation were simple. In more complicated situations (for example, diffraction by a solid of arbitrary shape) the calculations will be much more laborious (though Maxwell-type approximations could be exploited). It is possible that for the Neumann problem the formulation in terms of integral equations of the second kind is preferable for numerical work, but the general situation is not clear.

The approach that we have sketched to the numerical solution of mixed boundary value problems would have much more general applicability if we had efficient methods for numerical evaluation of integrals like (19), (21). At the moment we have to rely on tricks such as sophisticated transformations involving contour integrals when dealing with static-perturbation procedures for time-dependent problems in elasticity, for example. The subject of mixed boundary value problems would seem to be a classic case of a monumental expenditure of energy on ingenious analytical manipulations. Apart from the aesthetic pleasure derived from results like the D.S. Jones integral equation for diffraction by a disk, it would seem that adequate insight is given by the very simplest cases. The information that one obtains about more complicated geometries by laborious series expansions can be obtained more easily and more directly by numerical methods. Also numerical methods allow us to contemplate the solution of problems that one could not hope to tackle analytically.

In both classes of problems considered here, it would seem that a satisfactory analysis of the error involved in the numerical solution of the integral equations lies some way in the future.

I acknowledge gratefully the stimulus of collaborating with P. Linz and D.A. Spence on mixed boundary-value problems.

Most of the work reported here was carried out under Contract No.: DA-31-124-ARO-D-462. This paper was prepared under NSF Grant GY-9107, while on leave at Oberlin College, Oberlin, Ohio

REFERENCES

1. K. Atkinson, SIAM J. Num. Anal. 4 (1967), 337-348.
2. D. Greenspan and E. Silverman, Proc. I.E.E.E. 53 (1965), 1636.
3. J.L. Hess and A.M.O. Smith, Progress in Aero. Sci., 8 (1967), 1-138.
4. F.B. Hildebrand, Amer. Acad. Arts and Sci. 74 (1941), 287-295.
5. M.A. Jaswon, Proc. Roy. Soc. A 275 (1963), 23-32.
6. D.S. Jones, Comm. Pure Appl. Math. 9 (1956), 713-746.
7. R. Kussmaul, Computing 4 (1969), 246-273.
8. R. Kussmaul and P. Werner, Computing 3 (1968), 22-46.
9. P. Linz, MRC Tech. Summ. Report #826 (1967), University of Wisconsin Madison, Wis.
10. P. Linz, J. Engineering Math. 3 (1969), 245-249.
11. M.S. Lynn and W.P.Timlake, Num. Math. 11 (1968), 77-98.
12. M. Magnus and F. Oberhettinger, Special Functions of Math. Phys., Chelsea (1949).
13. J.C. Maxwell, ed., Electrical Researches of the Hon. Henry Cavendish, F.R.S. (1771-1781), Camb. Univ. Press (1879).
14. B. Noble, Proc. Symp. Int. Comp. Center, Rome, Birhauser Verlag (1960), 540-543.
15. B. Noble, MRC Tech. Summ. Rep. #730 (1966), University of Wisconsin, Madison, Wis.
16. B. Noble and D.A. Spence, MRC Tech. Summ. Rep. #1089 (1971), University of Wisconsin, Madison, Wis.
17. D.K. Reitan and T.J. Higgins, J. Appl. Phys. 22 (1951), 223-226.
18. H.A. Schenck, J. Acoust. Soc. Amer. 44 (1968), 41-58.
19. I.N. Sneddon, Mixed Boundary Value Problems in Potential Theory, North-Holland (1966).
20. I.N. Sneddon and M. Lowengrub, Crack Problems in the Classical Theory of Elasticity, Wiley (1969).
21. D.A. Spence, Proc. Roy. Soc. A 305 (1968), 55-80.

ON THE INVERSE EIGENVALUE PROBLEM FOR MATRICES AND RELATED PROBLEMS FOR DIFFERENCE AND DIFFERENTIAL EQUATIONS

M.R. Osborne

Abstract

The problem of estimating parameters $\alpha_1,\ldots,\alpha_k$ of the matrix valued function $M(\lambda,\underset{\sim}{\alpha})$ given eigenvalue data $\lambda_1,\ldots,\lambda_p$, $p \geq k$, is considered. Two algorithms are presented. The first reduces the estimation problem to an unconstrained minimisation and contains as special cases methods suggested by other authors. The second reduces the problem to one of minimisation subject to equality constraints. Examples are given to show that the behaviour of the solutions can be involved so that the application of numerical methods is probably of necessity tentative. The results of some numerical experiments are summarised.

1. Introduction

Perhaps the earliest computational algorithm for an inverse matrix eigenvalue problem was given by Downing and Householder [3]. They consider the problem of finding a diagonal matrix D such that the symmetric matrix A+D has prescribed eigenvalues. In this paper they stress the inherent difficulty of the problem and the lack of knowledge relating to conditions under which a solution is possible. Recently more progress has been made with this problem, and Hadeler [6] has given sufficient conditions for the existence of a solution and an algorithm based on successive approximations for its solution. His results are largely restricted to matrices which can be regarded as perturbations of a diagonal matrix and to well spaced eigenvalue data, and he motivates his paper by noting that the problem of determining $q(x)$ given the eigenvalues of the differential equation

$$\frac{d^2y}{dz^2} + (\lambda r(z) - q(z))y = 0 \qquad (1.1)$$

subject to appropriate boundary conditions becomes a problem of estimating certain elements of a matrix from eigenvalue data if finite difference methods are applied to equation (1.1).

Another recent paper (Andersson [1]) is directly concerned with a difference approximation to equation (1.1). Consider

$$\begin{bmatrix} s_{i+1} \\ y_{i+1} \end{bmatrix} = C_{i+1}(\lambda) \begin{bmatrix} s_i \\ y_i \end{bmatrix} \tag{1.2}$$

where

$$C_i = \begin{bmatrix} 1+h^2(q_i-\lambda r_i) & h(q_i-\lambda r_i) \\ h & 1 \end{bmatrix} , \tag{1.3}$$

$z_i = ih$, $q_i = q(z_i)$, and $r_i = r(z_i)$.

Applying this recurrence successively gives

$$\begin{bmatrix} s_n \\ y_n \end{bmatrix} = C_n C_{n-1} \ldots C_1 \begin{bmatrix} s_0 \\ y_0 \end{bmatrix} = S_n \begin{bmatrix} s_0 \\ y_0 \end{bmatrix} . \tag{1.4}$$

Writing

$$S_n = \begin{bmatrix} S_{11} & S_{12} \\ S_{21} & S_{22} \end{bmatrix} \tag{1.5}$$

Andersson proves the result that if $S_{11}(\lambda)$ and $S_{21}(\lambda)$ have all zeros real and simple, and if these zeros satisfy

$$\lambda_k^{(11)} < \lambda_k^{(21)} < \lambda_{k+1}^{(11)} , \quad k = 1,2,\ldots,n-1$$

then there are uniquely defined matrices $C_i(\lambda)$, and

$$A = \begin{bmatrix} a & b \\ 0 & 1/a \end{bmatrix} \quad \text{such that}$$

$$S_n = C_n C_{n-1} \ldots . C_1 A. \tag{1.6}$$

He also gives a recurrence for r_i and q_i given the zeros of S_{11} and S_{21}. These zeros are the eigenvalues corresponding to the boundary conditions $y_0 = s_n = 0$ and $y_0 = y_n = 0$.

Although these results appear comprehensive, they do not really get close to solving the problem which provided the original motivation. In particular, when the eigenvalues are the result of experimental observation the quantity of data available would be strictly limited, and it would also be subject to experimental error. Consider the example of a vibrating string. In this case the observations would consist of the frequencies of the first few modes, and the problem would be to determine the density (say). Another possible application is the determination of the velocity of sound in the ocean as a function of depth from observations of the speed of propagation of modes trapped in an underwater sound channel (see, for example, Ewing, Jardetzky and Press [4]). In this case $q(z) = \omega^2/c(z)^2$, and the data give $d\lambda/d\omega$ rather than λ.

<u>Remark.</u> In the case of the vibrating string (for example) it is quite likely that information would also be available on the eigenfunctions, and this permits an optimum solution to be given for the problem of determining $q(z)$. Let the first p eigenvalues and eigenfunctions be known. Then integrating equation (1.1) gives (assuming for simplicity that $r(x) = 1$)

$$\frac{dy_i(1)}{dz} - \frac{dy_i(0)}{dz} + \lambda_i \nu_i + \mu_i = 0, \quad i = 1,2,\ldots,p \tag{1.7}$$

where ν_i and μ_i are the Fourier coefficients for the expansion of 1 and $q(z)$ in terms of the eigenfunctions. Thus the leading coefficients in the Fourier expansion of q in terms of the eigenfunctions are available, and finite segments of this expansion have a well known best approximation property in the square norm.

It must also be stressed that the finite difference approximation can also be a source of error. In particular, the higher eigenvalues for a given discretization will bear little relation to the corresponding eigenvalues of the differential equation and hence little relation to the observed data.

A somewhat different approach has been given by Bellman [2]. Assume one boundary condition is $y(0) = 0$. Then the disposable scale factor can be fixed by setting $\frac{dy(0)}{dz} = K$. Assume also that $q(z)$ can be represented in the form $\Phi(z, \underset{\sim}{\alpha})$ where $\underset{\sim}{\alpha}$ is a vector of k disposable parameters. Let $y_i(z, \underset{\sim}{\alpha})$, $i=1,2,\ldots,p$ be obtained by integrating the resulting initial value problem for each given eigenvalue. Assuming that the correct terminal boundary condition is $y(1) = 0$, Bellman suggests that $\underset{\sim}{\alpha}$ be estimated by minimising

$$\sum_{i=1}^{p} y_i\,(1,\,\underset{\sim}{\alpha})^2 .$$

The specific algorithm recommended for this purpose is Quasilinearisation (the Gauss method), and an example is given in which q can be represented exactly by Φ. In this case $k = p = 2$ and satisfactory convergence is obtained.

This method has the advantage that if a good integration subroutine with automatic step length adjustment is used then problems due to the truncation error of the difference approximations should be avoided. Also, the use of an approximation to q with a finite number of parameters is a logical move considering the limited amount of data. However this approximation introduces a further source of error. Also, from the numerical point of view, neither simple shooting nor the Gauss method should be used uncritically.

2. A general approach

In this section we consider a mature valued function $M(\lambda,\underset{\sim}{\alpha})$. Our aim is to estimate the components $\alpha_1,\alpha_2,\ldots,\alpha_k$ of given values $\lambda_1,\lambda_2,\ldots,\lambda_p$ where $p \geqslant k$ for certain of the eigenvalues of M, and this will be done by minimising

$$\sum_{i=1}^{p} w_i\,\beta_i(\underset{\sim}{\alpha})^2$$

where w_i are certain weights and where $\beta_i(\underset{\sim}{\alpha})$ is defined by

$$M(\lambda_i,\underset{\sim}{\alpha})\underset{\sim}{v} = \beta_i(\underset{\sim}{\alpha})\underset{\sim}{x}_i \tag{2.1}$$

and

$$\underset{\sim}{s}_i^T\,\underset{\sim}{v} = K_i\;,\quad i = 1,2,\ldots,p. \tag{2.2}$$

Equation (2.2) is a scaling condition which ensures that as $\underset{\sim}{\alpha}$ varies to make $M(\lambda_i,\underset{\sim}{\alpha})$ approach a singularity $\beta_i(\underset{\sim}{\alpha}) \to 0$. Explicitly we have

$$\beta_i(\underset{\sim}{\alpha}) = \frac{K}{\underset{\sim}{s}_i^T\,M^{-1}\,\underset{\sim}{x}_i} \tag{2.3}$$

and

$$\frac{\partial\beta_i}{\partial\alpha_j} = \frac{K_i}{(\underset{\sim}{s}_i^T M^{-1}\underset{\sim}{x}_i)^2} \; \underset{\sim}{s}_i^T M^{-1} \frac{\partial M}{\partial\alpha_j} M^{-1} \underset{\sim}{x}_i \tag{2.4}$$

so that the implementation of an algorithm such as the Gauss method to minimise

$\sum_{i=1}^{p} w_i \, \beta_i(\underset{\sim}{\alpha})^2$ is quite straight forward.

From equations (2.3) and (2.4) it will be clear that there are similarities between our approach and inverse iteration (see, for example, the paper by J. H. Wilkinson in this volume), and this connexion will now be exploited to provide further analysis of the proposed method. To do this we assume that $\underset{\sim}{\alpha}$ is close to a solution of the inverse problem in the sense that there exist right and left eigenvectors $\underset{\sim}{u}_i$ and $\underset{\sim}{u}_i^*$ of $M(\lambda_i, \underset{\sim}{\alpha})$ normalised so their length is unity, such that the corresponding eigenvalue ϵ_i is small. Further we assume that all other eigenvalues are sufficiently well separated for $\beta_i(\underset{\sim}{\alpha})$ to be estimated to a good approximation by

$$\beta_i(\underset{\sim}{\alpha}) = \frac{K_i(\underset{\sim}{u}_i^{*T}\underset{\sim}{u}_i)}{(\underset{\sim}{u}_i^{*T}\underset{\sim}{x}_i)(\underset{\sim}{s}_i^T\underset{\sim}{u}_i)} \; \epsilon_i \tag{2.5}$$

Remark

(i) In the case where the model $M(\lambda,\underset{\sim}{\alpha})$ can be assumed to be adequate, and where errors e_i occur in the observations, then we have to have a first approximation

$$\epsilon_i = \frac{\underset{\sim}{u}_i^{*T}\frac{dM}{d\lambda}\underset{\sim}{u}_i}{\underset{\sim}{u}_i^{*T}\underset{\sim}{u}_i} \; e_i$$

If the e_i are independent normally distributed random variables with mean zero and standard deviation σ, then the principle of least squares indicates that the weights w_i should be chosen such that $\sum_{i=1}^{p} w_i\,\beta_i^2 = \sum_{i=1}^{p} e_i^2$. This gives

$$w_i = \left\{\frac{(\underset{\sim}{u}_i^{*T}\underset{\sim}{x}_i)(\underset{\sim}{s}_i^T\underset{\sim}{u}_i)}{K_i(\underset{\sim}{u}_i^{*T}\frac{dM}{d\lambda}\underset{\sim}{u}_i)}\right\}^2 .$$

(ii) From equation (2.5) it follows that $\underset{\sim}{s}_i$ and $\underset{\sim}{x}_i$ must be chosen so that neither of the scalar products $\underset{\sim}{u}_i^{*T} \underset{\sim}{x}_i$ nor $\underset{\sim}{s}_i^T \underset{\sim}{u}_i$ can vanish. Appropriate choices are $\underset{\sim}{x}_i = \underset{\sim}{u}_i$, $\underset{\sim}{s}_i = \underset{\sim}{u}_i^*$ if these are available. In practice it is frequently not difficult to construct good approximations to $\underset{\sim}{u}_i$ and $\underset{\sim}{u}_i^*$ and to update them during the progress of the computation.

The applicability of algorithms such as the Gauss method for minimising a sum of squares depends critically on the matrix $(\frac{\partial \beta}{\partial \alpha})$ with components $\frac{\partial \beta_i}{\partial \alpha_j}$ having its full rank. Consider, for example, the modification to the Gauss method in which the predicted correction is accepted provided it leads to a significant reduction in the sum of squares. Otherwise it is used as a search direction for a one dimensional minimisation of the sum of squares, and this minimum is taken as the new approximation. In this case we have the following result (Osborne [8]).

'Assume the sequence of iterates obtained using the modified Gauss method lies in a bounded region R. If we have in R that

(i) the smallest, nonzero, singular value of $(\frac{\partial \beta}{\partial \alpha}) \geq \delta > 0$, and

(ii) $$\max_{\underset{\sim}{t}, \|\underset{\sim}{t}\|=1} \left\| \sum_{i,j} \frac{\partial^2 \underset{\sim}{\beta}}{\partial t_i \partial t_j} t_i t_j \right\| \leq 2W,$$

and if at a certain stage we have $\|\underset{\sim}{\beta}\| \leq \frac{\delta^2}{4W}$, then the modified Gauss method converges. Further the full step method applied eventually, and the ultimate rate of convergence is geometric with common ratio $\leq \frac{1}{2}$. Second order convergence can only be demonstrated in general provided the system of equation $\underset{\sim}{\beta}(\underset{\sim}{\alpha}) = 0$ is compatible.'

If $\frac{\partial \beta_i}{\partial \alpha_j}$ is estimated in a similar fashion to that used above we then have

$$\frac{\partial \beta_i}{\partial \alpha_j} = \frac{K_i}{(\underset{\sim}{u}_i^{*T} \underset{\sim}{x}_i)(\underset{\sim}{x}_i^T \underset{\sim}{u}_i)} \underset{\sim}{u}_i^{*T} \frac{\partial M}{\partial \alpha_j} \underset{\sim}{u}_i \qquad (2.6)$$

<u>Example 1.</u> Let $M = A + D - \lambda I$ where D is diagonal, $D_{ii} = \alpha_i$, and $k = p = n$. In this case Newton's method can be applied. Noting that $\frac{\partial M}{\partial \alpha_j} = E_j$ where E_j has 1 in the jj position and zeros elsewhere we find using equations (2.5) and (2.6) that the Newton correction $\delta \underset{\sim}{\alpha}$ is given by the system of linear equations

$$Q \delta \underset{\sim}{\alpha} = -\underset{\sim}{W} \qquad (2.7)$$

where $$Q_{ij} = \frac{(\underset{\sim}{u}_i^*)_j(\underset{\sim}{u}_i^*)_j}{\underset{\sim}{u}_i^{*T}\ \underset{\sim}{u}_i}, \quad \text{and } W_i = \epsilon_i .$$

In the case A symmetric this corresponds to the iteration given by Downing and Householder [3]. The condition for the applicability of this algorithm is that the matrix Q have its full rank. That this condition need not hold is shown in the next example.

<u>Example 2.</u> Consider the finite difference approximation to equation (1.1) subject to the boundary conditions $y(0) = y(1) = 0$ defined by

$$\begin{aligned} M_{ij}(\lambda,\underset{\sim}{\alpha}) &= -2 + h^2\lambda r(z_i) + h^2\Phi(z_i,\underset{\sim}{\alpha}) \quad i = j, \\ &= 1, \ |i - j| = 1, \\ &= 0 \text{ otherwise} \end{aligned} \tag{2.8}$$

where $h = 1/(n+1)$ and $\Phi(z,\underset{\sim}{\alpha})$ is an approximation to $q(z)$. We note that M is symmetric so that $\underset{\sim}{u}_i = \underset{\sim}{u}_i^*$, and that $\frac{\partial M}{\partial \alpha_j}$ is independent of λ. Writing $(\frac{\partial \Phi}{\partial \alpha})$ for the matrix with components $\frac{\partial\Phi(z_i,\underset{\sim}{\alpha})}{\partial \alpha_j}$ we have from (2.6) that

$$\nabla\beta_i(\underset{\sim}{\alpha}) = \frac{K_i}{(\underset{\sim}{u}_i^T\ \underset{\sim}{x}_i)(\underset{\sim}{s}_i^T\ \underset{\sim}{u}_i)} \{(\underset{\sim}{u}_i)_1^2 \ldots (\underset{\sim}{u}_i)_n^2\} \ (\frac{\partial\Phi}{\partial\alpha}) \tag{2.9}$$

whence

$$(\frac{\partial\beta}{\partial\alpha}) = H\, Q\, (\frac{\partial\Phi}{\partial\alpha}) \tag{2.10}$$

where H is a diagonal matrix with $H_{ii} = \frac{K_i}{(\underset{\sim}{u}_i^T\ \underset{\sim}{x}_i)(\underset{\sim}{s}_i^T\ \underset{\sim}{u}_i)}$, and $Q_{ij} = (\underset{\sim}{u}_i)_j^2$. The close connexion between this result and that of the previous example should be noted. $(\frac{\partial\beta}{\partial\alpha})$ will have its full rank if the intersection of the sets $S=\{\underset{\sim}{x}:\underset{\sim}{x} = (\frac{\partial\Phi}{\partial\alpha})\underset{\sim}{y},\ \underset{\sim}{y} \in E_k,\ \|\underset{\sim}{y}\| = 1\}$ and $N = \{\underset{\sim}{x}:Q\underset{\sim}{x} = 0\}$ is empty. Again the matrix Q plays a key role. Presumably $(\frac{\partial\Phi}{\partial\alpha})$ will have rank k to ensure that the problem of approximating $q(z)$ by $\Phi(z,\underset{\sim}{\alpha})$ makes sense.

Consider the particular case $r = 1$, $q = 0$. Taking a finite difference grid consisting of two points ($h = \frac{1}{3}$) gives the algebraic eigenvalue problem

$$\begin{bmatrix} -2 + h^2\lambda & 1 \\ 1 & -2 + h^2\lambda \end{bmatrix} \underset{\sim}{y} = 0, \tag{2.11}$$

and this has the solutions

$$h^2\lambda_1 = 1, \quad \underset{\sim}{u}_1^T = \frac{1}{\sqrt{2}} \{1,1\} ,$$

$$h^2\lambda_2 = 3, \quad \underset{\sim}{u}_2^T = \frac{1}{\sqrt{2}} \{1,-1\} .$$

In this case we have

$$Q = \frac{1}{2} \begin{bmatrix} 1 & 1 \\ 1 & 1 \end{bmatrix}$$

which has rank 1. Taking $\Phi = \alpha_1 + \alpha_2 z$ it is readily seen that there is only one solution $\alpha_1 = \alpha_2 = 0$ corresponding to the exact eigenvalue data.

Consider now the perturbed data $h^2\lambda_1 = 1 + \epsilon_1$, $h^2\lambda_2 = 3 + \epsilon_2$. Here we find that

$$\left(\frac{h^2\alpha_2}{6}\right)^2 = \frac{\epsilon_1-\epsilon_2}{2} \left(\frac{\epsilon_1-\epsilon_2}{2} - 2\right) .$$

Thus these are a pair of roots (either real or imaginary roots are possible) unless $\epsilon_1 = \epsilon_2 = \epsilon$ in which case $\alpha_2 = 0$, $h^2\alpha_1 = -\epsilon$. This result illustrates well that the problem is potentially ill posed and further that the restrictions concerning the rank of Q derived in this and the previous example are natural.

Remark. If finite differences are used to implement Bellman's algorithm then it corresponds to the particular case of the above method in which $\underset{\sim}{s}_i = \underset{\sim}{e}_1$, $\underset{\sim}{x}_i = \underset{\sim}{e}_n$, $i = 1,2,\ldots,p$.

Example 3. Difficulties involved in trying to calculate higher eigenvalues using the difference approximation (2.8) can be reduced quite considerably (especially in the case $r = 1$) by using the difference approximation to equation (1.1)

$$y_{i+1} - 2 \cos (h\sqrt{\lambda r_i + q_i})y_i + y_{i-1} = 0 \tag{2.12}$$

This difference approximation has been frequently rediscovered. Its application to the eigenvalue problem is discussed in Osborne and Michaelson [10]. In this case we have

$$\nabla\beta_i(\underset{\sim}{\alpha}) = \frac{K_i}{(\underset{\sim}{u}_i^T \underset{\sim}{x}_i)(\underset{\sim}{s}_i^T \underset{\sim}{u}_i)} \rho_i(Q) \left(\frac{\partial\Phi}{\partial\alpha}\right) \tag{2.13}$$

where

$$Q_{ij} = \frac{\sin h\sqrt{\lambda_i r_j + \Phi(z_j,\underset{\sim}{\alpha})}}{h\sqrt{\lambda_i r_j + \Phi(z_j,\underset{\sim}{\alpha})}} (\underset{\sim}{u}_i)_j^2$$

As would be suspected there is a close connexion between this and the previous method in the case $h \to 0$ and a fixed quantity of eigenvalue data.

Example 4. Consider the general first order system

$$\frac{d\underset{\sim}{y}}{dz} = A(z,y,\underset{\sim}{\alpha})\underset{\sim}{y} , \tag{2.14}$$

subject to the boundary conditions

$$B_1 \underset{\sim}{y}(0) + B_2 \underset{\sim}{y}(1) = 0 . \tag{2.15}$$

Let points $z_i = ih$, $i = 1/n$ be given , and define $X_i(z)$ by

$$\frac{dX_i}{dz} = A X_i, \; X_i(z_{i-1}) = I , \tag{2.16}$$

In each interval (z_{i-1}, z_i) the solution to equation (2.14) can be written in the form $X_i\underset{\sim}{d}_i$, and the conditions of continuity at z_i and satisfaction of the boundary conditions give the problem

$$\begin{bmatrix} B_1 & & & B_2X_n(z_n,\lambda_i,\underset{\sim}{\alpha}) \\ X_1(z_1,\lambda_i,\underset{\sim}{\alpha}) & -I & & \\ & X_2(z_2,\lambda_i,\underset{\sim}{\alpha}) & -I & \\ & & \ddots & \\ & & X_{n-1}(z_{n-1},\lambda_i,\underset{\sim}{\alpha}) & -I \end{bmatrix} \begin{bmatrix} \underset{\sim}{d}_1^{(i)} \\ \\ \vdots \\ \\ \underset{\sim}{d}_n^{(i)} \end{bmatrix} = M(\lambda_i,\underset{\sim}{\alpha})\underset{\sim}{v}^{(i)} = 0 \tag{2.17}$$

Here the expression for $(\frac{\partial\beta}{\partial\alpha})$ takes a more complicated form. However, for the problem (1.1) it can be developed into a form similar to that of the preceeding examples by noting that for fixed λ as $h \to 0$

$$X_i = I + hA(z_{i-1},\lambda,\underset{\sim}{\alpha}) + 0(h^2) \tag{2.18}$$

where

$$A = \begin{bmatrix} 0 & 1 \\ -(\lambda r + \Phi) & 0 \end{bmatrix} .$$

In this case $\nabla\beta_i(\underset{\sim}{\alpha})$ takes the form of equation (2.13) with $Q_{ij} = (\underset{\sim}{d}^{*(i)}_{j+1})_2 \, (\underset{\sim}{d}^{(i)}_j)_1$.

The method outlined above is just an application of multiple shooting. The numerical advantages of this approach have been stressed by Osborne [9].

3. An alternative approach

One feature of the problem discussed in example 2 of the previous section is the possibility of complex solutions. This difficulty can often be removed by a slight change of approach which leads to a constrained minimisation problem. We note that β_i defined by equations (2.1) and (2.2) can be considered a function of λ as well as $\underset{\sim}{\alpha}$ and we pass the problem

$$\min \sum_{i=1}^{p} (\lambda_i - \mu_i)^2 \tag{3.1}$$

subject to

$$\beta_i(\mu_i, \underset{\sim}{\alpha}) = 0, \quad i = 1,2,\ldots,p. \tag{3.2}$$

Provided (for example) that $|\mu_i| \to \infty$, $i = 1,2,\ldots,p$, as $\|\underset{\sim}{\alpha}\| \to \infty$ this problem has at least one bounded solution for a given set of eigenvalue data. Effective methods are available for solving equality constrained problems - for example the penalty function methods of Fiacco and McCormick [5] and the improved penalty function approach of Powell [11]. However the condition that $(\frac{\partial\beta}{\partial\alpha})$ have its full rank is again important.

Consider, for example, $M(\mu,\underset{\sim}{\alpha})$ given by

$$M(\mu,\underset{\sim}{\alpha}) = \begin{bmatrix} \alpha_1+\mu & 1 \\ 1 & \alpha_2+\mu \end{bmatrix} .$$

The characteristic equation for M is

$$\mu^2 + (\alpha_1 + \alpha_2)\mu + \alpha_1\alpha_2 - 1 = 0$$

so that, defining $2y_1 = \alpha_1 + \alpha_2$, $2y_2 = \alpha_1 - \alpha_2$, the eigenvalues are

$$\mu_1 = -y_1 - \sqrt{1+y_2^2}, \quad \mu_2 = -y_1 + \sqrt{1+y_2^2} .$$

Let $\lambda_1 \leq \lambda_2$. Then the objective function is

$$R = (\lambda_1 + y_1 + \sqrt{1+y_2^2})^2 + (\lambda_2 + y_1 - \sqrt{1+y_2^2})^2$$

and the conditions for this to be stationary are

$$\frac{1}{2}\frac{\partial R}{\partial y_1} = 0 = \lambda_1 + \lambda_2 + 2y_1, \quad \text{and}$$

$$\frac{1}{2}\frac{\partial R}{\partial y_2} = 0 = (\lambda_1 - \lambda_2 + 2\sqrt{1+y_2^2})\frac{y_2}{\sqrt{1+y_2^2}} \quad .$$

Thus y_1 and y_2 are determined by

$$y_1 = -\frac{\lambda_1 + \lambda_2}{2}$$

$$y_2 = 0, \text{ or } \sqrt{\left(\frac{\lambda_1 - \lambda_2}{2}\right)^2 - 1} \quad .$$

The Hessian of R is positive definite provided $\frac{\partial^2 R}{\partial y_2^2}$ is positive. We have

$$\frac{\partial^2 R}{\partial y_2^2} = 2 + \frac{\lambda_1 - \lambda_2}{(1 + y_2^2)^{3/2}} \quad .$$

Thus the solution

$$y_1 = -\frac{\lambda_1 + \lambda_2}{2} \quad , \quad y_2 = 0$$

gives a minimum provided $\lambda_2 < \lambda_1 + 2$, while the solution

$$y_1 = -\frac{\lambda_1 + \lambda_2}{2} \, , \quad y_2^2 = \left(\frac{\lambda_1 - \lambda_2}{2}\right)^2 - 1$$

gives

$$\frac{\partial^2 R}{\partial y_2^2} = 2\left(1 - \sqrt{\frac{2}{\lambda_2 - \lambda_1}}\right)$$

so that it is a minimum for $\lambda_2 > \lambda_1 + 2$. As $R = 0$ when $\lambda_1 = \lambda_2 - 2$ it is a minimum in this case also.

Note (i) The condition $\lambda_2 \geq \lambda_1 + 2$ is necessary for the problem of example 2 of the previous section to have a real solution (a more elaborate treatment of this point is given by Hadeler [6]). In this case the method given in this section appears superior in the sense that it produces answers for a comprehensive range of the problem parameters. However the comparatively complex behaviour of the solutions to even this simple problem suggests that numerical calculations are likely to be difficult.

(ii) The two solutions corresponding to the $\pm$ values of y_2 correspond to interchanging α_1 and α_2. Multiple solutions of this kind correspond to additional symmetries in M and can be removed by imposing appropriate ordering conditions. For example we could have improved the constraint $\alpha_2 - \alpha_1 \geq 0$ in this case.

4. Numerical experience

Numerical experiments have been carried out principally to test the effects of error in the representation Φ on the performance of the method given in section 2. Two differential equations have been considered

$$\text{(i)}\ \frac{d^2y}{dz^2} + (\lambda + z)y = 0, \text{ and}$$

$$\text{(ii)}\ \frac{d^2y}{dz^2} + \left(\frac{\lambda}{16}(1+z^2)^2 + \frac{27z}{64}\left(1+\frac{z^2}{3}\right)(1+z^2)^2 + \frac{1-2z^2}{(1+z^2)^2}\right)y = 0$$

subject to the boundary condition $y(0) = y(1) = 0$, and in this case both equations have the same eigenvalues. The difference formula (2.12) has been used with $h = 1/25$, and the difference approximation eigenvalues were calculated for each differential equation to 15 significant figures to provide input data. To define β_i, $\underset{\sim}{x}_i$ was specified by estimating y by applying the WKB method to the differential equation with Φ defined by its initial parameter values, while $\underset{\sim}{s}_i$ was set to $\underset{\sim}{e}_{t_i}$ where t_i is the index of the component of maximum modulus of $\underset{\sim}{v}_i$ given by equation (2.1) in the first iteration. Φ was assumed to be a polynomial of degree k-1, and $\Phi = 0$ was taken as initial approximation to q. To minimise $\sum_{i=1}^{p} \beta_i^2$, the Marquardt method was used in the implementation described by Jennings and Osborne [7]. In this method the correction at each stage is obtained by solving

$$\left[\left(\frac{\partial\beta}{\partial\alpha}\right)^T\left(\frac{\partial\beta}{\partial\alpha}\right) + \delta I\right]\delta\underset{\sim}{\alpha} = -\left(\frac{\partial\beta}{\partial\alpha}\right)^T \underset{\sim}{\beta} \qquad (4.1)$$

where $\delta \geq 0$. The parameter δ serves to stabilise the computation, and the similarity to regularisation will be noted.

Runs were carried out for a range of values of k and p; 4 and 10 are typical values. In addition the eigenvalue data was truncated to 15, 12, 9 and 6 significant figures to provide perturbed data. For the first differential equation q can be represented exactly by Φ, and the numerical results were generally satisfactory. However, the solution for the six figure data gave rather a poor estimate of q. It is clear that the use of less accurate data would not have been satisfactory. For the differential equation (ii) the calculations were considerably more difficult and satisfactory convergence was only obtained with the nine figure data. The minimisation was difficult for the fifteeen figure data, but the final point was very close to that for the nine figure case. The representation Φ obtained agreed with q to within 5%. If anything, linearising k and p increased the difficulties. There was no indication of the coefficients of Φ converging, but the estimate of q was improved somewhat.

Note. For the differential equation (i) and an arbitrary coefficient function q we have the asymptotic estimate

$$\lambda_i = i^2\pi^2 - \int_0^1 q \; dz + O(1/i).$$

Thus truncating the eigenvalue data to a fixed number of significant figures causes relatively more information on q to be lost in the higher eigenvalue data. It is interesting that the sound propagation example would not suffer this difficulty.

REFERENCES

[1] L. E. Andersson: On the effective determination of the wave operator from given spectral data in the case of a difference equation corresponding to a Sturm-Liouville differential equation. J. Math. Anal. and Applic., 29 (1970), 467-497.

[2] R. E. Bellman, H. H. Kagiwada, R. E. Kalaba and R. Vasudevan: Quasilinearisation and the estimation of differential operators from eigenvalues. Comm. A.C.M., 11 (1968), 255-6.

[3] A. C. Downing and A. S. Householder: Some inverse characteristic value problems. J. Assoc. for Computing Machinery, 3 (1956), 203-207.

[4] W. M. Ewing, W. S. Jardetzky and F. Press: Elastic waves in layered media. McGraw-Hill, 1957.

[5] A. V. Fiacco and G. P. McCormick: Nonlinear Programming. Wiley, 1968.

[6] K. P. Hadeler: Ein inverses eigenvertproblem. Linear Algebra and Its Applications, 1 (1968), 83-101.

[7] L. S. Jennings and M. R. Osborne: Applications of orthogonal matrix transformations to the solution of systems of linear and nonlinear equations. Australian National University Computer Centre Tech. Rep. 37, 1970.

[8] M. R. Osborne: A class of methods for minimising a sum of squares. To be published.

[9] M. R. Osborne: On shooting methods for boundary value problems. J. Math. Anal. and Applic., 27 (1969), 417-433.

[10] M. R. Osborne and S. Michaelson: On the numerical solution of eigenvalue problems in which the eigenvalue parameter appears nonlinearly, with an application to differential equations. Computer J., 7 (1964), 66-71.

[11] M. J. D. Powell: A method for nonlinear constraints in minimisation problems. A.E.R.E. Tech. Rep. 310, Harwell, U.K., 1967.

THE DIFFERENTIAL CORRECTION ALGORITHM FOR RATIONAL L_∞ APPROXIMATION

M.J.D. Powell, I. Barrodale* and F.D.K. Roberts*
(* Mathematics Department, University of Victoria, Victoria, Canada)

SUMMARY*

Given a set of function values $f(x_t)$ $(t=1,2,\ldots,N)$, we consider the problem of calculating the rational function $R(x) = P(x)/Q(x)$ that minimizes the quantity

$$\max_t |f(x_t) - R(x_t)|,$$

where $P(x)$ and $Q(x)$ are polynomials of prescribed degrees. To solve this problem Cheney and Loeb [2] proposed a "differential correction algorithm", ODC say, but in a subsequent paper [3] they modified their algorithm, and now the modified algorithm, DC say, is nearly always used in place of ODC. The purpose of this paper is to direct attention back to the original algorithm, because in practice ODC seems to be much better.

The modified algorithm is usually preferred because it has been proved that it has sure convergence properties, see Cheney [1] for example. However now we show that the convergence of ODC is equally reliable. Moreover we prove that the rate of convergence of ODC is usually quadratic, but the rate of convergence of DC is only linear.

Some numerical examples are given to compare the two versions of the differential correction algorithm, and they confirm that ODC is faster and more accurate than DC.

References

[1] Cheney, E. W. "Introduction to approximation theory", McGraw-Hill (1966).

[2] Cheney, E. W. and Loeb, H. L. Numer. Math. 3, 72-75 (1961).

[3] Cheney, E. W. and Loeb, H. L. Numer. Math. 4, 124-127 (1962).

---

*The full text of this paper has been submitted for publication in a journal of the Society for Industrial and Applied Mathematics, so only a summary is given here.

RESOLUTION NUMERIQUE DE CERTAINS PROBLEMES HYPERBOLIQUES NON LINEAIRES. METHODE DE PSEUDO-VISCOSITE

P.A. Raviart

INTRODUCTION

On considère le problème de Cauchy-Dirichlet par l'équation hyperbolique non linéaire

$$(*)\qquad \frac{\partial^2 u}{\partial t^2} - \frac{\partial}{\partial x}\phi\left(\frac{\partial u}{\partial x}\right) = 0$$

où ϕ est une fonction de classe C^1 avec $\phi' > 0$. On sait que le problème précédent n'admet pas en général de solution globale qui soit une fois continûment dérivable, ceci quelle que soit la régularité des conditions initiales (phénomène d'ondes de choc).

Si on cherche à résoudre numériquement ce problème à l'aide du schéma aux différences finies explicite

$$(**)\qquad \frac{1}{k^2}(u_i^{n+1}-2u_i^n+u_i^{n-1}) - \frac{1}{h}\left[\phi\left(\frac{u_{i+1}^n-u_i^n}{h}\right) - \phi\left(\frac{u_i^n-u_{i-1}^n}{h}\right)\right] = 0$$

(h = pas d'espace, k = pas de temps, u_i^n = approximation de u(ih,nk)), on constate numériquement que ce schéma est inconditionnellement instable dès que la fonction ϕ n'est plus linéaire; les instabilités se développant à partir des points de discontinuité de $\frac{\partial u}{\partial t}$ et $\frac{\partial u}{\partial x}$. Nous sommes donc en présence du phénomène d'instabilité non linéaire. Un remède classique, dû à Von Neumann et Richtmyer cf. [6], consiste à introduire dans l'équation (*) un terme de pseudo-viscosité $-\varepsilon\frac{\partial q}{\partial x}$ ($\varepsilon > 0$ "petit") et à résoudre numériquement le problème de Cauchy-Dirichlet pour l'équation régularisée

$$(*)_\varepsilon \qquad \frac{\partial^2 u}{\partial t^2} - \frac{\partial}{\partial x}\phi\left(\frac{\partial u}{\partial x}\right) - \varepsilon \frac{\partial q}{\partial x} = 0.$$

Le but de cet article est de montrer rigoureusement comment l'adjonction de tels termes de pseudo-viscosité permet d'obtenir des schémas explicites stables sous des conditions de stabilité que nous préciserons; nous démontrerons la convergence de ces schémas dans un sens convenable. Nous étudierons deux choix simples du terme pseudo-viscosité qui permettent une analyse complète du problème :

(i) $q = \frac{\partial^2 u}{\partial x \partial t}$: pseudo-viscosité linéaire,

(ii) $q = \phi'\left(\frac{\partial u}{\partial x}\right) \frac{\partial^2 u}{\partial x \partial t}$: pseudo-viscosité quasi-linéaire.

Nous indiquerons comment l'analyse faite suggère d'autres choix de termes de pseudo-viscosité plus satisfaisants en pratique mais dont l'étude reste à faire.

Pour des considérations analogues mais techniquement différentes sur le schéma de Lax-Wendroff pour l'équation hyperbolique

$$\frac{\partial u}{\partial t} + u \frac{\partial u}{\partial x} = 0,$$

nous renvoyons à [5].

1. METHODE DE PSEUDO-VISCOSITE LINEAIRE

1.1. Notations - Théorème d'existence et d'unicité

Soit Ω l'intervalle ouvert $]0,1[$ de $\mathbb{R}$ de point générique x. Dans toute la suite, les fonctions définies sur Ω seront toujours à valeurs réelles. On introduit les espaces de Sobolev :

$$H^1_o(\Omega) = \{v|v, \frac{dv}{dx} \in L_2(\Omega),\ v(0) = v(1) = 0\},$$

$$H^2(\Omega) = \{v|v, \frac{dv}{dx}, \frac{d^2v}{dx^2} \in L_2(\Omega)\}$$

avec les normes

$$\|v\|_{H^1_o(\Omega)} = \left(\int_\Omega |\frac{dv}{dx}|^2 dx\right)^{1/2}.$$

$$\|v\|_{H^2(\Omega)} = \left(\int_\Omega \{|v|^2 + |\frac{dv}{dx}|^2 + |\frac{d^2v}{dx^2}|^2\} dx\right)^{1/2}.$$

On note $H^{-1}(\Omega)$ le dual fort de $H^1_o(\Omega)$.

Si X est un espace de Banach, on désigne par $C^o(0,T;X)$, $0 < T < \infty$, l'espace des fonctions continues sur $[0,T]$ à valeurs dans X, par $L_p(0,T;X)$, $1 \leq p < \infty$, l'espace des (classes de) fonctions $t \to v(t)$ définies sur $(0,T)$ fortement mesurables à valeurs dans X et telles que

$$\|v\|_{L_p(0,T;X)} = \left(\int_o^T \|v(t)\|^p_X dt\right)^{1/p} < \infty \quad .$$

Modification habituelle pour $p = \infty$. Si $v \in L_p(0,T;X)$, on note $\frac{\partial v}{\partial t}$ la dérivée de v au sens des distributions sur $]0,T[$ à valeurs dans X.

Soit ϕ une fonction une fois continûment différentiable de $\mathbb{R}$ dans $\mathbb{R}$ vérifiant

(1.1) $$\phi(0) = 0, \quad \phi'(\xi) \geq 0 \quad \forall \xi \in \mathbb{R} \quad .$$

On pose

(1.2) $$\Phi(\xi) = \int_o^\xi \phi(\eta) d\eta \qquad \forall \xi \in \mathbb{R} \quad .$$

La fonction Φ ainsi définie est alors ≥ 0 sur $\mathbb{R}$.

On peut maintenant énoncer le

<u>Théorème 1.1. Soient</u> u_o, u_1 <u>et</u> f <u>trois fonctions vérifiant</u> :

(1.3)
$$u_o \in H^2(\Omega) \cap H_o^1(\Omega),$$
$$u_1 \in L_2(\Omega),$$

(1.4)
$$f \in L_2(0,T;L_2(\Omega)).$$

<u>Alors, étant donné un nombre</u> $\varepsilon > 0$, <u>il existe une fonction</u> u_ε <u>et une seule telle que</u>

(1.5)
$$u_\varepsilon \in L_\infty(0,T;H^2(\Omega) \cap H_o^1(\Omega)),$$
$$\frac{\partial u_\varepsilon}{\partial t} \in L_2(0,T;H_o^1(\Omega)) \cap L_\infty(0,T;L_2(\Omega)),$$

(1.6)
$$\frac{\partial^2 u_\varepsilon}{\partial t^2} - \frac{\partial}{\partial x}\phi\left(\frac{\partial u_\varepsilon}{\partial x}\right) - \varepsilon\frac{\partial}{\partial t}\frac{\partial^2 u_\varepsilon}{\partial x^2} = f,$$

(1.7)
$$u_\varepsilon(0) = u_o,\ \frac{\partial u_\varepsilon}{\partial t}(0) = u_1.$$

<u>Remarque 1.1</u>. On déduit de (1.5) que $u_\varepsilon \in C^\circ(0,T;H_o^1(\Omega))$ (après modification éventuelle sur un ensemble de mesure nulle) de sorte que la 1ère condition (1.7) a un sens. D'autre part, toujours d'après (1.5), il est clair que $\frac{\partial u_\varepsilon}{\partial x} \in L_\infty(0,T;L_\infty(\Omega))$ d'où $\phi\left(\frac{\partial u_\varepsilon}{\partial x}\right) \in L_2(0,T;L_2(\Omega))$ par exemple; on déduit alors de (1.6) que $\frac{\partial^2 u_\varepsilon}{\partial t^2} \in L_2(0,T;H^{-1}(\Omega))$ d'où $\frac{\partial u_\varepsilon}{\partial t} \in C^\circ(0,T;L_2(\Omega))$ ce qui donne un sens à la 2ème condition (1.7) (cf. [3]).

<u>Démonstration du Théorème 1.1</u>. L'existence sera obtenue lors de l'étude de l'approximation par la méthode des différences finies (cf. [1], [2] pour des démonstrations plus directes). Prouvons donc l'unicité. On désigne par (,) le produit scalaire qui met $H^{-1}(\Omega)$ et $H_o^1(\Omega)$ en dualité. Soient u et v deux solutions; alors w = u-v satisfait aux équations

$$\begin{cases} \dfrac{\partial^2 w}{\partial t^2} - \dfrac{\partial}{\partial x}\left(\phi(\dfrac{\partial u}{\partial x}) - \phi(\dfrac{\partial v}{\partial x})\right) - \varepsilon \dfrac{\partial}{\partial t} \dfrac{\partial^2 w}{\partial x^2} = 0, \\ w(0) = \dfrac{\partial w}{\partial t}(0) = 0 \end{cases}$$

d'où par intégration en t

$$\begin{cases} \dfrac{\partial w}{\partial t}(t) - \varepsilon \dfrac{\partial^2 w}{\partial x^2}(t) - \int_0^t \dfrac{\partial}{\partial x}(\phi(\dfrac{\partial u}{\partial x}(\sigma)) - \phi(\dfrac{\partial v}{\partial x}(\sigma)))d\sigma = 0, \\ w(0) = 0. \end{cases}$$

On obtient alors pour $0 \leq s \leq T$

$$\int_0^s (\frac{\partial w}{\partial t}(t), w(t))dt + \varepsilon \int_0^s \|w(t)\|^2_{H^1_0(\Omega)} dt$$

$$+ \int_0^s \left(\int_0^t (\phi(\frac{\partial u}{\partial x}(\sigma)) - \phi(\frac{\partial v}{\partial x}(\sigma))\right)d\sigma, \frac{\partial w}{\partial x}(t))dt = 0.$$

On a d'abord la formule de Green

$$\int_0^s \left(\frac{\partial w}{\partial t}(t), w(t)\right)dt = \frac{1}{2}\|w(s)\|^2_{L_2(\Omega)} - \frac{1}{2}\|w(0)\|^2_{L_2(\Omega)} = \frac{1}{2}\|w(s)\|^2_{L_2(\Omega)}.$$

Ensuite, puisque $\dfrac{\partial u}{\partial x}, \dfrac{\partial v}{\partial x} \in L_\infty(0,T;L_\infty(\Omega))$ et que la fonction ϕ est de classe C^1, on peut écrire

$$\left| \int_0^s (\int_0^t (\phi \frac{\partial u}{\partial x}(\sigma)) - \phi(\frac{\partial v}{\partial x}(\sigma))d\sigma, \frac{\partial w}{\partial x}(t)\right)dt \right|$$

$$\leq C \int_0^s \left(\int_0^t \|w(\sigma)\|_{H^1_0(\Omega)} d\sigma\right) \|w(t)\|_{H^1_0(\Omega)} dt$$

$$\leq \frac{C}{\sqrt{2}} s \int_0^s \|w(t)\|^2_{H^1_0(\Omega)} dt.$$

On en déduit finalement

$$\|w(s)\|^2_{L_2(\Omega)} + (2\varepsilon - \sqrt{2}Cs) \int_0^s \|w(t)\|^2_{H^1_0(\Omega)} dt \leq 0.$$

Il en résulte que $w(s) = 0$ pour $0 \leq s \leq t_0 = \dfrac{\sqrt{2}\varepsilon}{C}$. En itérant le procédé on démontre que $w(s) = 0$ pour $t_0 \leq s \leq 2t_0$, etc... L'unicité est ainsi prouvée.

Dans la suite, nous prendrons $f = 0$ afin de simplifier un peu l'exposé mais ceci n'est nullement essentiel.

1.2. Le schéma aux différences finies. Notations

Comme on l'a déjà signalé, la méthode de pseudo-viscosité va consister à approcher la solution u_ε de (1.5), (1.6) et (1.7) à l'aide d'un schéma aux différences que l'on va maintenant décrire.

Soient I et N deux paramètres entiers > 0 destinés à tendre vers $+\infty$; on pose : $h = \frac{1}{I+1}$, $k = \frac{T}{N}$. On désignera par u_i^n une "approximation" de $u_\varepsilon(ih,nk)$, $i = 0,1,\ldots, I+1$, $n = 0,1,\ldots, N$. On considère le schéma aux différences finies explicite :

$$(1.8)\quad \begin{cases} \frac{1}{k^2}(u_i^{n+1}-2u_i^n+u_i^{n-1}) - \frac{1}{h}\left(\phi\left(\frac{u_{i+1}^n-u_i^n}{h}\right) - \phi\left(\frac{u_i^n-u_{i-1}^n}{h}\right)\right) - \\ \qquad - \frac{\varepsilon}{kh^2}\left((u_{i+1}^n-2u_i^n+u_{i-1}^n)-(u_{i+1}^{n-1}-2u_i^{n-1}+u_{i-1}^{n-1})\right) = 0, \end{cases}$$

$$i = 1, \ldots, I,\ n = 1, \ldots, N-1$$

$$(1.9)\quad u_i^o,\ u_i^1 \in R \text{ donnés pour } i = 0,1, \ldots, I+1,$$

$$(1.10)\quad u_o^n = u_{I+1}^n = 0,\ n = 0,1, \ldots, N.$$

Afin de mettre ce schéma sous une forme vectorielle agréable, on introduit un certain nombre de notations. On désigne par V_h l'espace des suites $v_h = \{v_i \in R\ ;\ i = 0,1, \ldots, I+1\}$ telles que $v_o = v_{I+1} = 0$. On munit V_h du produit scalaire

$$(1.11)\qquad (u_h,v_h)_h = h\sum_{i=1}^{I} u_i v_i,\quad u_h,v_h \in V_h$$

et on note $|\cdot|_h$ la norme correspondante. On définit une autre norme sur V_h soit

$$(1.12)\qquad \|v_h\|_h = \left(h\sum_{i=o}^{I}\left|\frac{v_{i+1}-v_i}{h}\right|^2\right)^{1/2},\quad v_h \in V_h.$$

On considère ensuite deux opérateurs A_h et B_h de V_h dans V_h donnés par

$$(1.13)\qquad (A_h v_h)_i = -\frac{1}{h^2}(v_{i+1}-2v_i+v_{i-1}),\ i = 1, \ldots, I.$$

$$(1.14)\qquad (B_h v_h)_i = -\frac{1}{h}\left[\phi\left(\frac{v_{i+1}-v_i}{h}\right) - \phi\left(\frac{v_i-v_{i-1}}{h}\right)\right],\ i = 1, \ldots, I.$$

On introduit enfin $V_{h,k}$ comme étant l'espace des suites $v_{h,k} = \{v_h^n \in V_h;\ n = 0,1, \ldots, N\}$.

Ceci posé, le schéma (1.8), (1.9), (1.10) consiste à calculer $u_{h,k} \in V_{h,k}$ solution de

$$(1.15)\qquad \begin{cases} \dfrac{1}{k^2}(u_h^{n+1}-2u_h^n+u_h^{n-1}) + B_h u_h^n + \dfrac{\varepsilon}{k} A_h(u_h^n-u_h^{n-1}) = 0,\ n = 1, \ldots, N-1, \\ u_h^o,\ u_h^1 \text{ donnés dans } V_h . \end{cases}$$

1.3. Stabilité du schéma. Majorations a priori

Dans ce N°, nous allons étudier la stabilité du schéma (1.15), c'est-à-dire chercher sous quelles conditions, appelées conditions de stabilité, il est possible d'obtenir un nombre suffisant de majorations a priori indépendantes de h et k sur la solution $u_{h,k}$ du schéma afin de pouvoir passer à la limite ultérieurement.

Commençons par donner quelques résultats préliminaires simples.

Lemme 1.1. On a pour tout $v_h \in V_h$

$$(1.16)\qquad (A_h v_h, v_h)_h = \|v_h\|_h^2,$$

$$(1.17)\qquad |A_h v_h|_h \leq \frac{2}{h}\|v_h\|_h.$$

Démonstration immédiate.

On pose ensuite

$$(1.18)\qquad \psi(\xi) = \frac{\phi(\xi)^2}{\Phi(\xi)} \quad \forall\ \xi \in R \qquad (^1)$$

puis si $u_h, v_h \in V_h$

(1) On vérifie aisément que $\Phi(\xi) = 0 \Longrightarrow \phi(\xi) = 0$. On a alors par convention $\psi(\xi) = 2\phi'(\xi)$ si $\Phi(\xi) = 0$

(1.19) $$w_{i+1/2} = \frac{1}{h}(u_{i+1}-u_i),\ z_{i+1/2} = \frac{1}{h}(v_{i+1}-v_i),\ i = 0,1,\ \ldots,\ I.$$

<u>Lemme 1.2</u>. <u>Si</u> $u_h, v_h \in V_h$, <u>on a</u>

(1.20) $$(B_h u_h, v_h)_h = h \sum_{i=o}^{I} \phi(w_{i+1/2}) z_{i+1/2},$$

(1.21) $$|B_h v_h|_h \leq \frac{2}{h} \max_{o \leq i \leq I} \psi(z_{i+1/2})^{1/2} \Big(h \sum_{i=o}^{I} \Phi(z_{i+1/2})\Big)^{1/2},$$

(1.22) $$(B_h u_h - B_h v_h, u_h - v_h)_h \leq \max_{o \leq i \leq I} \frac{\phi(w_{i+1/2})-\phi(z_{i+1/2})}{w_{i+1/2}-z_{i+1/2}} \| u_h - v_h \|_h^2$$ [1]

(1.23) $$|B_h u_h - B_h v_h|_h^2 \leq \frac{4}{h^2} \max_{o \leq i \leq I} \frac{\phi(w_{i+1/2})-\phi(z_{i+1/2})}{w_{i+1/2}-z_{i+1/2}} (B_h u_h - B_h v_h, u_h - v_h)_h$$

(1.24) $$2h \sum_{i=o}^{I} (\Phi(w_{i+1/2})-\Phi(z_{i+1/2}) - \frac{1}{2}(\phi(w_{i+1/2}) + \phi(z_{i+1/2}))(w_{i+1/2}-z_{i+1/2}))$$

$$\leq B_h u_h - B_h v_h, u_h - v_h)_h.$$

<u>Démonstration</u>. Prouvons par exemple (1.23) et (1.24). On a

$$|B_h u_h - B_h v_h|_h^2 = \frac{1}{h} \sum_{i=1}^{I} |(\phi(w_{i+1/2})-\phi(w_{i-1/2}))-(\phi(z_{i+1/2})-\phi(z_{i-1/2}))|^2 \leq$$

$$\leq \frac{4}{h} \sum_{i=o}^{I} |\phi(w_{i+1/2})-\phi(z_{i+1/2})|^2 \leq$$

$$\leq \frac{4}{h} \max_{o \leq i \leq I} \frac{\phi(w_{i+1/2})-\phi(z_{i+1/2})}{w_{i+1/2} - z_{i+1/2}} \sum_{i=o}^{I} (\phi(w_{i+1/2})-\phi(z_{i+1/2}))(w_{i+1/2}-z_{i+1/2}).$$

Or d'après (1.20)

---

[1] On pose $\frac{\phi(\xi)-\phi(\eta)}{\xi-\eta} = \phi'(\xi)$ si $\xi = \eta$.

$$(B_h u_h - B_h v_h, u_h - v_h)_h = h \sum_{i=o}^{I} (\phi(w_{i+1/2}) - \phi(z_{i+1/2}))(w_{i+1/2} - z_{i+1/2})$$

d'où (1.23). Pour vérifier (1.24), on remarque que

$$\Phi(\xi)-\Phi(\eta) - \frac{1}{2}(\phi(\xi)+\phi(\eta))(\xi-\eta) = - \int_\eta^\xi (t - \frac{1}{2}(\xi+\eta))\phi'(t)dt \quad \forall \xi, \eta \in R .$$

Or, puisque $\phi'(t)$ est $\geq 0 \quad \forall t \in R$, on peut écrire

$$- \int_\eta^\xi (t - \frac{1}{2} (\xi+\eta))\phi'(t)dt = - \int_\eta^\xi (t-\eta)\phi'(t)dt + \frac{1}{2} \int_\eta^\xi (\xi-\eta)\phi'(t)dt \leq$$

$$\leq \frac{1}{2} (\phi(\xi)-\phi(\eta))(\xi-\eta) .$$

L'inégalité (1.24) en résulte immédiatement d'après (1.20).

Si $u_{h,k} \in V_{h,k}$ est la solution de (1.15), on pose

(1.25) $$w^n_{i+1/2} = \frac{1}{h}(u^n_{i+1}-u^n_i), \ i = 0,1, \dots, I, \ n = 0,1, \dots, N.$$

On peut alors donner le premier résultat de majoration.

<u>Théorème 1.2</u>. <u>Soit</u> $r \in \{1, \dots, N-1\}$; <u>on suppose qu'il existe deux constantes</u> $\lambda, \mu \in \,]0,1[$ <u>telles que</u>

(1.26) $$\frac{k^2}{2h^2} \max_{o \leq i \leq I} \psi(w^n_{i+1/2}) \leq (1-\lambda)^2(1-\mu)^2, \ n = 0,1, \dots, r,$$

(1.27) $$\frac{\varepsilon}{2\lambda} \frac{k}{h^2} + \frac{k}{2\varepsilon} \max_{o \leq i \leq I} \frac{\phi(w^n_{i+1/2})-\phi(w^{n-1}_{i+1/2})}{w^n_{i+1/2}-w^{n-1}_{i+1/2}} \leq (1-\lambda)(1-\mu), \ n = 1, \dots, r;$$

<u>alors la solution</u> $u_{h,k} \in V_{h,k}$ <u>du schéma aux différences finies</u> (1.15) <u>vérifie l'inégalité de l'énergie suivante pour</u> $n = 0,1, \dots, r$:

(1.28) $$\begin{cases} \frac{1}{k^2}|u^{n+1}_h-u^n_h|^2_h + 2h \sum_{i=o}^{I} \Phi(w^n_{i+1/2}) + \frac{2\varepsilon}{k} \sum_{m=1}^{n} \| u^m_h-u^{m-1}_h \|^2_h \leq \\ \leq \frac{2-\mu}{\mu} (\frac{1}{k^2} |u^1_h-u^o_h|^2_h + 2h \sum_{i=o}^{I} \Phi(w^o_{i+1/2})). \end{cases}$$

<u>Démonstration</u>. On adapte la méthode donnée dans [4] Chapitre II : on prend le produit scalaire dans V_h de l'équation (1.15) avec $u^{n+1}_h-u^{n-1}_h$ et on somme de $n = 1$ à $n = r$ ($1 \leq r \leq N-1$); on obtient (en supprimant les indices h pour allé-

ger l'écriture

$$\frac{1}{k^2}\sum_{n=1}^{r}(u^{n+1}-2u^n+u^{n-1},u^{n+1}-u^{n-1}) + \sum_{n=1}^{r}(Bu^n,u^{n+1}-u^{n-1}) +$$
$$+\frac{\varepsilon}{k}\sum_{n=1}^{r}(A(u^n-u^{n-1}),\ u^{n+1}-u^{n-1}) = 0.$$

On remarque d'abord que

$$\sum_{n=1}^{r}(u^{n+1}-2u^n+u^{n-1},u^{n+1}-u^{n-1}) = |u^{r+1}-u^r|^2 - |u^1-u^o|^2. \tag{1.30}$$

Ensuite, on a d'après (1.20) et (1.25)

$$(Bu^n,u^{n+1}-u^{n-1}) = h\sum_{i=o}^{I}\phi(w^n_{i+1/2})(w^{n+1}_{i+1/2}-w^{n-1}_{i+1/2}) .$$

Or

$$\sum_{n=1}^{r}\phi(w^n_{i+1/2})(w^{n+1}_{i+1/2}-w^{n-1}_{i+1/2}) = 2\Phi(w^r_{i+1/2}) + \phi(w^{r+1}_{i+1/2})(w^{r+1}_{i+1/2}-w^r_{i+1/2}) -$$
$$- 2\Phi(w^o_{i+1/2}) - \phi(w^o_{i+1/2})(w^1_{i+1/2}-w^o_{i+1/2}) -$$
$$- 2\sum_{n=1}^{r}(\Phi(w^n_{i+1/2}) - \Phi(w^{n-1}_{i+1/2}) -$$
$$- \frac{1}{2}(\phi(w^n_{i+1/2})+\phi(w^{n-1}_{i+1/2}))(w^n_{i+1/2}-w^{n-1}_{i+1/2})).$$

En multipliant par h et en sommant de i = 0 à i = I, on trouve donc

$$\sum_{n=1}^{r}(Bu^n,u^{n+1}-u^{n-1}) = 2h\sum_{i=o}^{I}\Phi(w^r_{i+1/2}) + (Bu^r,u^{r+1}-u^r) -$$
$$- 2h\sum_{i=o}^{I}\Phi(w^o_{i+1/2}) + (Bu^o,\ u^1-u^o) - \tag{1.31}$$
$$- 2h\sum_{n=1}^{r}\sum_{i=o}^{I}(\Phi(w^n_{i+1/2})-\Phi(w^{n-1}_{i+1/2}) - \frac{1}{2}(\phi(w^n_{i+1/2}) +$$
$$+ \phi(w_{i+1/2}))(w^n_{i+1/2}-v^{n-1}_{i+1/2})).$$

Enfin, on a d'après (1.16)

$$(A(u^n-u^{n-1}),u^{n+1}-u^{n-1}) = 2\,\|u^n-u^{n-1}\|^2 + (A(u^n-u^{n-1}),u^{n+1}-2u^n+u^{n-1}). \tag{1.32}$$

En reportant les expressions (1.30), (1.31) et (1.32) dans l'équation (1.29), on obtient

$$(1.33)\quad \begin{aligned} &\frac{1}{k^2}|u^{r+1}-u^r|^2 + 2h\sum_{i=o}^{I}\Phi(w^r_{i+1/2}) + (Bu^r,u^{r+1}-u^r) + \frac{2\varepsilon}{k}\sum_{n=1}^{r}\|u^n-u^{n-1}\|^2 = \\ &= \frac{1}{k^2}|u^1-u^o|^2 + 2h\sum_{i=o}^{I}\Phi(w^o_{i+1/2}) + (Bu^o,\ u^1-u^o) + \\ &+ 2h\sum_{n=1}^{r}\sum_{i=o}^{I}(\Phi(w^n_{i+1/2})-\Phi(w^{n-1}_{i+1/2})-\frac{1}{2}(\phi(w^n_{i+1/2})+\phi(w^{n-1}_{i+1/2}))(w^n_{i+1/2}-w^{n-1}_{i+1/2})) - \\ &- \frac{\varepsilon}{k}\sum_{n=1}^{r}(A(u^n-u^{n-1}),\ u^{n+1}-2u^n+u^{n-1}),\quad r = 1,\ \ldots,\ N-1. \end{aligned}$$

Pour obtenir une inégalité de l'énergie, il convient d'abord de majorer "judicieusement" les deux derniers termes de l'équation (1.33). On a d'après (1.24)

$$(1.34)\quad \begin{aligned} &2h\sum_{n=1}^{r}\sum_{i=o}^{I}(\Phi(w^n_{i+1/2})-\Phi(w^{n-1}_{i+1/2}) - \frac{1}{2}(\phi(w^n_{i+1/2}) + \phi(w^{n-1}_{i+1/2}))(w^n_{i+1/2}-w^{n-1}_{i+1/2})) \\ &\qquad \leq \sum_{n=1}^{r}(Bu^n-Bu^{n-1},\ u^n-u^{n-1}). \end{aligned}$$

Estimons d'autre part le terme

$$-\frac{\varepsilon}{k}\sum_{n=1}^{r}(A(u^n-u^{n-1}),\ u^{n+1}-2u^n+u^{n-1}).$$

Pour cela, on prend le produit scalaire de (1.15) avec $u^{n+1}-2u^n+u^{n-1}$, d'où

$$\frac{1}{k^2}|u^{n+1}-2u^n+u^{n-1}|^2 = -\frac{\varepsilon}{k}(A(u^n-u^{n-1}),\ u^{n+1}-2u^n+u^{n-1}) - $$
$$- (Bu^n,u^{n+1}-2u^n+u^{n-1})$$

et

$$\frac{1}{k^2}|u^{n+1}-2u^n+u^{n-1}|^2 \leq \frac{\varepsilon^2}{4\lambda}|A(u^n-u^{n-1})|^2 + \frac{\lambda}{k^2}|u^{n+1}-2u^n+u^{n-1}|^2 - $$
$$- (Bu^n,u^{n+1}-2u^n+u^{n-1})$$

où λ est une constante > 0 arbitraire. On en déduit pour $0 < \lambda < 1$

$$\frac{1}{k^2}|u^{n+1}-2u^n+u^{n-1}|^2 \leq \frac{\varepsilon^2}{4\lambda(1-\lambda)}|A(u^n-u^{n-1})|^2 - \frac{1}{1-\lambda}(Bu^n,u^{n+1}-2u^n+u^{n-1})$$

d'où

$$-\frac{\varepsilon}{k}(A(u^n-u^{n-1}),\ u^{n+1}-2u^n+u^{n-1}) \le \frac{\varepsilon^2}{4\lambda(1-\lambda)}|A(u^n-u^{n-1})|^2 - $$
$$-\frac{\lambda}{1-\lambda}(Bu^n,u^{n+1}-2u^n+u^{n-1}).$$

Comme

$$\sum_{n=1}^{r}(Bu^n,u^{n+1}-2u^n+u^{n-1}) = (Bu^r,u^{r+1}-u^r)-(Bu^o,u^1-u^o) - $$
$$-\sum_{n=1}^{r}(Bu^n-Bu^{n-1},u^n-u^{n-1}),$$

on obtient finalement

$$(1.35)\quad -\frac{\varepsilon}{k}\sum_{n=1}^{r}(A(u^n-u^{n-1}),\ u^{n+1}-2u^n+u^{n-1}) \le \frac{\varepsilon^2}{4\lambda(1-\lambda)}\sum_{n=1}^{r}|A(u^n-u^{n-1})|^2 - $$
$$-\frac{\lambda}{1-\lambda}\{(Bu^r,u^{r+1}-u^r)-(Bu^o,u^1-u^o)-\sum_{n=1}^{r}(Bu^n-Bu^{n-1},u^n-u^{n-1})\}.$$

En reportant les majorations (1.34) et (1.35) dans l'équation (1.33), on trouve pour $0 < \lambda < 1$

$$\frac{1}{k^2}|u^{r+1}-u^r|^2 + 2h\sum_{i=o}^{I}\Phi(w^r_{i+1/2}) + \frac{1}{1-\lambda}(Bu^r,u^{r+1}-u^r) + \frac{2\varepsilon}{k}\sum_{n=1}^{r}\|u^n-u^{n-1}\|^2 \le$$
$$(1.36)\quad \le \frac{1}{k^2}|u^1-u^o|^2 + 2h\sum_{i=o}^{I}\Phi(w^o_{i+1/2}) + \frac{1}{1-\lambda}(Bu^o,u^1-u^o) + \frac{\varepsilon^2}{4\lambda(1-\lambda)}\sum_{n=1}^{r}|A(u^n-u^{n-1})|^2 +$$
$$+\frac{1}{1-\lambda}\sum_{n=1}^{r}(Bu^n-Bu^{n-1},u^n-u^{n-1}).$$

Il est maintenant possible de montrer que l'inégalité (1.36) entraîne l'inégalité de l'énergie cherchée sous les conditions de stabilité (1.26) et (1.27), ceci pour n = r. On peut en effet écrire d'après (1.21)

$$|(Bu^r,u^{r+1}-u^r)| \le \frac{\sqrt{2}}{h}(\max_{o\le i\le I}\psi(w^r_{i+1/2})^{1/2})(2h\sum_{i=o}^{I}\Phi(w^r_{i+1/2}))^{1/2}|u^{r+1}-u^r| \le$$
$$\le \frac{1}{\sqrt{2}}\frac{k}{h}(\max_{o\le i\le I}\psi(w^r_{i+1/2})^{1/2})(\frac{1}{k^2}|u^{r+1}-u^r|^2 + 2h\sum_{i=o}^{I}\Phi(w^r_{i+1/2})).$$

D'autre part (1.17) donne

$$|A(u^n-u^{n-1})|^2 \le \frac{4}{h^2}\|u^n-u^{n-1}\|^2$$

et (1.22) entraîne

$$(Bu^n-Bu^{n-1},u^n-u^{n-1}) \leq \max_{0\leq i\leq I} \frac{\phi(w^n_{i+1/2})-\phi(w^{n-1}_{i+1/2})}{w^n_{i+1/2}-w^{n-1}_{i+1/2}} \|u^n-u^{n-1}\|^2.$$

On obtient donc à partir de (1.36)

$$(1-\frac{1}{\sqrt{2(1-\lambda)}}\frac{k}{h}\max_{0\leq i\leq I}\psi(w^r_{i+1/2})^{1/2})(\frac{1}{k^2}|u^{r+1}-u^r|^2+2h\sum_{i=o}^{I}\Phi(w^r_{i+1/2}))+$$

$$(1.37)\quad +(1-\frac{\varepsilon}{2\lambda(1-\lambda)}\frac{k}{h^2}-\frac{1}{1-\lambda}\frac{k}{2\varepsilon}\max_{0\leq i\leq I}\frac{\phi(w^n_{i+1/2})-\phi(w^{n-1}_{i+1/2})}{w^n_{i+1/2}-w^{n-1}_{i+1/2}})\frac{2\varepsilon}{k}\sum_{n=1}^{r}\|u^n-u^{n-1}\|^2\leq$$

$$\leq (1+\frac{1}{\sqrt{2(1-\lambda)}}\frac{k}{h}\max_{0\leq i\leq I}\psi(w^o_{i+1/2})^{1/2})(\frac{1}{k^2}|u^1-u^o|^2+2h\sum_{i=o}^{I}\phi(w^o_{i+1/2})).$$

Il est désormais immédiat de vérifier que (1.37) entraîne la majoration (1.25) sous les conditions de stabilité (1.26) et (1.27) (avec n remplacé par r).

Remarque 1.2. Considérons le cas où $\varepsilon = 0$ et supposons, pour simplifier, que la fonction ϕ soit deux fois continûment dérivable sur $\mathbb{R}$. Alors la méthode précédente ne permet d'obtenir une inégalité de l'énergie du type (1.28) avec $\varepsilon = 0$ sous une condition de stabilité de nature hyperbolique de la forme (1.26) que si la fonction ϕ est linéaire. En effet, prenons l'équation (1.33) où nous faisons $\varepsilon = 0$. Pour obtenir une telle estimation, il faudrait que l'on ait

$$(1.38)\qquad \Phi(\xi)-\Phi(\eta)-\frac{1}{2}(\phi(\xi)+\phi(\eta))(\xi-\eta)\leq 0 \qquad \xi,\eta\in\mathbb{R}\,.$$

Or on vérifie aisément que, lorsque la fonction ϕ est de classe C^2, on a (1.38) si et seulement si la fonction ϕ est linéaire (auquel cas, l'expression précédente est d'ailleurs identiquement nulle). Le terme de pseudo viscosité a donc pour rôle d'absorber le terme perturbateur

$$2h\sum_{n=1}^{r}\sum_{i=o}^{I}(\Phi(w^n_{i+1/2})-\Phi(w^{n-1}_{i+1/2})-\frac{1}{2}(\phi(w^n_{i+1/2})+\phi(w^{n-1}_{i+1/2}))(w^n_{i+1/2}-w^{n-1}_{i+1/2}))$$

dont le signe est quelconque. Cette analyse permet ainsi de comprendre l'apparition du phénomène d'instabilité non linéaire et le mécanisme stabilisateur du terme de pseudo-viscosité.

Remarque 1.3. La condition de stabilité (1.26) est de nature hyperbolique classique, la condition de stabilité (1.27) contient une condition de nature parabolique due à la présence du terme de pseudo-viscosité $- \varepsilon \frac{\partial}{\partial t} \frac{\partial^2 u}{\partial x^2}$ et une condition exprimant que le pas de temps k doit être choisi assez petit devant ε . Notons que (1.27) interdit de choisir la constante λ trop voisine de 0 dans (1.26) tandis que la constante μ peut être prise arbitrairement petite.

Il convient maintenant de prouver que, h étant fixé, on peut toujours choisir k assez petit pour que les conditions de stabilité (1.26) et (1.27) soient vérifiées. On se fixe donc le pas d'espace h et les constantes λ et $\mu \in]0,1[$. On fait l'hypothèse

(1.39) $\left\{ \begin{array}{l} \text{il existe une constante } C > 0 \text{ indépendante de h et k telle que} \\ \frac{1}{k^2} |u_h^1 - u_h^o|_h^2 + 2h \sum_{i=o}^{I} \Phi(w^o_{i+1/2}) \leq C^2 \end{array} \right.$

et on pose

$$L(h,k) = \max_{\Phi(\xi) \leq \frac{2-\mu}{\mu} \frac{C^2}{2h}} |\xi| + \frac{2k}{h^{3/2}} \sqrt{\frac{2-\mu}{\mu}}\, C,$$

(1.40) $$M(h,k) = \max_{|\xi| \leq L(h,k)} \psi(\xi),$$

$$N(h,k) = \max_{|\xi| \leq L(h,k)} \phi'(\xi).$$

<u>Théorème 1.3</u>. <u>On fait l'hypothèse</u> (1.39). <u>Si</u> k <u>vérifie</u>

(1.41) $$\begin{aligned} &\frac{k^2}{2h^2} M(h,k) \leq (1-\lambda)^2 (1-\mu)^2, \\ &\frac{\varepsilon}{2\lambda} \frac{k}{h^2} + \frac{k}{2\varepsilon} N(h,k) \leq (1-\lambda)(1-\mu), \end{aligned}$$

<u>les conditions de stabilité</u> (1.26) <u>et</u> (1.27) <u>ont lieu pour</u> r = 1, ..., N-1 .

<u>Démonstration</u>. Nous allons raisonner par récurrence. On a tout d'abord d'après (1.39)

$$2h \sum_{i=o}^{I} \Phi(w^o_{i+1/2}) \leq C^2 < \frac{2-\mu}{\mu} C^2$$

d'où

$$\max_{o\leq i\leq I} \Phi(w^o_{i+1/2}) \leq \frac{2-\mu}{\mu}\,\frac{C^2}{2h}$$

et

$$\max_{o\leq i\leq I} |w^o_{i+1/2}| \leq L(h,k) \quad .$$

On déduit alors de la première condition (1.41)

$$\frac{k^2}{2h^2} \max_{o\leq i\leq I} \psi(w^o_{i+1/2}) \leq \frac{k^2}{2h^2} M(h,k) \leq (1-\lambda)^2(1-\mu)^2.$$

Ceci signifie que les conditions de stabilité ont lieu pour "r = o". Supposons alors que ces conditions de stabilité aient lieu pour r (r $\geq$ 0); on déduit de (1.28)

$$\frac{1}{k^2} |u^{r+1}_h - u^r_h|^2_h + 2h \sum_{i=o}^{I} \Phi(w^r_{i+1/2}) \leq \frac{2-\mu}{\mu} C^2.$$

Ceci entraîne

$$\max_{o\leq i\leq I} \Phi(w^r_{i+1/2}) \leq \frac{2-\mu}{\mu}\,\frac{C^2}{2h}$$

puis

$$\begin{aligned}\max_{o\leq i\leq I} |w^{r+1}_{i+1/2}| &\leq \max_{o\leq i\leq I} |w^r_{i+1/2}| + \max_{o\leq i\leq I} |w^{r+1}_{i+1/2} - w^r_{i+1/2}| \leq \\ &\leq \max_{o\leq i\leq I} |w^r_{i+1/2}| + \frac{2}{h} \max_{1\leq i\leq I} |u^{r+1}_i - u^r_i| \leq \\ &\leq \max_{o\leq i\leq I} |w^r_{i+1/2}| + \frac{2k}{h^{3/2}} |u^{r+1}_h - u^r_h|_h \leq \\ &\leq \max_{o\leq i\leq I} |w^r_{i+1/2}| + \frac{2k}{h^{3/2}} \sqrt{\frac{2-\mu}{\mu}}\, C.\end{aligned}$$

On trouve ainsi

$$\max_{o\leq i\leq I} |w^r_{i+1/2}|,\ \max_{o\leq i\leq I} |w^{r+1}_{i+1/2}| \leq L(h,k)$$

d'où

$$\frac{k^2}{2h^2} \max_{o\leq i\leq I} \psi(w^r_{i+1/2}) \leq \frac{k^2}{2h^2} M(h,k) \leq (1-\lambda)^2(1-\mu)^2,$$

$$\frac{\varepsilon}{2\lambda}\,\frac{k}{h^2} + \frac{k}{2\varepsilon} \max_{o\leq i\leq I} \frac{\phi(w^{r+1}_{i+1/2}) - \phi(w^r_{i+1/2})}{w^{r+1}_{i+1/2} - w^r_{i+1/2}} \leq \frac{\varepsilon}{2\lambda}\,\frac{k}{h^2} + \frac{k}{2\varepsilon} N(h,k) \leq (1-\lambda)(1-\mu).$$

Les conditions de stabilité ont encore lieu pour r+1, d'où le théorème.

Soit $k_o = k_o(h,\lambda,\mu,\varepsilon)$ le plus grand des nombres $k > 0$ vérifiant (1.41) ; un tel nombre k_o existe. Alors, il est clair que, pour $k \leq k_o$, les conditions de stabilité (1.26) et (1.27) sont toujours remplies.

Remarque 1.4. En pratique, on choisit un pas de temps k satisfaisant aux conditions (1.26) et (1.27) pour r = 1. A chaque temps rk, on vérifie que les conditions de stabilité

$$\frac{k^2}{2h^2} \max_{o\leq i\leq I} \psi(w^r_{i+1/2}) < (1-\lambda)^2,$$

$$\frac{\varepsilon}{2\lambda}\,\frac{k}{h^2} + \frac{k}{2\varepsilon} \max_{o\leq i\leq I} \frac{\phi(w^r_{i+1/2})-\phi(w^{r-1}_{i+1/2})}{w^r_{i+1/2}-w^{r-1}_{i+1/2}} < 1-\lambda$$

ont lieu afin de pouvoir calculer u^{r+1}_h. Sinon, on diminue k de façon à pouvoir continuer le calcul. Une variante du Théorème 1.3 montre alors que l'on peut ainsi atteindre n'importe quel temps T donné à l'avance.

La majoration (1.28) étant insuffisante pour passer à la limite, nous allons chercher une estimation supplémentaire.

Théorème 1.4. La solution $u_{h,k} \in V_{h,k}$ du schéma aux différences finies (1.15) vérifie l'inégalité suivante pour $r = 1, \ldots, N-1$:

$$|A_h u^r_h|^2_h \leq \nu \left\{ |A_h u^o_h|^2_h + \frac{1}{\varepsilon^2 k^2} |u^1_h - u^o_h|^2_h + \frac{1}{\varepsilon^2 k^2} |u^{r+1}_h - u^r_h|^2_h + \right.$$
(1.42)
$$\left. + \frac{1}{\varepsilon k} \sum_{n=1}^{r} \|u^n_h - u^{n-1}_h\|^2_h \right\}$$

où ν est une constante > 0 indépendante de h, k et ε.

Démonstration. On prend le produit scalaire dans V_h de l'équation (1.15) avec $\frac{2k}{\varepsilon} A_h u^n_h$ et on somme de n = 1 à n = r ($1 \leq r \leq N-1$); on obtient (en supprimant les indices h)

$$\frac{2}{\varepsilon k} \sum_{n=1}^{r} (u^{n+1}-2u^n+u^{n-1}, Au^n) + \frac{2k}{\varepsilon} \sum_{n=1}^{r} (Bu^n, Au^n) + \tag{1.43}$$
$$+ 2 \sum_{n=1}^{r} (A(u^n-u^{n-1}), Au^n) = 0.$$

On a tout d'abord

$$(1.44)\qquad \sum_{n=1}^{r}(u^{n+1}-2u^n+u^{n-1},Au^n) = (u^{r+1}-u^r,Au^r)-(u^1-u^\circ,Au^\circ) - \sum_{n=1}^{r}\|u^n-u^{n-1}\|^2.$$

D'autre part,

$$(1.45)\qquad (Bu^n,Au^n) = \frac{1}{h}\sum_{i=1}^{I}(\phi(w^n_{i+1/2})-\phi(w^n_{i-1/2}))(w^n_{i+1/2}-w^n_{i-1/2}) \geq 0.$$

Enfin

$$2(A(u^n-u^{n-1}),\ Au^n) = |Au^n|^2 - |Au^{n-1}|^2 + |A(u^n-u^{n-1})|^2$$

d'où

$$(1.46)\qquad 2\sum_{n=1}^{r}(A(u^n-u^{n-1}),Au^n) \geq |Au^r|^2 - |Au^\circ|^2.$$

On déduit de (1.43), ..., (1.46)

$$|Au^r|^2 + \frac{2}{\varepsilon k}(u^{r+1}-u^r,Au^r) \leq |Au^\circ|^2 + \frac{2}{\varepsilon k}(u^1-u^\circ,Au^\circ) + \frac{2}{\varepsilon k}\sum_{n=1}^{r}\|u^n-u^{n-1}\|^2.$$

Le résultat découle alors de l'inégalité

$$\frac{2}{\varepsilon k}|(u^{r+1}-u^r,Au^r)| \leq \frac{1}{2}|Au^r|^2 + \frac{2}{\varepsilon^2k^2}|u^{r+1}-u^r|^2.$$

1.4. Convergence du schéma

Introduisons quelques notations. Si $v_h \in V_h$, on définit les fonctions p_hv_h et q_hv_h sur Ω par

$$(1.47)\qquad \begin{aligned} p_hv_h(x) &= v_i + \frac{1}{h}(v_{i+1}-v_i)(x-ih),\ x\in [ih,(i+1)h],\ i = 0,1,\ldots,I, \\ q_hv_h(x) &= v_i,\ x\in\ ](i-\tfrac{1}{2})h,\ (i+\tfrac{1}{2})h[\ \cap\ \Omega,\ i = 0,1,\ldots,I+1. \end{aligned}$$

Il est clair que $p_hv_h \in H^1_o(\Omega)$ et que

$$\|p_hv_h\|_{L_2(\Omega)} \leq |v_h|_h.$$

Si $v \in H^1_o(\Omega)$, on définit $r_hv \in V_h$ par

$$(1.48)\qquad (r_hv)_i = v(ih),\ i = 0,1,\ldots,I+1.$$

On voit alors aisément que

$$(1.49)\qquad \lim_{h\to o} p_h r_h v = v \text{ dans } H_o^1(\Omega) \text{ fort,} \qquad \lim_{h\to o} q_h r_h v = v \text{ dans } L_2(\Omega) \text{ fort.}$$

De même, si $v_{h,k} \in V_{h,k}$, on définit sur $[0,T[$ les fonctions

$$(1.50)\qquad \begin{aligned} p_{h,k}v_{h,k}(t) &= p_h v_h^n + \frac{1}{k}(p_h v_h^{n+1} - p_h v_h^n)(t-nk), \quad t \in [nk,(n+1)k], \\ p_k q_h v_{h,k}(t) &= q_h v_h^n + \frac{1}{k}(q_h v_h^{n+1} - q_h v_h^n)(t-nk), \quad t \in [nk,(n+1)k],\ n=o,1,\dots,N-1. \\ q_k p_h v_{h,k}(t) &= p_h v_h^n, \quad t \in [nk,(n+1)k[\ . \end{aligned}$$

Nous pouvons alors énoncer le résultat de convergence.

Théorème 1.5. On fait les hypothèses :

$$(1.51)\qquad \left\{ \begin{aligned} &\text{il existe une constante } C > 0 \text{ indépendante de } h \text{ et } k \text{ telle que} \\ &\frac{1}{k^2}|u_h^1 - u_h^o|_h^2 + 2h\sum_{i=o}^{I}\Phi(w_{i+1/2}^o) + |A_h u_h^o|_h^2 \le C^2, \end{aligned} \right.$$

$$(1.52)\qquad \left\{ \begin{aligned} &\lim_{h\to o} q_h u_h^o = u_o \\ &\lim_{h,k\to o} \frac{1}{k}(q_h u_h^1 - q_h u_h^o) = u_1 \end{aligned} \right. \quad \text{dans } L_2(\Omega) \text{ fort.}$$

Alors, si h et k tendent vers 0 de façon que les conditions de stabilité (1.26) et (1.27) aient lieu pour r = 1, ..., N-1 avec une constante μ indépendante de h et k [1], la solution $u_{h,k} \in V_{h,k}$ du schéma (1.15) vérifie

$$(1.53)\qquad \begin{aligned} &\lim_{h,k\to o} p_{h,k}u_{h,k} = u_\varepsilon \text{ dans } L_2(0,T;H_o^1(\Omega)) \text{ fort et dans } L_\infty(0,T;H_o^1(\Omega)) \text{ faible-étoile,} \\ &\lim_{h,k\to o} \frac{\partial}{\partial t} p_{h,k}u_{h,k} = \frac{\partial u_\varepsilon}{\partial t} \text{ dans } L_2(0,T;H_o^1(\Omega)) \text{ faible et dans } L_\infty(0,T;L_2(\Omega)) \text{ faible étoile} \ . \end{aligned}$$

---

(1) Ceci est loisible en vertu de (1.51) et du Théorème 1.3 .

<u>Démonstration</u>. On désigne par K une constante > 0 arbitraire indépendante de h,k et ε. Il sera commode de prolonger $p_h v_h (v_h \in V_h)$ par 0 en dehors de Ω et de poser

$$D_h p_h v_h(x) = \frac{1}{h}(p_h v_h(x+\frac{h}{2})-p_h v_h(x-\frac{h}{2})).$$

1/ Ceci précisé, on déduit de (1.51) et du Théorème 1.2 que, si les conditions de stabilité sont vérifiées pour r = 1, ..., N-1 avec une constante μ indépendante de h et k, on a

$$\| \frac{\partial}{\partial t} p_k q_h u_{h,k} \|_{L_\infty(0,T;L_2(\Omega))} = \max_{0 \le r \le N-1} \frac{1}{k} |u_h^{r+1}-u_h^r|_h \le K,$$

$$\| \frac{\partial}{\partial t} p_{h,k} u_{h,k} \|_{L_\infty(0,T;L_2(\Omega))} \le \max_{0 \le r \le N-1} \frac{1}{k} |u_h^{r+1}-u_h^r|_h \le K,$$

$$\int_0^T \| \frac{\partial}{\partial t} p_{h,k} u_{h,k} \|^2_{H_0^1(\Omega)} dt = \frac{1}{k} \sum_{n=1}^{N} \| u_h^n - u_h^{n-1} \|_h^2 \le$$

$$\le \frac{1}{k} \sum_{n=1}^{N-1} \| u_h^n - u_h^{n-1} \|_h^2 + \frac{4k}{h^2} \frac{1}{k^2} |u_h^N - u_h^{N-1}|_h^2 \le \frac{K}{\varepsilon} .$$

On a d'après (1.51)

$$|A_h u_h^o|_h \le C$$

ce qui entraîne

$$\| q_h u_h^o \|_{L_2(\Omega)}, \| p_h u_h^o \|_{H_0^1(\Omega)} \le K.$$

On obtient alors

$$\| p_k q_h u_{h,k} \|_{L_\infty(0,T;L_2(\Omega))} \le K,$$

$$\| p_{h,k} u_{h,k} \|_{L_\infty(0,T;H_0^1(\Omega))}, \| q_k p_h u_{h,k} \|_{L_\infty(0,T;H_0^1(\Omega))} \le \frac{K}{\sqrt{\varepsilon}} .$$

Enfin, on déduit de (1.51) et des Théorèmes 1.2 et 1.4 :

$$\| D_h \frac{\partial}{\partial x} q_k p_h u_{h,k} \|_{L_\infty(0,T;L_2(\Omega))} = \max_{0 \le r \le N-1} |A_h u_h^r|_h \le \frac{K}{\varepsilon},$$

$$\| D_h \frac{\partial}{\partial x} p_{h,k} u_{h,k} \|_{L_\infty(0,T;L_2(\Omega))} = \max_{0 \le r \le N} |A_h u_h^r|_h \le$$

$$\le \max_{0 \le r \le N-1} |A_h u_h^r|_h + |A_h(u_h^N - u_h^{N-1})|_h \le$$

$$\le \max_{0 \le r \le N-1} |A_h u_r^r|_h + \frac{4k}{h^2} \frac{1}{k} |u_h^N - u_h^{N-1}|_h \le \frac{K}{\varepsilon} .$$

2/ Il en résulte que, de la suite (h,k) vérifiant les conditions de stabilité, on peut extraire une sous-suite notée encore (h,k) telle que

$$(1.54)\quad \lim_{h,k\to o} p_{h,k}u_{h,k} = \lim_{h,k\to o} q_k p_h u_{h,k} = u^* \text{ dans } L_\infty(0,T;H_o^1(\Omega)) \text{ faible-étoile,}$$

$$(1.55)\quad \lim_{h,k\to o} p_k q_h u_{h,k} = u^* \text{ dans } L_\infty(0,T;L_2(\Omega)) \text{ faible-étoile,}$$

$$(1.56)\quad \lim_{h,k\to o} \frac{\partial}{\partial t} p_{h,k}u_{h,k} = \frac{\partial u^*}{\partial t} \text{ dans } L_2(0,T;H_o^1(\Omega)) \text{ faible et dans } L_\infty(0,T;L_2(\Omega)) \text{ faible-étoile,}$$

$$(1.57)\quad \lim_{h,k\to o} \frac{\partial}{\partial t} p_k q_h u_{h,k} = \frac{\partial u^*}{\partial t} \text{ dans } L_\infty(0,T;L_2(\Omega)) \text{ faible-étoile,}$$

$$(1.58)\quad \lim_{h,k\to o} D_h \frac{\partial}{\partial x} p_{h,k}u_{h,k} = \lim_{h,k\to o} D_h \frac{\partial}{\partial x} q_k p_h u_{h,k} = \frac{\partial^2 u^*}{\partial x^2} \text{ dans } L_\infty(0,T;L_2(\Omega)) \text{ faible-étoile.}$$

On déduit alors de (1.56), (1.58) et d'une variante semi-discrète du théorème de compacité de Kondrachov-Sobolev que, quitte à extraire une nouvelle sous-suite, on a

$$(1.59)\quad \lim_{h,k\to o} \frac{\partial}{\partial x} p_{h,k}u_{h,k} = \frac{\partial u^*}{\partial x} \text{ dans } L_2(0,T;L_2(\Omega)) \text{ fort.}$$

Il est facile de vérifier que ceci implique

$$\lim_{h,k\to o} \frac{\partial}{\partial x} q_k p_h u_{h,k} = \frac{\partial u^*}{\partial x} \text{ dans } L_2(0,T;L_2(\Omega)) \text{ fort.}$$

Mais (1.54) et (1.58) entraînent que $\frac{\partial}{\partial x} q_k p_h u_{h,k}$ reste borné dans $L_\infty(0,T;L_\infty(\Omega))$ et que $\frac{\partial u^*}{\partial x} \in L_\infty(0,T;L_\infty(\Omega))$; on trouve donc

$$\left\| \phi\left(\frac{\partial}{\partial x} q_k p_h u_{h,k}\right) - \phi\left(\frac{\partial u^*}{\partial x}\right) \right\|_{L_2(0,T;L_2(\Omega))} \leq K(\varepsilon) \left\| \frac{\partial}{\partial x}(q_k p_h u_{h,k} - u^*) \right\|_{L_2(0,T;L_2(\Omega))}$$

d'où

$$(1.60)\quad \lim_{h,k\to o} \phi\left(\frac{\partial}{\partial x} q_k p_h u_{h,k}\right) = \phi\left(\frac{\partial u^*}{\partial x}\right) \text{ dans } L_2(0,T;L_2(\Omega)) \text{ fort.}$$

3/ Montrons que $u^* = u_\varepsilon$. On a par construction

$$u^* \in L_\infty(0,T;H^2(\Omega) \cap H_o^1(\Omega)),$$

$$\frac{\partial u^*}{\partial t} \in L_2(0,T;H_o^1(\Omega)) \cap L_\infty(0,T;L_2(\Omega)).$$

Il reste à prouver que

$$\frac{\partial^2 u^*}{\partial t^2} - \frac{\partial}{\partial x}\phi\left(\frac{\partial u^*}{\partial x}\right) - \varepsilon \frac{\partial}{\partial t}\frac{\partial^2 u^*}{\partial x^2} = 0,$$

$$u^*(0) = u_o, \quad \frac{\partial u^*}{\partial t}(0) = u_1.$$

Soit alors $v \in H_o^1(\Omega)$ et $\chi \in C^1(0,T)$ avec $\chi(T) = 0$. On déduit de (1.15)

$$\frac{1}{k}\sum_{n=1}^{N-1}(u_h^{n+1}-2u_h^n+u_h^{n-1},r_hv)_h\chi(nk) + k\sum_{n=1}^{N-1}(B_hu_h^n,r_hv)_h\chi(nk) +$$

$$+ \varepsilon\sum_{n=1}^{N-1}(A_h(u_h^n-u_h^{n-1}),r_hv)_h\chi(nk) = 0.$$

Or

$$\frac{1}{k}\sum_{n=1}^{N-1}(u_h^{n+1}-2u_h^n+u_h^{n-1},r_hv)_h\chi(nk) =$$

$$= -\frac{1}{k}\sum_{n=1}^{N}(u_h^n-u_h^{n-1},r_hv)_h(\chi(nk)-\chi(n-1)k)) - \frac{1}{k}(u_h^1-u_h^o,r_hv)_h\chi(0) =$$

$$= -\int_o^T\left(\frac{\partial}{\partial t}p_kq_hu_{h,k}(t),q_hr_hv\right)\frac{1}{k}(\chi_k(t+k)-\chi_k(t))dt - \frac{1}{k}(q_hu_h^1u_h^o,q_hr_hv)\chi(0).$$

avec

$$\chi_k(t) = \chi(nk), \; nk \leq t < ((n+1)k), \; n = 0,1, \ldots, N-1.$$

D'autre part,

$$k\sum_{n=1}^{N-1}(B_hu_h^n,r_hv)_h\chi(nk) = \int_k^T\left(\phi\left(\frac{\partial}{\partial x}q_kp_hu_{h,k}(t)\right), \frac{\partial}{\partial x}p_hr_hv\right)\chi_k(t)dt,$$

$$\varepsilon\sum_{n=1}^{N-1}(A_h(u_h^n-u_h^{n-1}),r_hv)_k\chi(nk) = \varepsilon\int_o^{T-k}\left(\frac{\partial}{\partial t}\frac{\partial}{\partial x}p_{h,k}u_{h,k}(t), \frac{\partial}{\partial x}p_hr_hv\right)\chi_k(t)dt.$$

On obtient finalement

$$(1.61)\quad \begin{aligned} & -\int_0^T\left(\frac{\partial}{\partial t}\, p_k q_h u_{h,k}(t), q_h r_h v\right)\frac{1}{k}\left(\chi_k(t+k)-\chi_k(t)\right)dt \;+ \\ & +\int_k^T\left(\phi\left(\frac{\partial}{\partial x}\, q_k p_h u_{h,k}(t)\right), \frac{\partial}{\partial x}\, p_h r_h v\right)\chi_k(t)dt \;+ \\ & +\varepsilon\int_0^{T-k}\left(\frac{\partial}{\partial t}\frac{\partial}{\partial x}\, p_{h,k}u_{h,k}(t), \frac{\partial}{\partial x}\, p_h r_h v\right)\chi_k(t)dt = \\ & = \frac{1}{k}\left(q_h u_h^1 - q_h u_h^o, q_h r_h v\right)\chi(0). \end{aligned}$$

En passant à la limite grâce à (1.49), (1.52), (1.56), (1.57) et (1.59), on trouve

$$(1.62)\quad \begin{aligned} & -\int_0^T\left(\frac{\partial u^*}{\partial t}(t), v\right)\chi'(t)dt + \int_0^T\left(\phi\left(\frac{\partial u^*}{\partial x}(t)\right), \frac{\partial v}{\partial x}\right)\chi(t)dt \;+ \\ & +\varepsilon\int_0^T\left(\frac{\partial}{\partial t}\frac{\partial u^*}{\partial x}(t), \frac{\partial v}{\partial x}\right)\chi(t)dt = (u_1, v)\chi(0) \quad \forall\, v\in H_o^1(\Omega). \end{aligned}$$

En prenant $\chi \in \mathcal{D}(]0,T[)$, espace des fonctions indéfiniment dérivables à support compact dans $]0,T[$, on en déduit que u^* vérifie

$$(1.63)\quad \frac{\partial^2 u^*}{\partial t^2} - \frac{\partial}{\partial x}\,\phi\left(\frac{\partial u^*}{\partial x}\right) - \varepsilon\frac{\partial}{\partial t}\frac{\partial^2 u^*}{\partial x^2} = 0$$

au sens des distributions sur $]0,T[$ à valeurs dans $H^{-1}(\Omega)$. A partir de (1.62) et (1.63), il est facile de voir que

$$(1.64)\quad \frac{\partial u^*}{\partial t}(0) = u_1 .$$

On vérifierait de manière analogue que

$$(1.65)\quad u^*(0) = u_o .$$

Ainsi $u^* = u_\varepsilon$.

4/ De l'unicité de la solution u_ε, on déduit que c'est toute la famille $u_{h,k}$ qui converge au sens indiqué. En particulier

$p_{h,k}u_{h,k} \to u_\varepsilon$ dans $L_2(0,T;H_o^1(\Omega))$ fort et dans $L_\infty(0,T;H_o^1(\Omega))$ faible-étoile,

$\frac{\partial}{\partial t}\, p_{h,k}u_{h,k} \to \frac{\partial u_\varepsilon}{\partial t}$ dans $L_2(0,T;H_o^1(\Omega))$ faible et dans $L_\infty(0,T;L_2(\Omega))$ faible-étoile.

Ceci achève la démonstration du théorème.

Il convient maintenant de choisir u_h^o et u_h^1 de façon que les hypothèses (1.51) et (1.52) soient vérifiées. Il est aisé de prouver que le choix

$$(1.66)\qquad \begin{aligned} & u_i^o = u_o(ih),\ i = 0,1,\ \ldots,\ I+1, \\ & \frac{1}{k}(u_i^1-u_i^o) = \frac{1}{h}\int_{(i-1/2)h}^{(i+1/2)h} u_1(x)dx,\ i = 1,\ \ldots,\ I, \end{aligned}$$

répond à la question. Si u_1 est une fonction continue sur Ω, on peut prendre également

$$\frac{1}{k}(u_i^1-u_i^o) = u_1(ih),\ i = 1,\ \ldots,\ I,$$

etc ...

<u>Remarque 1.5</u>. La démonstration précédente prouve l'existence d'une fonction u_ε satisfaisant à (1.5), (1.6) et (1.7). Notons que les majorations a priori obtenues au cours de cette démonstration dépendent de ε, ce qui ne permet pas de passer à la limite lorsque h,k,ε tendent vers 0.

<u>Remarque 1.6</u>. Sous les conditions d'application du Théorème 1.5, on peut démontrer un résultat de convergence plus fort que (1.53). On a en effet

$$(1.67)\qquad \begin{aligned} & \lim_{h,k\to o} p_{h,k}u_{h,k} = u_\varepsilon \text{ dans } L_\infty(0,T;H_o^1(\Omega)) \text{ fort}, \\ & \lim_{h,k\to o} \frac{\partial}{\partial t} p_{h,k}u_{h,k} = \frac{\partial u_\varepsilon}{\partial t} \text{ dans } L_2(0,T;H_o^1(\Omega)) \cap L_\infty(0,T;L_2(\Omega)) \text{ fort}. \end{aligned}$$

La démonstration est très technique et n'est pas reproduite ici.

2. METHODE DE PSEUDO-VISCOSITE QUASI-LINEAIRE

On a vu (cf. Remarque 1.2) que le phénomène d'instabilité non linéaire était dû à la présence du terme perturbateur

$$2h\sum_{n=1}^{r}\sum_{i=o}^{I}(\Phi(w^n_{i+1/2})-\Phi(w^{n-1}_{i+1/2})-\frac{1}{2}(\phi(w^n_{i+1/2})+\phi(w^{n-1}_{i+1/2}))(w^n_{i+1/2}-w^{n-1}_{i+1/2})),$$

terme que l'on majore par $(B_hu^n_h-B_hu^{n-1}_h,u^n_h-u^{n-1}_h)_h$. Il est donc naturel de songer à remplacer dans l'équation (1.15) le terme de pseudo-viscosité $\frac{\varepsilon}{k}A_h(u^n_h-u^{n-1}_h)$ par $\frac{\varepsilon}{k}(B_hu^n_h-B_hu^{n-1}_h)$, c'est-à-dire de remplacer dans l'équation (1.6) $-\varepsilon\frac{\partial}{\partial t}\frac{\partial^2u_\varepsilon}{\partial x^2}$ par $-\varepsilon\frac{\partial}{\partial t}\frac{\partial}{\partial x}\phi\left(\frac{\partial u_\varepsilon}{\partial x}\right)$. La 2ème partie de ce travail va donc être consacrée à ce cas.

2.1. <u>Théorème d'existence et d'unicité</u>. <u>Le schéma aux différences finies</u>

<u>Pour simplifier</u>, nous allons faire dans toute la suite l'hypothèse suivante :

(2.1) $\qquad \phi'(\xi)\geq\alpha>0 \qquad \forall\xi\in R$.

Il est immédiat de voir que (2.1) entraîne

(2.2) $\qquad (\phi(\xi)-\phi(\eta))(\xi-\eta)\geq\alpha|\xi-\eta|^2 \qquad \forall\,\xi,\eta\in R$

(2.3) $\qquad \Phi(\xi)\geq\frac{\alpha}{2}|\xi|^2 \qquad \forall\,\xi\in R$.

<u>Théorème 2.1</u>. <u>Soient</u> u_o,u_1 <u>et</u> f <u>trois fonctions vérifiant</u>

(2.4) $$\begin{cases} u_o\in H^2(\Omega)\cap H^1_o(\Omega),\\ u_1\in L_2(\Omega),\end{cases}$$

(2.5) $\qquad f\in L_2(0,T;L_2(\Omega))$.

<u>On fait l'hypothèse</u> (2.1). <u>Alors, étant donné un nombre</u> $\varepsilon>0$, <u>il existe une fonction</u> u_ε <u>et une seule telle que</u>

$$(2.6)\quad \begin{cases} u_\varepsilon \in L_\infty(0,T;H^2(\Omega)\cap H_o^1(\Omega)), \\ \dfrac{\partial u_\varepsilon}{\partial t} \in L_2(0,T;H_o^1(\Omega))\cap L_\infty(0,T;L_2(\Omega)), \end{cases}$$

$$(2.7)\quad \frac{\partial^2 u_\varepsilon}{\partial t^2} - \frac{\partial}{\partial x}\phi\left(\frac{\partial u_\varepsilon}{\partial x}\right) - \varepsilon \frac{\partial}{\partial t}\frac{\partial}{\partial x}\phi\frac{\partial u_\varepsilon}{\partial x} = f,$$

$$(2.8)\quad u_\varepsilon(0) = u_o,\ \frac{\partial u_\varepsilon}{\partial t}(0) = u_1.$$

On vérifie comme au N° 1.1 que les conditions initiales (2.8) ont un sens. On démontrera l'existence d'une solution u_ε lors de l'étude de l'approximation par la méthode des différences finies. L'unicité s'obtient de la même manière qu'au N° 1.1. Dans la suite, nous prendrons $f = 0$ pour simplifier.

On approche la solution u_ε à l'aide du schéma aux différences finies explicite

$$(2.9)\quad \begin{aligned} &\frac{1}{k^2}(u_i^{n+1}-2u_i^n+u_i^{n-1}) - \frac{1}{h}\left(\phi\left(\frac{u_{i+1}^n-u_i^n}{h}\right)-\phi\left(\frac{u_i^n-u_{i-1}^n}{h}\right)\right) - \\ &- \frac{\varepsilon}{kh}\left[\left(\phi\left(\frac{u_{i+1}^n-u_i^n}{h}\right)-\phi\left(\frac{u_i^n-u_{i-1}^n}{h}\right)\right) - \left(\phi\left(\frac{u_{i+1}^{n-1}-u_i^{n-1}}{h}\right)-\phi\left(\frac{u_i^{n-1}-u_{i-1}^{n-1}}{h}\right)\right)\right] = 0, \end{aligned}$$

$$i = 1, \ldots, I,\ n = 1, \ldots, N-1,$$

$$(2.10)\quad u_i^o, u_i^1 \in R \text{ donnés pour } i = 0,1,\ \ldots,\ I+1,$$

$$(2.11)\quad u_o^n = u_{I+1}^n = 0,\ n = 0,1,\ \ldots,\ N.$$

Avec les notations introduites au N°1.2, ce schéma peut s'écrire sous la forme suivante : trouver $u_{h,k} \in V_{h,k}$ solution de

$$(2.12)\quad \begin{cases} \dfrac{1}{k^2}(u_h^{n+1}-2u_h^n+u_h^{n-1}) + B_h u_h^n + \dfrac{\varepsilon}{k}(B_h u_h^n - B_h u_h^{n-1}) = 0,\ n = 1,\ \ldots,\ N-1, \\ u_h^o, u_h^1 \text{ donnés dans } V_h. \end{cases}$$

2.2. Stabilité du schéma. Majorations a priori

Comme au N° 1.3, nous allons chercher un nombre suffisant de majorations a priori sur la solution $u_{h,k}$ du schéma afin de pouvoir passer à la limite. Pour les obtenir, nous allons légèrement modifier les techniques données au N°1.3.

Théorème 2.2. Soit $r \in \{1, \ldots, N-1\}$; on suppose qu'il existe une constante $\mu \in \,]0,1[$ telle que

$$\left(\frac{k^2}{2h^2} + (1-\mu)\frac{\varepsilon k}{h^2}\right) \max_{o\leq i\leq I} \psi(w^n_{i+1/2}) \leq (1-\mu)^2, \quad n = 0,1, \ldots, r, \tag{2.13}$$

$$\frac{k}{2\varepsilon} + \left(\frac{k^2}{h^2} + \frac{2\varepsilon k}{h^2}\right) \max_{o\leq i\leq I} \frac{\phi(w^n_{i+1/2})-\phi(w^{n-1}_{i+1/2})}{w^n_{i+1/2}-w^{n-1}_{i+1/2}} \leq 1-\mu, \quad n = 1, \ldots, r \,; \tag{2.14}$$

alors la solution $u_{h,k} \in V_{h,k}$ du schéma aux différences finies (2.12) vérifie l'inégalité de l'énergie suivante pour $n = 1, \ldots, r$:

$$\begin{aligned} &\frac{1}{k^2}|u_h^{n+1}-u_h^n|_h^2 + 2h\sum_{i=o}^{I} \Phi(w^n_{i+1/2}) + \frac{2\varepsilon}{k}\sum_{m=1}^{n} (B_h u_h^m - B_h u_h^{m-1}, u_h^m - u_h^{m-1})_h \leq \\ &\leq \frac{2-\mu}{\mu}\left(\frac{1}{k^2}|u_h^1-u_h^o|_h^2 + 2h\sum_{i=o}^{I}\Phi(w^o_{i+1/2})\right). \end{aligned} \tag{2.15}$$

Démonstration. A partir de (2.12), on obtient pour $1\leq r\leq N-1$ (en supprimant les indices h) :

$$\begin{aligned} &\frac{1}{k^2}\sum_{n=1}^{r} (u^{n+1}-2u^n+u^{n-1}, u^{n+1}-u^{n-1}) + \sum_{n=1}^{r} (Bu^n, u^{n+1}-u^{n-1}) + \\ &+ \frac{\varepsilon}{k}\sum_{n=1}^{r} (Bu^n - Bu^{n-1}, u^{n+1}-u^{n-1}) = 0. \end{aligned} \tag{2.16}$$

On écrit

$$\begin{aligned} (Bu^n-Bu^{n-1}, u^{n+1}-u^{n-1}) &= 2(Bu^n-Bu^{n-1}, u^n-u^{n-1}) + \\ &+ (Bu^n-Bu^{n-1}, u^{n+1}-2u^n+u^{n-1}). \end{aligned}$$

On trouve en utilisant à nouveau l'équation (2.12)

$$(Bu^n-Bu^{n-1},u^{n+1}-2u^n+u^{n-1}) = -k^2(Bu^n-Bu^{n-1},Bu^n)-\varepsilon k|Bu^n-Bu^{n-1}|^2 =$$

$$= -\frac{k^2}{2}|Bu^n|^2 + \frac{k^2}{2}|Bu^{n-1}|^2 - k(\varepsilon+\frac{k}{2})|Bu^n-Bu^{n-1}|^2.$$

On en déduit donc

$$\frac{\varepsilon}{k}\sum_{n=1}^{r}(Bu^n-Bu^{n-1},u^{n+1}-u^{n-1}) = \frac{2\varepsilon}{k}\sum_{n=1}^{r}(Bu^n-Bu^{n-1},u^n-u^{n-1}) - \tag{2.17}$$

$$-\frac{\varepsilon k}{2}|Bu^r|^2 + \frac{\varepsilon k}{2}|Bu^o|^2 - \varepsilon(\varepsilon+\frac{k}{2})\sum_{n=1}^{r}|Bu^n-Bu^{n-1}|^2.$$

En utilisant (1.30), (1.31) et (2.17), l'équation (2.16) se met sous la forme

$$\frac{1}{k^2}|u^{r+1}-u^r|^2 + 2h\sum_{i=o}^{I}\Phi(w^r_{i+1/2}) + (Bu^r,u^{r+1}-u^r) - \frac{\varepsilon k}{2}|Bu^r|^2 +$$

$$+\frac{2\varepsilon}{k}\sum_{n=1}^{r}(Bu^n-Bu^{n-1},u^n-u^{n-1}) - \varepsilon(\varepsilon+\frac{k}{2})\sum_{n=1}^{r}|Bu^n-Bu^{n-1}|^2 -$$

$$-2h\sum_{n=1}^{r}\sum_{i=o}^{I}(\Phi(w^n_{i+1/2})-\Phi(w^{n-1}_{i+1/2}) - \frac{1}{2}(\phi(w^n_{i+1/2}) + \phi(w^{n-1}_{i+1/2}))(w^n_{i+1/2}-w^{n-1}_{i+1/2})) =$$

$$= \frac{1}{k^2}|u^1-u^o|^2 + 2h\sum_{i=o}^{I}\Phi(w^o_{i+1/2}) + (Bu^o,u^1-u^o) - \frac{\varepsilon k}{2}|Bu^o|^2 .$$

On déduit de (1.21)

$$|(Bu^r,u^{r+1}-u^r)| \leq \frac{2k}{h}\big(\max_{o\leq i<I}\psi(w^r_{i+1/2})^{1/2}\big)\big(h\sum_{i=o}^{I}\Phi(w^r_{i+1/2})\big)^{1/2}\frac{1}{k}|u^{r+1}-u^r| \leq$$

$$\leq (1-\eta)\frac{1}{k^2}|u^{r+1}-u^r|^2 + \frac{1}{1-\eta}\frac{k^2}{2h^2}(\max_{o\leq i\leq I}\psi(w^r_{i+1/2}))\,2h\sum_{i=o}^{I}\Phi(w^r_{i+1/2})$$

où η est une constante < 1 arbitraire, puis

$$\frac{\varepsilon k}{2}|Bu^r|^2 \leq \frac{\varepsilon k}{h^2}(\max_{o\leq i\leq I}\psi(w^r_{i+1/2}))\,2h\sum_{i=o}^{I}\Phi(w^r_{i+1/2})$$

d'où

$$\frac{1}{k^2}|u^{r+1}-u^r|^2 + 2h\sum_{i=o}^{I}\Phi(w^r_{i+1/2}) + (Bu^{r+1},u^{r+1}-u^r) - \frac{\varepsilon k}{2}|Bu^r|^2 \geq$$

$$\geq \eta\frac{1}{k^2}|u^{r+1}-u^r|^2 + (1-(\frac{\varepsilon k}{h^2} + \frac{1}{1-\eta}\frac{k^2}{2h^2})(\max_{o\leq i\leq I}\psi(w^r_{i+1/2}))2h\sum_{i=o}^{I}\Phi(w^r_{i+1/2}).$$

En choisissant $\eta = \eta^r$ avec

$$(2.19)\quad \eta^r = 1 - \frac{\varepsilon k}{2h^2} \max_{o\leq i\leq I} \psi(w^r_{i+1/2}) - \left(\frac{\varepsilon^2 k^2}{4h^4} \max_{o\leq i\leq I} \psi(w^r_{i+1/2})^2 + \frac{k^2}{2h^2} \max_{o\leq i\leq I} \psi(w^r_{i+1/2})\right)^{1/2},$$

ce qui est loisible, on trouve

$$(2.20)\quad \begin{aligned} &\frac{1}{k^2}|u^{r+1}-u^r|^2 + 2h\sum_{i=o}^{I} \phi(w^r_{i+1/2}) + (Bu^r,u^{r+1}-u^r) - \frac{\varepsilon k}{2}|Bu^r|^2 \geq \\ &\geq \eta^r \left(\frac{1}{k^2}|u^{r+1}-u^r|^2 + 2h\sum_{i=o}^{I} \phi(w^r_{i+1/2})\right). \end{aligned}$$

On obtient d'autre part à partir de (1.23)

$$(2.21)\quad |Bu^n-Bu^{n-1}|^2 \leq \frac{4}{h^2} \max_{o\leq i\leq I} \frac{\phi(w^n_{i+1/2})-\phi(w^{n-1}_{i+1/2})}{w^n_{i+1/2}-w^{n-1}_{i+1/2}} (Bu^n-Bu^{n-1},u^n-u^{n-1}).$$

Enfin, on voit facilement que

$$(2.22)\quad \frac{1}{k^2}|u^1-u^o|^2 + 2h\sum_{i=o}^{I} \phi(w^o_{i+1/2}) + (Bu^o,u^1-u^o) \leq (2-\eta^o)\left(\frac{1}{k^2}|u^1-u^o|^2 + 2h\sum_{i=o}^{I} \phi(w^o_{i+1/2})\right).$$

En reportant les majorations (1.34), (2.19), (2.20), (2.21) et (2.22) dans l'équation (2.18), on a

$$(2.23)\quad \begin{aligned} &\eta^r\left(\frac{1}{k^2}|u^{r+1}-u^r|^2 + 2h\sum_{i=o}^{I} \phi(w^r_{i+1/2})\right) + \\ &\left(\frac{2\varepsilon}{k} - 1 - \frac{4\varepsilon}{h^2}\left(\varepsilon + \frac{k}{2}\right) \max_{\substack{o\leq i\leq I\\ 1\leq n\leq r}} \frac{\phi(w^n_{i+1/2})-\phi(w^{n-1}_{i+1/2})}{w^n_{i+1/2}-w^{n-1}_{i+1/2}}\right) \sum_{n=1}^{r} (Bu^n-Bu^{n-1},u^n-u^{n-1}) \leq \\ &\leq (2-\eta^o)\left(\frac{1}{k^2}|u^1-u^o|^2 + 2h\sum_{i=o}^{I} \phi(w^o_{i+1/2})\right). \end{aligned}$$

Les conditions de stabilité (2.13) et (2.14) entraînent respectivement

$$\eta^r,\ \eta^o \geq \mu$$

$$\frac{2\varepsilon}{k} - 1 - \frac{4\varepsilon}{h^2}\left(\varepsilon + \frac{k}{2} \max_{\substack{o\leq i\leq I\\ 1\leq n\leq r}} \frac{\phi(w^n_{i+1/2})-\phi(w^{n-1}_{i+1/2})}{w^n_{i+1/2}-w^{n-1}_{i+1/2}}\right) \geq 2\mu \frac{\varepsilon}{k}.$$

On en déduit l'inégalité de l'énergie (2.15) (avec n remplacé par r).

Corollaire. Sous les conditions d'application du théorème précédent, on a :

$$(2.24)\quad \begin{aligned} &\frac{1}{k^2}|u_h^{n+1}-u_h^n|_h^2 + \alpha \| u_h^n \|_h^2 + 2\alpha \frac{\varepsilon}{k} \sum_{m=1}^{n} \| u_h^m - u_h^{m-1} \|_h^2 \leq \\ &\leq \frac{2-\mu}{\mu} \left(\frac{1}{k^2}|u_h^1-u_h^o|_h^2 + 2h \sum_{i=o}^{I} \Phi(w_{i+1/2}^o)\right), \quad n = 1, \ldots, r. \end{aligned}$$

Démonstration. On a d'après (2.2) et (2.3)

$$(Bu^m - Bu^{m-1}, u^m - u^{m-1}) \geq \alpha \| u^m - u^{m-1} \|^2,$$

$$2h \sum_{i=o}^{I} \Phi(w_{i+1/2}^n) \geq \alpha \| u^n \|^2.$$

Alors (2.24) est une conséquence immédiate de (2.15).

Remarque 2.1. La condition de stabilité (2.13) couple une condition de nature hyperbolique et une condition de nature parabolique naturelles. La condition de stabilité (2.14) couple deux conditions analogues et une condition exprimant que le pas de temps k doit être choisi assez petit devant ε. Notons que la constante μ peut être encore prise arbitrairement petite.

Remarque 2.2. Grâce à la forme particulièrement adéquate du terme de pseudo-viscosité, on a pu utiliser une technique de majoration plus fine que celle développée dans la démonstration du théorème 1.2. En effet, on pourrait ici appliquer la même méthode de majoration qu'au N°I mais cela conduirait à des conditions de stabilité beaucoup plus restrictives.

Il est facile de prouver un résultat analogue au théorème 1.3. Il existe donc un nombre $k_o = k_o(h,\mu,\varepsilon)$ tel que, pour $k \leq k_o$, on ait (2.13) et (2.14) pour $r = 1, \ldots, N-1$.

Remarque 2.3. En pratique, à chaque temps rk, on vérifie que les conditions de stabilité

$$\left(\frac{k^2}{2h^2} + \frac{\varepsilon k}{h^2}\right) \max_{o\leq i\leq I} \psi(w^r_{i+1/2}) < 1,$$

$$\frac{k}{2\varepsilon} + \left(\frac{k^2}{h^2} + \frac{2\varepsilon k}{h^2}\right) \max_{o\leq i\leq I} \frac{\phi(w^r_{i+1/2}) - (w^{r-1}_{i+1/2})}{w^r_{i+1/2} - w^{r-1}_{i+1/2}} < 1,$$

ont lieu afin de calculer u_h^{r+1}. Sinon, on diminue le pas de temps k de façon à pouvoir continuer le calcul. On peut ainsi atteindre tout temps T donné arbitrairement.

Cherchons maintenant une majoration a priori supplémentaire.

Théorème 2.3. La solution $u_{h,k} \in V_{h,k}$ du schéma aux différences finies (2.12) vérifie l'inégalité suivante pour $r = 1, \ldots, N-1$:

$$|B_h u_h^r|_h^2 \leq \nu \left\{ |B_h u_h^o|_h^2 + \frac{1}{\varepsilon^2 k^2} |u_h^1 - u_h^o|_h^2 + \frac{1}{\varepsilon^2 k^2} |u_h^{r+1} - u_h^r|_h^2 + \right.$$
$$\left. + \frac{1}{\varepsilon k} \sum_{n=1}^{r} (B_h u_h^n - B_h u_h^{n-1}, u_h^n - u_h^{n-1})_h \right\}$$

où ν est une constante > 0 indépendante de h,k et ε.

Démonstration. On prend le produit scalaire de l'équation (2.12) avec $\frac{2k}{\varepsilon} B_h u_h^n$ et on somme de n = 1 à n = r ($1\leq r\leq N-1$); On obtient

$$\frac{2}{\varepsilon k} \sum_{n=1}^{r} (u^{n+1} - 2u^n + u^{n-1}, Bu^n) + \frac{2k}{\varepsilon} \sum_{n=1}^{r} |Bu^n|^2 +$$
$$+ 2 \sum_{n=1}^{r} (Bu^n - Bu^{n-1}, Bu^n) = 0.$$

Un raisonnement en tout point identique à la démonstration du théorème 1.4 montre alors que

$$|Bu^r|^2 + \frac{2}{\varepsilon k}(u^{r+1} - u^r, Bu^r) + \frac{2k}{\varepsilon} \sum_{n=1}^{r} |Bu^n|^2 \leq$$
$$\leq |Bu^o|^2 + \frac{2}{\varepsilon k} (u^1 - u^o, Bu^o) + \frac{2}{\varepsilon k} \sum_{n=1}^{r} (Bu^n - Bu^{n-1}, u^n - u^{n-1}).$$

Le résultat est maintenant immédiat.

Corollaire. On a pour $r = 1, \ldots, N-1$ l'inégalité

$$|A_h u_h^r|_h^2 \leq \frac{\nu}{\alpha^2} \{|B_h u_h^o|_h^2 + \frac{1}{\varepsilon^2 k^2} |u_h^1 - u_h^o|_h^2 + \frac{1}{\varepsilon^2 k^2} |u_h^{r+1} - u_h^r|_h^2 + $$

$$+ \frac{1}{\varepsilon k} \sum_{n=1}^{r} (B_h u_h^n - B_h u_h^{n-1}, u_h^n - u_h^{n-1})_k\}.$$

Démonstration. En effet

$$|Bu^r|^2 = \frac{1}{h} \sum_{i=1}^{I} |\phi(w_{i+1/2}^r) - \phi(w_{i-1/2}^r)|^2.$$

En appliquant l'hypothèse (2.1), on trouve

$$|Bu^r|^2 \geq \frac{\alpha^2}{h} \sum_{i=1}^{I} |w_{i+1/2}^r - w_{i-1/2}^r|^2 = \alpha^2 |Au^r|^2 .$$

L'inégalité (2.26) résulte alors de (2.25).

Remarque 2.4. Les théorèmes de majoration 2.2 et 2.3 sont valables sous la seule hypothèse $\phi'(\xi) \geq 0 \quad \forall \xi \in R$.

2.3. Convergence du schéma

Théorème 2.4. On fait les hypothèses :

$$(2.27) \quad \begin{cases} \text{il existe une constante } C > 0 \text{ indépendante de } h \text{ et } k \text{ telle que} \\ \frac{1}{k^2} |u_h^1 - u_h^o|_h^2 + 2h \sum_{i=o}^{I} \Phi(w_{i+1/2}^o) + |B_h u_h^o|_h^2 \leq C^2, \end{cases}$$

$$(2.28) \quad \begin{cases} \lim_{h \to o} q_h u_h^o = u_o \\ \lim_{h \to o} \phi(\frac{\partial}{\partial x} p_h u_h^o) = \phi(\frac{\partial u_o}{\partial x}) \quad \text{dans } L_2(\Omega) \text{ fort.} \\ \lim_{h,k \to o} \frac{1}{k} (q_h u_h^1 - q_h u_h^o) = u_1 \quad . \end{cases}$$

Alors si h et k tendent vers 0 de façon que les conditions de stabilité (2.13) et (2.14) aient lieu pour $r = 1, \ldots, N-1$ avec une constante μ indépendante de h et k, la solution $u_{h,k} \in V_{h,k}$ du schéma (2.12) vérifie

$$(2.29)\quad \begin{aligned}&\lim_{h,k\to o} p_{h,k}u_{h,k} = u_\varepsilon \text{ \underline{dans} } L_2(0,T;H_o^1(\Omega)) \text{ \underline{fort et dans} } L_\infty(0,T;H_o^1(\Omega)) \text{ \underline{faible-étoile},}\\ &\lim_{h,k\to o} \frac{\partial}{\partial t} p_{h,k}u_{h,k} = \frac{\partial u_\varepsilon}{\partial t} \text{ \underline{dans} } L_2(0,T;H_o^1(\Omega)) \text{ \underline{faible et dans} } L_\infty(0,T;L_2(\Omega)) \text{ \underline{faible-étoile},}\end{aligned}$$

__où__ u_ε __est la solution de__ (2.6), (2.7), 2.8) __avec__ $f = 0$.

__Démonstration.__ La démonstration est analogue à celle du théorème 1.5. D'après le théorème 2.2 et son corollaire, le corollaire du théorème 2.3 et (2.27), on déduit que, de la suite (h,k) vérifiant les conditions de stabilité, on peut extraire une sous-suite notée encore (h,k) telle que

$$(2.30)\quad \begin{aligned}&\lim_{h,k\to o} p_{h,k}u_{h,k} = u^* \text{ dans } L_2(0,T;H_o^1(\Omega)) \text{ fort et dans } L_\infty(0,T;H_o^1(\Omega)) \text{ faible-étoile,}\\ &\lim_{h,k\to o} \phi\left(\frac{\partial}{\partial x} q_k p_h u_{h,k}\right) = \phi\left(\frac{\partial u^*}{\partial x}\right) \text{ dans } L_2(0,T;L_2(\Omega)) \text{ fort,}\\ &\lim_{h,k\to o} \frac{\partial}{\partial t} p_{h,k}u_{h,k} = \frac{\partial u^*}{\partial t} \text{ dans } L_2(0,T;H_o^1(\Omega)) \text{ faible et dans } L_\infty(0,T;L_2(\Omega)) \text{ faible-étoile}\\ &\lim_{h,k\to o} \frac{\partial}{\partial t} p_k q_h u_{h,k} = \frac{\partial u^*}{\partial t} \text{ dans } L_\infty(0,T;L_2(\Omega)) \text{ faible-étoile,}\\ &\lim_{h,k\to o} D_h \frac{\partial}{\partial x} p_{h,k}u_{h,k} = \frac{\partial^2 u^*}{\partial x^2} \text{ dans } L_\infty(0,T;L_2(\Omega)) \text{ faible-étoile.}\end{aligned}$$

On a alors

$$u^* \in L_\infty(0,T;H^2(\Omega) \cap H_o^1(\Omega))$$

$$\frac{\partial u^*}{\partial t} \in L_2(0,T;H_o^1(\Omega)) \cap L_\infty(0,T;L_2(\Omega)).$$

Montrons que

$$(2.31)\quad \frac{\partial^2 u^*}{\partial t^2} - \frac{\partial}{\partial x}\phi\left(\frac{\partial u^*}{\partial x}\right) - \varepsilon \frac{\partial}{\partial t}\frac{\partial}{\partial x}\phi\left(\frac{\partial u^*}{\partial x}\right) = 0.$$

Soit $v \in H_o^1(\Omega)$ et $\chi \in C^1(0,T)$ avec $\chi(T) = 0$; on peut écrire

$$\frac{1}{k}\sum_{n=1}^{N-1}(u_h^{n+1}-2u_h^n+u_h^{n-1},r_hv)_h\chi(nk) + k\sum_{n=1}^{N-1}(B_hu_h^n,r_hv)_h\chi(nk) +$$
$$+\ \varepsilon\sum_{n=1}^{N-1}(B_hu_h^n-B_hu_h^{n-1},r_hv)_h\chi(nk) = 0$$

d'où en utilisant des formules de sommation par parties

$$-\frac{1}{k}\sum_{n=1}^{N}(u_h^n-u_h^{n-1},r_hv)_h\ (\chi(nk)-\chi(n-1)k)) + k\sum_{n=1}^{N-1}(B_hu_h^n,r_hv)_h\chi(nk) -$$
$$-\ \varepsilon\sum_{n=1}^{N}(B_hu_h^{n-1},r_hv)_h\ (\chi(nk)-\chi((n-1)k)) =$$
$$= \frac{1}{k}\ (u_h^1-u_h^o,r_hv)_h\chi(0) + (B_hu_h^o,r_hv)_h\chi(0)$$

c'est-à-dire

$$-\int_o^T(\frac{\partial}{\partial t}\ p_kq_hu_{h,k}(t),q_hr_hv)\ \frac{1}{k}(\chi_k(t+k)-\chi_k(t))dt +$$
$$+\int_k^T(\phi(\frac{\partial}{\partial x}\ q_kp_hu_{h,k}(t)),\ \frac{\partial}{\partial x}\ p_hr_hv)\chi_k(t)dt -$$
$$-\ \varepsilon\int_o^T(\phi(\frac{\partial}{\partial x}\ q_kp_hu_{h,k}(t),\ \frac{\partial}{\partial x}\ p_hr_hv)\ \frac{1}{k}(\chi_k(t+k)-\chi_k(t))dt =$$
$$= \frac{1}{k}(q_hu_h^1-q_hu_h^o,q_hr_hv)\chi(0) + (\frac{\partial}{\partial x}\ p_hu_h^o,\frac{\partial}{\partial x}\ p_hr_hv)\chi(0).$$

En passant à la limite grâce à (2.28) et (2.30), on trouve

$$(2.32)\quad -\int_o^T(\frac{\partial u^*}{\partial t}(t),v)\chi'(t)dt + \int_o^T(\phi(\frac{\partial u^*}{\partial x}(t)),\frac{\partial v}{\partial x})\chi(t)dt - \varepsilon\int_o^T(\phi(\frac{\partial u^*}{\partial x}(t)),\frac{\partial v}{\partial x})\chi'(t)dt =$$
$$= (u_1,v)\chi(0) + (\frac{\partial u_o}{\partial x},\frac{\partial v}{\partial x})\chi(0) \qquad v\in H_o^1(\Omega).$$

En prenant $\chi \in \mathcal{D}\ (]0,T[)$, on en déduit que u^* vérifie (2.31) au sens distributions sur $]0,T[$ à valeurs dans $H^{-1}(\Omega)$. A partir de (2.31) et (2.32), on vérifie aisément

$$(2.33)\qquad u^*(0) = u_o,\ \frac{\partial u^*}{\partial t}\ (0) = u_1.$$

Ainsi $u^* = u_\varepsilon$ et c'est toute la famille $u_{h,k}$ qui converge au sens indiqué.

Les hypothèses (2.27) et (2.28) sont satisfaites lorsque l'on choisit u_h^o et u_h^1 donnés par (1.66). Ce point est facile à voir.

Remarque 2.5. Supposons ϕ convexe par exemple. Puisque

$$\Phi(\xi)-\Phi(\eta) - \frac{1}{2}(\phi(\xi)+\phi(\eta))(\xi-\eta) = - \int_{\frac{\xi+\eta}{2}}^{\xi}(t - \frac{1}{2}(\xi+\eta))(\phi'(t)-\phi'(\xi+\eta-t))dt,$$

on obtient alors

$$(2.34)\quad \Phi(\xi)-\Phi(\eta) - \frac{1}{2}(\phi(\xi)+\phi(\eta))(\xi-\eta) \begin{cases} \leq 0 \text{ si } \xi \geq \eta, \\ \geq 0 \text{ si } \xi < \eta. \end{cases}$$

Lorsque $\xi < \eta$, on peut majorer l'expression précédente par $\frac{1}{2}(\phi(\xi)-\phi(\eta))(\xi-\eta)$ (cf. démonstration de (1.24)). Il est donc naturel de songer à remplacer dans (2.9) le terme de pseudo-viscosité par

$$(2.35)\quad - \frac{\varepsilon}{h}(q_{i+1/2}^{n-1/2} - q_{i-1/2}^{n-1/2})$$

avec

$$(2.36)\quad q_{i+1/2}^{n-1/2} = \begin{cases} \frac{1}{h}(\phi(w_{i+1/2}^n)-\phi(w_{i+1/2}^{n-1})) & \text{si } w_{i+1/2}^n < w_{i+1/2}^{n-1}, \\ 0 & \text{si } w_{i+1/2}^n \geq w_{i+1/2}^{n-1}. \end{cases}$$

Ceci revient à remplacer dans l'équation (2.6) le terme $- \varepsilon \frac{\partial}{\partial t}\frac{\partial}{\partial x}\phi(\frac{\partial u}{\partial x})$ par $- \varepsilon \frac{\partial q}{\partial x}$ avec

$$(2.37)\quad q = \begin{cases} \phi'(\frac{\partial u}{\partial x})\frac{\partial^2 u}{\partial x \partial t} \text{ si } \frac{\partial^2 u}{\partial x \partial t} < 0, \\ 0 \text{ si } \frac{\partial^2 u}{\partial x \partial t} \geq 0. \end{cases}$$

Malheureusement, l'auteur ne connaît aucun théorème d'existence ni d'unicité pour les solutions du problème

$$(2.38)\qquad \begin{aligned} &\frac{\partial^2 u}{\partial t^2} - \frac{\partial}{\partial x}\phi\left(\frac{\partial u}{\partial x}\right) - \varepsilon \frac{\partial q}{\partial x} = 0 \\ &u(0) = u_o,\ \frac{\partial u}{\partial t}(0) = u_1 \end{aligned}$$

où q est donné par (1.37). Néanmoins, on peut démontrer à l'aide des méthodes de la partie I l'analogue du théorème 1.2 (ou du théorème 2.2). On ne sait pas obtenir l'analogue du théorème 1.4 (ou du théorème 2.3), ce qui permettrait de passer à la limite.

De la même façon, supposons ϕ de classe C^2; on peut écrire

$$\Phi(\xi)-\Phi(\eta) - \frac{1}{2}(\phi(\xi)+\phi(\eta))(\xi-\eta) = \frac{1}{2}\int_\eta^\xi (t-\xi)(t-\eta)\phi''(t)dt \quad .$$

Si, de plus, la fonction ϕ est <u>convexe</u>, on peut écrire

$$(2.39)\quad \Phi(\xi)-\Phi(\eta) - \frac{1}{2}(\phi(\xi)+\phi(\eta))(\xi-\eta) \begin{cases} \leq 0 & \text{si } \xi \geq \eta , \\ \leq -\frac{1}{8}(\phi'(\xi)-\phi'(\eta))(\xi-\eta)^2 & \text{si } \xi < \eta. \end{cases}$$

Il est donc encore naturel de remplacer dans (2.9) le terme de pseudo-viscosité par (2.35) avec

$$(2.40)\quad q_{i+1/2}^{n-1/2} = \begin{cases} -\frac{1}{k^2}(\phi'(w_{i+1/2}^{n})-\phi'(w_{i+1/2}^{n-1}))(w_{i+1/2}^{n}-w_{i+1/2}^{n-1}) & \text{si } w_{i+1/2}^{n} < w_{i+1/2}^{n-1}, \\ 0 & \text{si } w_{i+1/2}^{n} \geq w_{i+1/2}^{n-1}. \end{cases}$$

Ceci revient à prendre dans l'équation (2.6) $- \varepsilon \frac{\partial q}{\partial x}$ comme terme de pseudo-viscosité avec

$$(2.41)\qquad q = \begin{cases} -\phi''\left(\frac{\partial u}{\partial x}\right)\left(\frac{\partial^2 u}{\partial x \partial t}\right)^2 & \text{si } \frac{\partial^2 u}{\partial x\partial t} < 0, \\ 0 & \text{si } \frac{\partial^2 u}{\partial x\partial t} \geq 0. \end{cases}$$

Là encore, on ne connaît rien sur l'existence et l'unicité des solutions de (2.38) avec q donné par (2.41). On peut pourtant démontrer un résultat analogue au théorème 1.2.

On constate numériquement que les formes (2.37) et (2.41) du terme de pseudo-viscosité sont les mieux adaptées (cf. [6], Chapitre 12 pour des considérations voisines).

3. CONCLUSION

On <u>conjecture</u> que, moyennant des hypothèses raisonnables sur la fonction ϕ, le problème

$$(3.1)\quad \begin{cases} \dfrac{\partial^2 u}{\partial t^2} \quad \dfrac{\partial}{\partial x}\,\phi\left(\dfrac{\partial u}{\partial x}\right) = f \quad \text{dans } \Omega \times]0,T[, \\ u(0,t) = u(1,t) = 0 \;,\; t \in]0,T[, \\ u(x,o) = u_o(x),\; \dfrac{\partial u}{\partial t}(x,0) = u_1(x) \;,\; x \in \Omega, \\ u \text{ vérifie une condition d'entropie convenable,} \end{cases}$$

admet une solution unique u et que la solution u_ε de (1.5), (1.6), (1.7) et la solution u_ε de (2.6), (2.7), (2.8) convergent dans un sens convenable vers u lorsque ε tend vers 0. Alors, si on se <u>fixe</u> ε "petit", on peut considérer que dans chaque partie I et II, on résoud numériquement un problème dont la solution est "voisine" de la solution du problème initial (3.1). C'est en sens que l'on peut dire que les méthodes de différences finies (1.8), (1.9), (1.10) et (2.9), (2.10), (2.11) fournissent sous les conditions de stabilité données des approximations de la solution u de (3.1). Notons en outre que rien ne permet de croire à la convergence des schémas aux différences donnés lorsque h,k <u>et</u> ε tendent vers 0.

En pratique, h est fixé "petit"; on prend alors habituellement $\varepsilon = \eta h$ où η est une constante numérique à déterminer au mieux. Les conditions de stabilité données précédemment ne font plus intervenir les pas h et k que par le rapport $\frac{k}{h}$.

R E F E R E N C E S

[1] J.M. GREENBERG. On the existence, uniqueness and stability of solutions of the equation $\rho_o \mathcal{X}_{tt} = E(\mathcal{X}_x)\mathcal{X}_{xx} + \lambda \mathcal{X}_{xxt}$, Journal of Math. Anal. and Appl., 25 (1969), 575-591.

[2] J.M. GREENBERG; R.C. Mac CAMY et V.J. MIZEL. On the existence, uniqueness and stability of solutions of the equation $\sigma'(u_x)u_{xx}+\lambda u_{xtx} = \rho_o u_{tt}$, Journal of Math. Mech., 17 (1968), 707-728.

[3] J.L. LIONS et E. MAGENES. Problèmes aux limites non homogènes, vol 1, Paris, Dunod, 1968.

[4] P.A. RAVIART. Sur l'approximation de certaines équations d'évolution linéaires et non linéaires, Journal Math. pures et appl., 46 (1967), 11-183.

[5] P.A. RAVIART. Sur la résolution numérique de l'équation $\frac{\partial u}{\partial t}+u\frac{\partial u}{\partial x} -\varepsilon\frac{\partial}{\partial x}(|\frac{\partial u}{\partial t}|\frac{\partial u}{\partial x})$, Journal of Diff. Eq., 8 (1970), 56-94.

[6] R.D. RICHTMYER et K.W. MORTON. Difference methods for initial value problems, New-York, Interscience, 1967.

STABILITY OF DISCRETIZATIONS ON INFINITE INTERVALS

Hans J. STETTER

I. INTRODUCTION

In the theory of discretization methods for systems of ordinary differential equations (SODE) the most fundamental results are due to Dahlquist ([1]) and Henrici ([2], [3]). In view of the fact that a general analysis of discretization methods is not feasible (except in simple examples) they designed an asymptotic theory which describes the situation as the step h tends to zero. This theory has proved to be an extremely valuable tool in understanding the behavior of discretizations for sufficiently small steps and even produces quantitative results for this case. (What is "sufficiently small", however!) It has been further elaborated through the asymptotic expansion results of Gragg ([4]) and has been applied to increasingly many classes of discretization methods (hybrid, non-polynomial approximant, variable coefficient, cyclic, etc.).

This asymptotic theory exclusively refers to discretization of SODE on fixed finite intervals. That even its basic convergence results may become meaningless when the interval increases has become well-known by the analysis of the midpoint and Simpson rules (see, e.g., [2]). Here, for a given SODE with decreasing solutions

for any fixed interval T we may choose the step h such that the discretization error is arbitrarily small, but

for any fixed step h we may choose the interval T such that this error is arbitrarily large!

More generally, the different growth with t of the various coefficients in the asymptotic expansion of the solution of a discretization may make the concept of "lowest order in h" meaningless, a concept which is the principal tool of the asymptotic theory for $h \to 0$.

This suggest the design of a second type of asymptotic theory which describes the behavior of the discretization and its solution for fixed h as the length T of the interval of integration increases beyond bound. Again we are really not interested in the limiting case proper but we hope that the theory for $T \to \infty$ will also explain the phenomena which occur for large T. Naturally, "large" is a relative term (as "small" is for h) since it strongly depends on the formulation of our problems. In fact we may characterize the notorious stiff problems as SODE where the long-interval effects occur very soon.

This paper represents an attempt to outline such an asymptotic theory of discretization methods for SODE on the infinite interval $[0,\infty)$. It fits the few existing results into the framework of this theory and gives some new results, furthermore it points out various unsolved problems. Proofs are only sketched or completely omitted due to the limited amount of space for this presentation; a full elaboration will be found in the forthcoming book ([5]) by this author.

II. STABILITY THEORY FOR SODE

The discretization of a SODE should display the same sensitivity, or rather insensitivity, to perturbations as the differential equations themselves. We will therefore begin our analysis by recalling a few basic facts from the perturbation theory of SODE on infinite intervals, a theory which is commonly called stability theory of differential equations. We quote no specific references for these well-known results but refer the uninitiated reader to text-books on the subject, e.g. [6].

We compare the solution $\tilde{y}$ of the perturbed SODE

$$\tilde{y}(0) = y_0 + \delta_0, \qquad \tilde{y}' = f(t,\tilde{y}) + \delta(t), \qquad t \geq 0,$$

to the solution y of the unperturbed SODE

$$(2.1) \qquad y(0) = y_0, \qquad y' = f(t,y), \qquad t \geq 0,$$

and obtain the fundamental SODE

$$(2.2) \qquad \begin{array}{ll} e(0) = \delta_0, & e' = g(t,e) + \delta(t), \qquad t \geq 0 \\ \text{with} \quad g(t,e) := f(t,y(t)+e) - f(t,y(t)) & \end{array}$$

for the difference $e := \tilde{y} - y$. It is always assumed that (2.2) possesses the unique solution $e \equiv 0$ for vanishing perturbations $(\delta_0,\delta(t))$; $e \equiv 0$ is called the "equilibrium" of (2.2).

<u>Def. 2.1:</u> (2.2) - and also (2.1) - is called <u>totally stable</u> (under persistent perturbation) if it is possible for each $\rho > 0$ ($\rho \le r$) to find $\bar{\delta}_0 > 0$ and $\bar{\delta}_1 > 0$ such that the solution e of (2.2) exists and satisfies*) $\|e(t)\| < \rho$, $t \ge 0$, whenever

$$(2.3) \qquad \|\delta_0\| < \bar{\delta}_0 \qquad \text{and} \qquad \|\delta(t)\| < \bar{\delta}_1 , \qquad t \ge 0 .$$

Obviously, (2.1) is <u>properly posed</u> (w.r.t. uniform convergence on the infinite interval $[0,\infty)$) if and only if it is totally stable. In the following we will exclusively consider totally stable SODE.

In place of (2.2) one often considers the family of SODE, $t_0 \ge 0$,

$$(2.4) \qquad e(t_0) = \delta_0 , \qquad e' = g(t,e), \qquad t \ge t_0, \qquad g(t) \text{ from (2.2);}$$

the solution of (2.4) will be denoted by $\bar{e}(t;\delta_0,t_0)$.

<u>Def. 2.2:</u> (2.4) - and also (2.1) - is called <u>uniformly asymptotically stable</u> if, for $\|\delta_0\| < r$, $\bar{e}(t;\delta_0,t_0)$ exists and satisfies, uniformly in t_0,

$$(2.5) \qquad \|\bar{e}(t;\delta_0,t_0)\| \le \varphi(\|\delta_0\|)\ \sigma(t-t_0)$$

where φ and σ are functions $R \to R$ independent of t_0 satisfying $\varphi(0) = 0$, $\varphi\uparrow$, and $\lim_{t\to\infty} \sigma(t) = 0$.

(2.5) bounds the effect of a "displacement" at t_0 of the solution of (2.1). It is well-known that uniform asymptotic stability implies total stability.

<u>Def. 2.3:</u> (2.4) - and also (2.1) - is called <u>exponentially stable</u> if $\bar{e}(t;\delta_0,t_0)$ exists for $\|\delta_0\| < r$ and satisfies, uniformly in t_0,

$$(2.6) \qquad \|\bar{e}(t;\delta_0,t_0)\| \le a\ \|\delta_0\|\ \exp(-\mu(t-t_0)), \qquad \mu > 0 .$$

Obviously (2.6) is a special case of (2.5) more suitable for a quantitative analysis. If g is essentially linear in e (see (2.10) below) then (2.5) implies (2.6) so that the assumption of exponential stability means only a slight loss of generality. In the following we will use the basic domain

*) The norms are arbitrary but fixed vector norms in R^s where s is the number of equations in the SODE (2.1). Norms of mappings $R^s \to R^s$ are the associated l.u.b. norms.

$$B := \{(t,e) : t \in [0,\infty), \quad e \in R^s, \quad \|e\| < r\}, \qquad \text{with some } r > 0.$$

<u>Theorem 2.1</u> (Liapunov function): Assume that $g \in C^{(1)}(B)$ with uniformly bounded derivatives. (2.4) is exponentially stable if and only if there exists a mapping $v\colon R\times B \to R$ satisfying uniformly in B

(2.7) $$a_1\|e\|^2 \le v(t,e) \le a_2\|e\|^2 ,$$

with positive a_1,a_2,a_3 .

(2.8) $$\frac{d}{d\tau} v(\tau,\bar{e}(\tau;e,t))\Big|_{\tau=t} \le -a_3\|e\|^2 ,$$

v may then be chosen such that also $\|\frac{\partial v}{\partial e}(t,e)\| \le a_4\|e\|$, $a_4 > 0$.
v is called a Liapunov function for (2.4) resp. (2.1).

<u>Theorem 2.2</u> (Linearization): Assume that, uniformly in t,

(2.9) $$g(t,e) = g(t)e + o(\|e\|) \qquad \text{for } \|e\| \to 0$$

where $g(t) \in L(R^s \to R^s)$. Then (2.4) is exponentially stable if and only if

(2.10) $$e(t_o) = \delta_o, \qquad e' = g(t)e, \qquad t \ge t_o ,$$

is exponentially stable.

<u>Def. 2.4:</u> For a given norm in $L(R^s \to R^s)$

$$\mu[A] := \lim_{h\to 0} \frac{1}{h}[\ \|I + hA\| - 1\ ]$$

is called the associated <u>logarithmic norm</u>. ($\mu[A]$ may be negative!)

<u>Theorem 2.3</u> (logarithmic norm): If g in (2.9) satisfies

$$\mu[g(t)] \le -\mu < 0 , \qquad t \ge 0 ,$$

then (2.9) is exponentially stable.

III. STABILITY THEORY FOR DIFFERENCE EQUATIONS

We will now obtain results for difference equations on $[0,\infty)$ which are analogous to those of section II. For reasons of simplicity we will only regard difference equations on equidistant grids $G_h := \{t_\nu := \nu h, \ \nu=0,1,2,\ldots\}$, $h > 0$ is the stepsize; but all considerations may be extended to grids with variable steps.

Values of functions $\eta: G_h \to R^s$ will be denoted by $\eta_\nu := \eta(t_\nu)$. Furthermore, we will denote k-tuples of such values by

$$\underline{\eta}_\nu := (\eta_\nu, \eta_{\nu-1}, \ldots, \eta_{\nu-k+1}), \quad \text{with} \quad \|\underline{\eta}_\nu\| := \max_{\varkappa=0(1)k-1} \|\eta_{\nu-\varkappa}\| .$$

We are interested in solutions of systems of difference equations (SΔE) of order k

$$(3.1) \qquad \underline{\eta}_{k-1} = \underline{\eta}^{(o)}, \qquad \eta_\nu = \varphi_h(t_\nu; \eta_{\nu-1}, \ldots, \eta_{\nu-k}) = \varphi_h(t_\nu, \underline{\eta}_{\nu-1}), \quad \nu \geq k,$$

and their behavior under perturbation. Let $\tilde{\eta}$ be the solution of

$$\underline{\tilde{\eta}}_{k-1} = \eta^{(o)} + \underline{\delta}_{k-1} , \qquad \tilde{\eta}_\nu = \varphi_h(t_\nu, \underline{\tilde{\eta}}_{\nu-1}) + \delta_\nu , \qquad \nu \geq k ,$$

then $\varepsilon := \tilde{\eta} - \eta$ satisfies the SΔE (comp. (2.2))

$$(3.2) \qquad \begin{aligned} &\underline{\varepsilon}_{k-1} = \underline{\delta}_{k-1} , \qquad \varepsilon_\nu = \psi_h(t_\nu, \underline{\varepsilon}_{\nu-1}) + \delta_\nu , \qquad \nu \geq k, \\ &\text{with} \quad \psi_h(t_\nu, \underline{\varepsilon}_\nu) := \varphi_h(t_\nu, \underline{\eta}_\nu + \underline{\varepsilon}_\nu) - \varphi_h(t_\nu, \underline{\eta}_\nu) . \end{aligned}$$

The explicit appearance of (3.1) is not meant to imply that our SΔE are necessarily explicit. An originally implicit SΔE may be <u>formally</u> written like (3.1) if it possesses a unique solution within a suitable domain. For (3.2), we will assume that the domains of uniqueness are, with a suitable $r > 0$,

$$B_h := \{(t_\nu, \underline{\varepsilon}_\nu): t_\nu \in G_h, \ \underline{\varepsilon}_\nu \in R^{ks}, \ \|\underline{\varepsilon}_\nu\| < r\} .$$

Throughout this section the stepsize h will have an arbitrary but <u>fixed value</u> which enters into all results as a parameter.

<u>Def. 3.1:</u> (3.2) - and also (3.1) - is called <u>totally stable</u> (under persistent perturbation) if it is possible, for each $\rho > 0$ ($\rho \leq r$), to find $\bar{\delta}_o > 0$ and $\bar{\delta}_1 > 0$ such that the solution of (3.2) exists and satisfies $\|\varepsilon_\nu\| < \rho'$ for all $\nu \geq 0$ whenever

$$(3.3) \qquad \|\underline{\delta}_{k-1}\| < \bar{\delta}_o \quad \text{and} \quad \|\delta_\nu\| < \bar{\delta}_1 \quad \text{for all } \nu \geq k .$$

Again, the total stability and the properly-posedness of (3.1) are equivalent properties.

(It is interesting to compare Def. 3.1 to the definition of "stability" in the asymptotic theory of difference equations (3.1) on fixed finite intervals of length T for $h \to 0$. We need only replace the phrase "for all ν" by "for $\nu = 0(1)T/h$ and all $h > 0$" to obtain the stability concept of the Dahlquist-Henrici theory.)

We now consider the case of a single perturbation at t_{ν_0}:

(3.4) $$\underline{\varepsilon}_{\nu_0} = \underline{\delta}_{\nu_0}, \qquad \varepsilon_\nu = \psi_h(t_\nu, \underline{\varepsilon}_{\nu-1}), \qquad \nu > \nu_0 ;$$

we denote the solution of (3.4) by $\bar{\varepsilon}_\nu(\underline{\delta}_{\nu_0}, t_{\nu_0})$. One may now define uniform asymptotic stability in a completely analogous manner as in Def. 2.2, but we proceed immediately to the quantitative concept of exponential stability.

<u>Def. 3.2:</u> (3.4) - and also (3.1) - is called <u>exponentially</u> stable if $\bar{\varepsilon}_\nu(\underline{\delta}_{\nu_0}, t_{\nu_0})$ exists for $\|\underline{\delta}_{\nu_0}\| < r$ and satisfies, uniformly in ν_0,

(3.5) $$\|\bar{\varepsilon}_\nu(\underline{\delta}_{\nu_0}, t_{\nu_0})\| \leq a \|\underline{\delta}_{\nu_0}\| \exp(-\mu(t_\nu - t_{\nu_0})), \qquad \mu > 0 .$$

<u>Theorem 3.1</u> (total stability): If (3.1) is exponentially stable it is also totally stable.

<u>Proof:</u> If (3.1) is linear the assertion follows from (3.5) by linear superposition. In the non-linear case one has to use Theorem 3.2.

<u>Theorem 3.2</u> (Liapunov function): (3.4) is exponentially stable if and only if there exists a mapping $v_h: R \times R^{ks} \to R$ satisfying

(3.6) $$a_1\|\underline{\varepsilon}_\nu\|^2 \leq v_h(t_\nu, \underline{\varepsilon}_\nu) \leq a_2\|\underline{\varepsilon}_\nu\|^2 ,$$

(3.7) $$\Delta_h v_h(t_\nu, \underline{\varepsilon}_\nu) := \frac{1}{h}\,[v_h(t_{\nu+1}, \underline{\bar{\varepsilon}_{\nu+1}(\underline{\varepsilon}_\nu, t_\nu)}) - v_h(t_\nu, \underline{\varepsilon}_\nu)] \leq -a_3\|\underline{\varepsilon}_\nu\|^2,$$

with positive constants a_1, a_2, a_3, uniformly in B_h. v_h may be chosen such that in B_h

(3.8) $$|v_h(t_\nu, \underline{\varepsilon}_\nu^1) - v_h(t_\nu, \underline{\varepsilon}_\nu^2)| \leq a_4 \max(\|\underline{\varepsilon}_\nu^1\|, \|\underline{\varepsilon}_\nu^2\|)\, \|\underline{\varepsilon}_\nu^1 - \underline{\varepsilon}_\nu^2\| ,$$

with $a_4 > 0$. v_h is called a Liapunov function for (3.1) resp. (3.4).

<u>Proof:</u> Necessity: By an analogous construction as in the analytic case (comp., e.g., [6]):

$$v_h(t_\nu, \underline{\varepsilon}_\nu) := h \sum_{\lambda=\nu}^{\nu+N} \|\bar{\varepsilon}_\lambda(\underline{\varepsilon}_\nu, t_\nu)\|^2 \quad \text{with a suitable N.}$$

Sufficiency: Follows from the validity of the difference inequality

$$\Delta_h v_h(t_\nu,\underline{\varepsilon}_\nu) \le -\frac{a_3}{a_2} v_h(t_\nu,\underline{\varepsilon}_\nu) .$$

Theorem 3.3 (neighboring difference equations): In (3.4), let

$$\psi_h(t_\nu,\underline{\varepsilon}_{\nu-1}) = \psi_h^o(t_\nu,\underline{\varepsilon}_{\nu-1}) + \chi_h(t_\nu,\underline{\varepsilon}_{\nu-1}) \tag{3.9}$$

and assume that both ψ_h^o and χ_h are uniformly Lipschitz-continous w.r.t. their second argument in B_h. Then (3.4) is exponentially stable if and only if the SΔE

$$\underline{\varepsilon}_{\nu_o} = \underline{\delta}_{\nu_o} , \qquad \varepsilon_\nu = \psi_h^o(t_\nu,\underline{\varepsilon}_{\nu-1}) , \qquad \nu > \nu_o , \tag{3.10}$$

is exponentially stable, provided that the Lipschitz constant M of χ_h is sufficiently small*).

Corollary 3.4 (Linearization): Assume that, uniformly in t_ν,

$$\psi_h(t_\nu,\underline{\varepsilon}_{\nu-1}) = \psi_h^o(t_\nu)\underline{\varepsilon}_{\nu-1} + o(\|\underline{\varepsilon}_{\nu-1}\|) \text{ for } \|\underline{\varepsilon}_{\nu-1}\| \to 0 \tag{3.11}$$

where $\psi_h^o \in L(R^{ks} \to R^s)$. Then (3.4) is exponentially stable if and only if

$$\underline{\varepsilon}_{\nu_o} = \underline{\delta}_{\nu_o} , \qquad \varepsilon_\nu = \psi_h^o(t_\nu)\underline{\varepsilon}_{\nu-1} , \qquad \nu > \nu_o , \tag{3.12}$$

is exponentially stable.

IV. STRONG EXPONENTIAL STABILITY

We are interested only in those SΔE which appear as discretizations of SODE; here we would like to be sure that the discretizations of a SODE share its stability properties.

Def. 4.1: A discretization method is called <u>strongly exponentially stable</u> if it produces an exponentially stable SΔE whenever it is applied to an exponentially stable SODE, with an arbitrary step h.

---

*) An explicit condition on M can be given in terms of the constants a_3 and a_4 for the Liapunov function of the SΔE which is assumed as exponentially stable.

Strong exponential stability would be a desirable property for a discretization method which is to be used over infinite (or very long) **intervals**. However, it turns out that there are no discretization methods which are strongly exponentially stable without some restriction. Such restrictions may concern the admissible class of SODE or the size of h or both.

Def. 4.2: A discretization method is called strongly exponentially stable w.r.t. a class K of SODE if it produces an exponentially stable SΔE whenever it is applied to an exponentially stable SODE from K.

Def. 4.3: A discretization method is called strongly exponentially stable for sufficiently small h [w.r.t. K] if for each exponentially stable SODE [from K] there exists a $h_o > 0$ such that the SΔE produced by the discretization method is exponentially stable for $h \le h_o$.

These restrictions present the following problems:

a) Find meaningful classes of SODE w.r.t. which there exist strongly exponentially stable discretization methods.

b) Find criteria which permit the determination of h_o for a given SODE and a given discretization method.

The only successful solution of these problems has been achieved for the class J_c of linear SODE with constant coefficients (= SODE (2.1) with a constant Jacobian $\frac{\partial}{\partial y} f(t,y(t))$). Here the well-known regions of absolute stability*) provide the criterion on the admissible size of h, at least for constant **h**. For a SODE $\in J_c$, with Jacobian matrix g, denote the eigenvalues of hg by $\lambda_\sigma(hg)$, $\sigma=1(1)s$. Then

(4.1) $$\lambda_\sigma(hg) \in H_o, \qquad \sigma = 1(1)s,$$

implies the exponential stability of the discretization.

*) Let $\zeta_\varkappa(h)$ be the zeros of the characteristic polynomial for the linear difference equation with constant coefficients which a discretization method produces for $y' = y$ and step h. Then

$$H_o := \{h \in \mathbb{C}: |\zeta_\varkappa(h)| < 1 \text{ for all } \varkappa\} \subset \mathbb{C}$$

is the region of absolute stability for this discretization method.

From the various definitions which refer to the character of the domain H_o we mention only:

A discretization method is called A-stable if $H_o \supset \{z \in \mathbb{C}, \mathrm{Re}\, z < 0\}$.

The regions of absolute stability have been used extensively to judge the stability properties of discretization methods (e.g. predictor-corrector methods and the like), and many researchers have tried to construct methods with large H_o. However, the range of validity of any assertion based upon regions of absolute stability for discretizations of SODE not from J_c is still unknown! A few such assertions the validity of which is commonly assumed are the following:

Let $\lambda_\sigma(hg(t))$ be the "local" eigenvalues of the Jacobian $g(t) := f_y(t,y(t))$ along the true solution. If all $\lambda_\sigma(hg(t)) \in H_o$ for all $t \geq 0$ then the discretization is exponentially stable.

Consider discretization methods 1 and 2 and assume $H_o^{(1)} \subset H_o^{(2)}$. If method 1 produces an exponentially stable SΔE for some SODE so does method 2.

If a discretization method is A-stable it produces an exponentially stable SΔE for each exponentially stable SODE irrespective of the stepsize used (i.e. it is strongly exponentially stable).

All these three - and many other "intuitively correct" - assertions are false ! Thus, the stability regions of discretization methods, although a valuable tool for the assessment of their basic merits w.r.t. strong exponential stability, are not sufficient to characterize the stability behavior of discretizations to SODE on infinite intervals.

V. RESULTS FOR ONE-STEP METHODS

(We employ the term "one-step method" in its usual sense, it includes, e.g., all RK-methods (explicit and implicit). If a variable stepsize is permitted, the choice of the local stepsize in dependence on a parameter h is part of the "method".)

<u>Theorem 5.1:</u> Consider a totally stable SODE (2.1) with solution y and assume that $g(t,e)$ of (2.2) satisfies a Lipschitz condition w.r.t. t and e uniformly in t for $\|e\| < r$. Let $\eta(h)$ be the solution of the discretization produced by some consistent one-step method, with stepsize (parameter) h. For each $\varepsilon > 0$ ($\varepsilon < r$) there exists $h_o(\varepsilon) > 0$ such that

$$(5.1) \qquad \|\eta_\nu(h) - y(t_\nu)\| < \varepsilon \quad \text{for all} \quad t_\nu \in G_h \,, \quad \text{if} \quad h \le h_o(\varepsilon) \,.$$

Furthermore, the SΔE produced by the one-step method is also totally stable.

<u>Proof</u> (idea): The linear interpolant $\tilde{y}(t)$ of the η_ν is a continuous, piecewise differentiable function which satisfies a SODE which is a perturbation of (2.1). Thus $e(t) := \tilde{y}(t) - y(t)$ satisfies (2.2), the estimate for $\|\delta(t)\|$ is proportional to h and to the Lipschitz constant of the SODE.

If η is replaced by the solution $\tilde{\eta}$ of a perturbed SΔE this adds only another term to the estimate for $\|\delta(t)\|$ which must be uniformly small.

Theorem 5.1 shows that one-step methods provide suitable means to determine uniformly good approximations on $[0,\infty)$ to a wide variety of SODE which are properly posed on this infinite interval. Furthermore, such uniformly good approximations may be computed numerically in spite of the persistent perturbation by round-off errors.

The requirement of a uniform Lipschitz condition for the SODE is essential: Consider $y' = -ty$, $y(t_o) = \delta_o$, with solution

$$y(t) = \delta_o \exp(t_o^2 - t^2)/2 \le \delta_o e^{1/2} \exp(-(t - t_o)) \,.$$

The Euler method produces $\eta_\nu = (1 - ht_{\nu-1})\eta_{\nu-1}$ which cannot be totally stable as $|1 - ht| \to \infty$ for $t \to \infty$. (Obviously we have to exclude SODE which become arbitrarily stiff as $t \to \infty$. See also section VII.)

<u>Theorem 5.2:</u> Consider an exponentially stable SODE with differentiable f and a uniform Lipschitz constant on f_y in the vicinity of the solution y. For each consistent one-step method there exists a $h_o > 0$ such that the discretization of the SODE, with $h < h_o$, is also

exponentially stable.

From the detailed proofs of Theorem 5.1 and 5.2 one may obtain bounds for h_0 from a knowledge of the stability data of the SODE. In many cases, these bounds ask for steps h which are ineconomically small for practical computations. Remembering our original idea of strong exponential stability (see Def. 4.1) we therefore wonder whether there are special one-step methods which achieve exponential stability for arbitrary h at least w.r.t. large classes of SODE (see Def. 4.2).

In 1963, Dahlquist [7] has shown that the implicit trapezoidal rule is strongly exponentially stable w.r.t. the class of SODE which satisfy his condition C'_λ, see [7, p.36]. A meaningful subclass of this class consists of those SODE for which the linearization (2.10) possesses a time-independent quadratic Liapunov function, see Theorem 5.4 below. Even then his result seems to hold only for constant step-size (which is all he considers) as it is possible to construct counter-examples with SODE which satisfy (5.2) below when variable (periodic) steps are permitted. These and similar counter-examples prove that A-stability does not imply strong exponential stability.

For the implicit Euler method $\eta_\nu = \eta_{\nu-1} + hf(t_\nu,\eta_\nu)$ one may exhibit a few meaningful classes of SODE which discretize into exponentially stable SΔE for arbitrary (variable) steps:

Theorem 5.3: For a SODE, if g(t) in (2.9) is a normal matrix and satisfies

$$(5.2) \qquad \max_\sigma \operatorname{Re} \lambda_\sigma(g(t)) \leq -\mu < 0 \qquad \text{for all} \quad t \geq 0$$

then the implicit Euler discretization of the SODE is exponentially stable for arbitrary steps.

Theorem 5.4: For a SODE, if the linearized variational equation (2.10) possesses a Liapunov function of the form $v(t_\nu,\varepsilon) = \varepsilon^T G\,\varepsilon$, where G is a constant (positive-definite) symmetric matrix, then the implicit Euler discretization of the SODE is exponentially stable for arbitrary steps.

In both cases it is not even necessary to require the uniform Lipschitz-boundedness of g in (2.2) so that the implicit Euler method is strongly exponentially stable w.r.t. certain SODE which become arbitrarily stiff as $t \to \infty$, like the one discussed after Theorem 5.1. (This pleasant property is not shared by the trapezoidal rule.)

For more general one-step methods and SODE the following approach is suggested by our Theorem 3.3 on neighboring S∆E: The linearized variational equation (2.1) of the given SODE is compared to a SODE with a piecewise constant function $g(t)$. The exponential stability of the discretization of this "comparison-SODE" may be judged by the use of stability regions which may be supplemented by interior "level curves" which indicate the size of the exponent μ in (3.5). (The domains H_μ enclosed by these curves are sometimes called regions of relative or μ-exponential stability.) The actual variation of $g(t)$ for the given SODE can then be weighed against the exponential decay of a perturbation in the discretization of the "comparison-SODE" for a chosen step h by the analysis on which Theorem 3.3 is based. This approach may lead to concise criteria for the admissible stepsize in the numerical integration over long intervals of realistic SODE by one-step methods.

VI. RESULTS FOR MULTISTEP METHODS

In this section we consider k-step methods of the form

$$(6.1) \qquad \sum_{\varkappa=0}^{k} \alpha_\varkappa \eta_{\nu-k+\varkappa} = h\,\Psi(t_\nu, \eta_\nu, \eta_{\nu-1}, \dots, \eta_{\nu-k}; h)$$

where Ψ satisfies a Lipschitz-condition w.r.t. the η-arguments. (6.1) includes the usual predictor-corrector methods. Methods with variable coefficients $\alpha_\varkappa$ have not been included in the analysis although these seem to have very promising stability properties (see e.g., Brunner [8], [9] and Lambert [10]).

It is clear that the stability behavior of multistep discretizations on infinite intervals is far more complicated than that of one-step discretizations. However, for sufficiently small h some of the inherent difficulties may be overcome.

<u>Theorem 6.1:</u> Consider a totally stable SODE as in Theorem 5.1. Let $\eta(h)$ be the solution of the discretization with stepsize h, produced by a consistent k-step method (6.1). If the polynomial $\rho(z) = \sum_{\varkappa=0}^{k} \alpha_\varkappa z^\varkappa$ has no zero outside the closed unit disk and no zero except 1 on the unit circle, then for each $\varepsilon > 0$ ($\varepsilon < r$) there exists $h_0(\varepsilon)$ such that (5.1) holds.

Furthermore, the SΔE produced by the k-step method is also totally stable.

<u>Proof</u> (idea): Let $\rho(z) = (z-1)\,\bar{\rho}(z) = (z-1) \sum_{\varkappa=0}^{k-1} \bar{\alpha}_\varkappa z^\varkappa$ and set

(6.2) $$\bar{\eta}_\nu := \sum_{\varkappa=0}^{k-1} \bar{\alpha}_\varkappa \, \eta_{\nu-(k-1)+\varkappa} \, .$$

Then

(6.3) $$\bar{\eta}_\nu = \bar{\eta}_{\nu-1} + h\,\Psi(t_\nu, \eta_\nu, \ldots, \eta_{\nu-k}; h)$$

so that $\bar{\eta}_\nu$ obeys essentially a first order SΔE and the approach of the proof of Theorem 5.1 may be used. As the characteristic polynomial $\bar{\rho}$ of (6.2), interpreted as a difference equation for the η_ν in terms of the $\bar{\eta}_\nu$, has all its zeros inside the open unit disk the representation of the η_ν through the $\bar{\eta}_\nu$ in the arguments of Ψ in (6.3) introduces no effect which might grow with t.

Theorem 6.1 shows that multistep methods without essential roots except the principal root 1 (in the terminology of Henrici [2]) may be safely used for the integration of totally stable SODE over $[0,\infty)$ if a sufficiently small step is used. Presumably these multistep methods are even strongly exponentially stable for sufficiently small h w.r.t. SODE with a uniform Lipschitz condition on f_y, compare Theorem 5.2.

The possibility to use relatively large steps without loss of exponential stability seems very limited with methods of the form (6.1) and $k \geq 2$ as there are only few such methods which are A-stable. Perhaps the variable coefficient methods mentioned above may be able to fill that gap.

The general approach outlined at the end of section V may also be applicable to multistep methods.

VII. STIFF EQUATIONS

In conclusion, I would like to make a few remarks concerning "stiff" SODE. In our present context we may characterize them as SODE where "long interval effects occur on short intervals"! Thus the developed stability theory of SΔE on infinite intervals should be a valuable tool in understanding the peculiar difficulties which arise in the numerical integration of stiff SODE.

At the present stage, the main problems are:

(i) Which discretization methods should be used?

A number of method have been suggested, notably by Gear [11] and Osborne [12]. The use of successive Richardson extrapolation with the implicit trapezoidal rule (in analogy to the well-known Gragg-Bulirsch-Stoer extrapolation method for "normal" SODE) was discouraged by Dahlquist [7] on the basis of the observation that

$$(7.1)\qquad \lim_{hg\to-\infty} \frac{4}{3}\left(\frac{1+hg/2}{1-hg/2}\right)^2 - \frac{1}{3}\left(\frac{1+hg/2}{1-hg/2}\right) = \frac{5}{3} > 1 .$$

But the corresponding limit for an odd number of different subdivisions in one extrapolation step is <u>smaller</u> than 1, or the odd power in (7.1) may be made to disappear by the choice of subdivisions into even numbers of steps only.

It has been shown by this author that the extrapolation method on the basis of the implicit trapezoidal rule (or the implicit midpoint rule) is an $A(\alpha)$-stable method when it is considered as a rather complicated one-step method. Here $\tan\alpha \approx 4.3115$ so that we have almost A-stability; furthermore the propagation factors become very small except for extremely large $|hg|$. Thus this method should be a powerful competitor with the existing "stiff methods".

(ii) How should the implicit equations be solved?

All reasonable methods for stiff SODE have to be implicit, thus the effective solution of the arising implicit "algebraic" equations presents a serious problem. Simple contraction iteration is not applicable since it reintroduces explicitness. Modifications of Newton iteration seems to be the best approach; here the choice of a good initial guess for the iteration becomes essential because of the large iteration effort (computation of Jacobians).

(iii) How should the stepsizes be controlled?

A powerful stepsize control strategy becomes the more important the larger steps one is able to use due to the strong stability of the discretization method employed. With some stiff SODE one may typically increase the stepsize by a factor like 100 after the "bad" transients have died away. If this possibility is not immediately recognized by the algorithm it may waste huge amounts of computation.

Here, the extrapolation method with its parallel computations over the same basic steps gives considerable insight into the situation and may permit the design of superior stepsize control mechanisms.

BIBLIOGRAPHY

[1] G. DAHLQUIST: Convergence and stability in the numerical integration of ordinary differential equations, Math. Scand. 4 (1956) 33-53.

[2] P. HENRICI: Discrete variable methods in ordinary differential equations, Wiley, New York, 1962.

[3] P. HENRICI: Error propagation for difference methods, Wiley, New York, 1963.

[4] W.B. GRAGG: Repeated extrapolation to the limit in the numerical solution of ordinary differential equations, Ph.D. thesis, UCLA, 1964.

[5] H.J. STETTER: Analysis of discretization methods, Springer-Verlag, Berlin-Heidelberg-New York, to appear.

[6] W. HAHN: Stability of motion, Springer-Verlag, Berlin-Heidelberg-New York, 1967.

[7] G. DAHLQUIST: A special stability problem for linear multistep methods, BIT 3 (1963) 27-43.

[8] H. BRUNNER: Stabilisierung optimaler Differenzenverfahren zur numerischen Integration gewöhnlicher Differentialgleichungen, Ph.D. Thesis, ETH Zürich, 1969.

[9] H. BRUNNER: Marginal stability and stabilization in the numerical integration of ordinary differential equations, Math. Comp. 24 (1970) 635-646.

[10] J.D. LAMBERT: Linear multistep methods with mildly varying coefficients, Math. Comp. 24 (1970) 81-93.

[11] C.W. GEAR: The automatic integration of stiff ordinary differential equations, Proceed. IFIP Congress 1968, 195-199.

[12] M.R. OSBORNE: A new method for the integration of stiff systems of ordinary differential equations, Proceed. IFIP Congress 1968, 200-204.

Added in print:

After the completion of the above manuscript it was pointed out to me that Theorem 6.1 has been (essentially) obtained in I.BABUSKA, M.PRÁGER, E.VITÁSEK: Numerical processes in differential equations, Interscience, London - New York - Sydney, 1966, as Theorem 3.3. The method of proof is nearly the same.

A RATIONAL BASIS FOR FUNCTION APPROXIMATION

Eugene L. Wachspress

PART I : The Rational Basis

1. Wedge Base Functions

Approximation of continuous functions of one variable has been analysed extensively and a broad mathematical theory has been developed. Less satisfactory results have been obtained for multi-variable function approximation. We shall consider here a particular continuous approximation to a function whose values are specified on the "threads" of a convex spider web in a plane

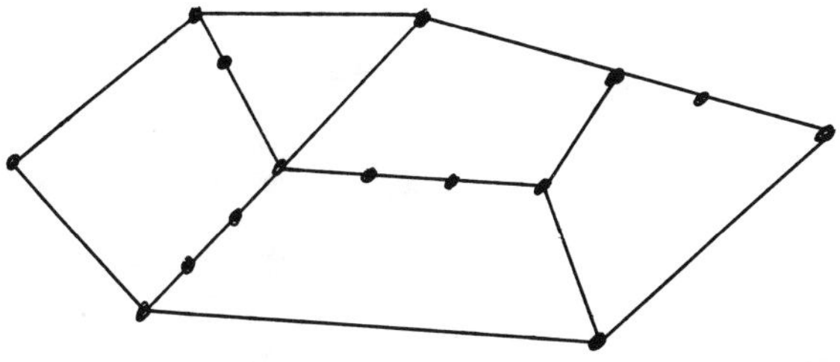

• denotes a data point.

By convex we mean that each polygon in the web is convex. We exclude the case where a data point is a vertex of one polygon and a side point of another. For a triangular web with no data points other than at the web nodes, it is common practice to use a piecewise linear approximation. Areal (or barycentric) coordinates [1] provide a convenient basis for such an approximation. For example, the areal coordinate associated with vertex 1 of triangle 123 is the "wedge" function shown in the following diagram:

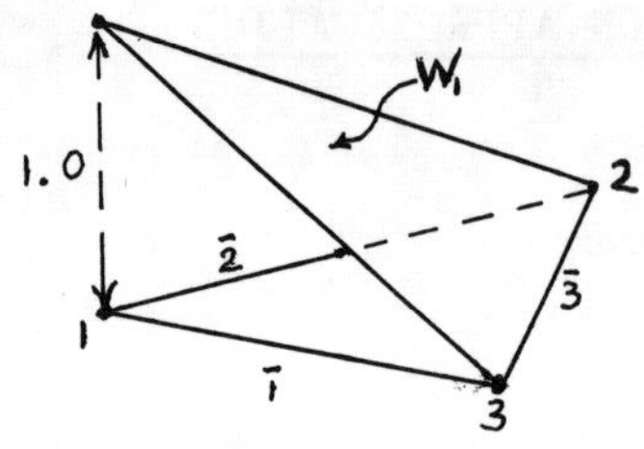

$$\bar{s} = \{x,y \mid a_s x + b_s y + c_s = 0\}$$

$$\underline{s} = \frac{a_s x + b_s y + c_s}{(a_s^2 + b_s^2)^{\frac{1}{2}}} \qquad (1)$$

$$\underline{s}_i = \underline{s}(x_i, y_i)$$

$$W_1(x,y) = \frac{\underline{3}}{\underline{3}_1}$$

(Note that $|\underline{s}|$ is the distance of (x,y) from line $\bar{s}$.)

Thus the values of u_i at the vertices $(i = 1,2,3)$ are fitted by linear approximation

$$u(x,y) = \sum_{i=1}^{3} u_i \, W_i(x,y) \quad . \qquad (2)$$

This linear approximation generalises to a bilinear over a parallelogram;

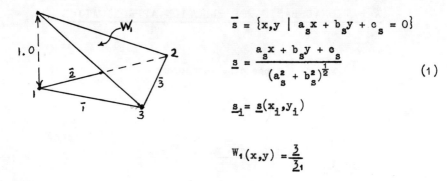

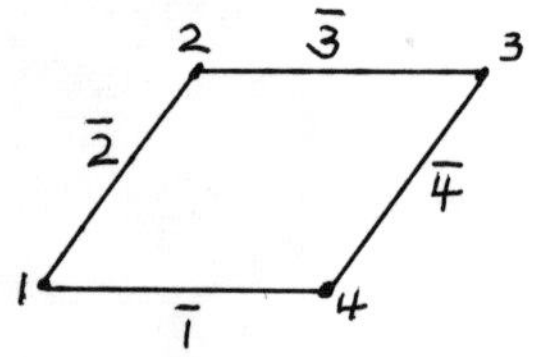

$$W_1 = \frac{\underline{3}\,\underline{4}}{(\underline{3}_1\,\underline{4}_1)} \text{ , etc}$$

$$u(x,y) = \sum_{i=1}^{4} u_i \, W_i(x,y) \quad . \qquad (3)$$

For both triangular and parallelogram elements, the approximation is linear on the sides. (The bilinear approximation over the parallelogram is linear on its sides. For example on $\bar{1}$ we note that $|\underline{3}|$ is the constant distance from $\bar{3}$ to $\bar{1}$ so that $W_1 = \frac{\underline{4}}{\underline{4}_1}$ on $\bar{1}$.)

This linearity assures continuity of the composite "patch-work" approximation over the web. For quadrilaterals which are not parallelograms or for n-gons with $n > 4$, it can be established that polynomial wedge functions will not yield this linearity property essential for continuity across web threads.

This difficulty has been resolved in the past by adding threads to the web to reduce all of the polygons to triangles or parallelograms. This is not always desirable, and is not acceptable if we append the requirement that the approximation be regular interior to the polygons of the web:

$$u(x,y) \in C^{\infty} \text{ (polygon interior).}$$

Derivatives normal to the web threads need not exist. The approximation may be "creased" along the threads.

We shall be concerned with a generalisation of the triangle wedges (areal coordinates) to convex polygons, thereby retaining regularity interior to each polygon. Having constructed the polygon wedges we will return to the problem of fitting data on the web threads. We have already noted that data at nodes only may be fitted over an n-gon by $u(x,y) = \sum_{i=1}^{n} u_i W_i(x,y)$. We have not yet discussed "side point" data even for triangular webs.

Let p (for "poly") denote the perimeter and g (for "gon") the interior of a convex n-gon. Let pg ("polygon") denote p $\cup$ g. Let i denote the vertex at $\overline{i} \bullet \overline{i+1}$, i.e. the point of intersection of sides $\overline{i}$ and $\overline{i+1}$.

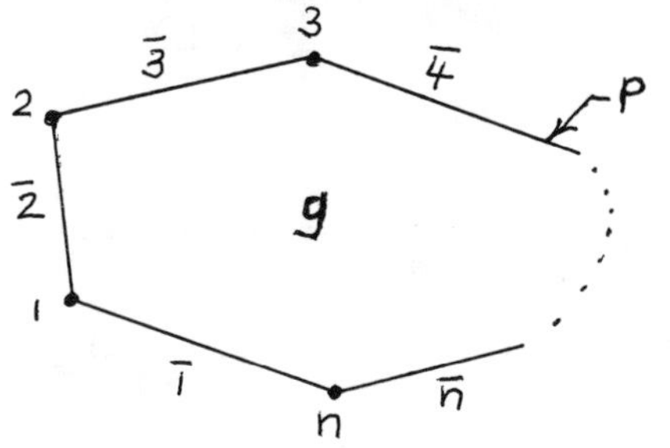

The index i is mod n in the ensuing development .

Polynomials do not suffice as wedge functions for the general n-gon. We shall introduce a particularly convenient rational basis. Wedge functions having the following properties will be constructed:

<u>Wedge Function Properties for an n-gon</u>

P 1. W_i is a rational function for $i = 1,2,\ldots,n$.

P 2. $W_i \in C^{\infty}(pg)$.

P 3. $W_i(x_i, y_i) = 1$.

P 4. $W_i(x,y) = 0$ for $(x,y) \in \bar{j}$ and $\bar{j} \neq \bar{i}, \overline{i+1}$.

(Wedge W_i vanishes on all sides "opposite" vertex i in the n-gon.)

P 5. W_i is linear on sides of p "adjacent" to vertex i (i.e. on sides $\bar{j} = \bar{i}, \overline{i+1}$).

P 6. $W_i(x,y) > 0$ for $(x,y) \in g$.

P 7. $\sum_{i=1}^{n} W_i(x,y) = 1$ for $(x,y) \in pg$.

Wedge Construction for an n-gon

We note the following:

Lemma

If lines $\bar{a}$, $\bar{b}$ and $\bar{c}$ meet at a common point then $\frac{\underline{a}}{\underline{b}}$ is constant on $\bar{c}$, $\frac{\underline{a}}{\underline{c}}$ is constant on $\bar{b}$ and $\frac{\underline{b}}{\underline{c}}$ is constant on $\bar{a}$. (The ratios are not defined at the intersection point.)

Proof.

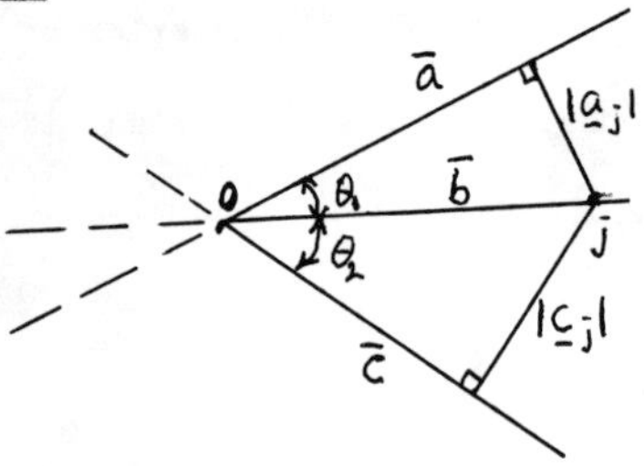

For j on $\bar{b}$,

$$|\underline{a}_j| \sin \theta_2 = |\underline{c}_j| \sin \theta_1$$

$$\frac{|\underline{a}_j|}{|\underline{c}_j|} = \frac{\sin \theta_1}{\sin \theta_2}$$

sign $\left(\frac{\underline{a}_j}{\underline{c}_j} \right)$ is constant on $\bar{b}$ except at the point of intersection, where $\frac{\underline{a}_j}{\underline{c}_j}$ is not defined. Both $\underline{a}_j$ and $\underline{c}_j$ change sign as j crosses 0. The same argument applies to the other two cases.

We direct our attention to vertex 1 of a quadrilateral, no two of whose sides are parallel

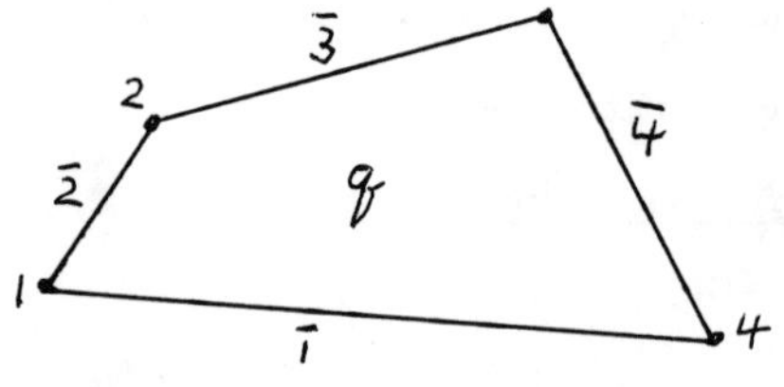

We seek a wedge of the form $W_1(x,y) = k_1 \left(\frac{\underline{3}\,\underline{4}}{\underline{5}} \right)$, where k_1 is the normalisation $k_1 = \left(\frac{\underline{5}_1}{\underline{3}_1\,\underline{4}_1}\right)$. Linear form $\underline{5}$ must not vanish in pg. It is easily verified that if $\bar{5}$ does not intersect the pg, then W_1 satisfies properties P1-P4 and P6. Property P5, linearity along the adjacent sides, is the critical property which yields a unique $\bar{5}$. We choose $\bar{5}$ so that $\frac{\underline{3}}{\underline{5}}$ is a constant on $\bar{1}$ and $\frac{\underline{4}}{\underline{5}}$ is a constant on $\bar{2}$. Invoking the lemma:

$\bar{1}.\bar{3}.\bar{5} \neq \phi \Rightarrow$

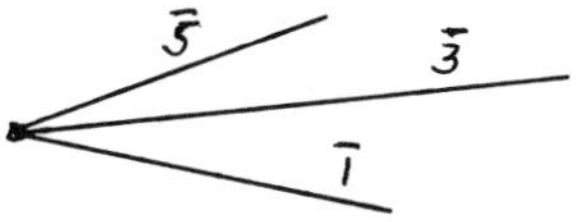

$\bar{2}.\bar{4}.\bar{5} \neq \phi \Rightarrow$

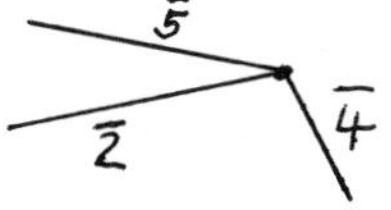

Hence, $\bar{5}$ is the line determined by the points $\bar{1}.\bar{3}$ and $\bar{2}.\bar{4}$:

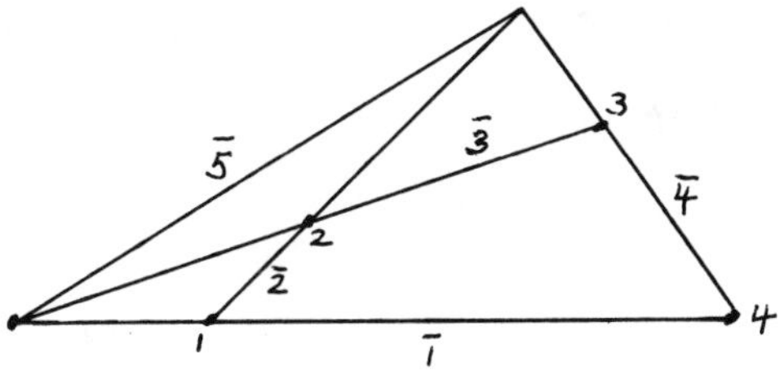

$$W_1(x,y) = \frac{\underline{5}_1}{\underline{3}_1\,\underline{4}_1} \cdot \frac{\underline{3}\,\underline{4}}{\underline{5}}$$

For any convex quadrilateral, $\bar{5}$ does not have any point in common with the quadrilateral. It is the "exterior diagonal" of the "complete quadrilateral" [1].

For a parallelogram, we define $\underline{5}(x,y) = 1$. In this case the wedge is bilinear.

If two sides are parallel and the other two are not, we define $\bar{5}$ as follows:

$\bar{1} \parallel \bar{3} \Rightarrow \bar{5} \parallel \bar{1}$ and $\bar{5}$ passes through $\bar{2}.\bar{4}$

$\bar{2} \parallel \bar{4} \Rightarrow \bar{5} \parallel \bar{2}$ and $\bar{5}$ passes through $\bar{1}.\bar{3}$:

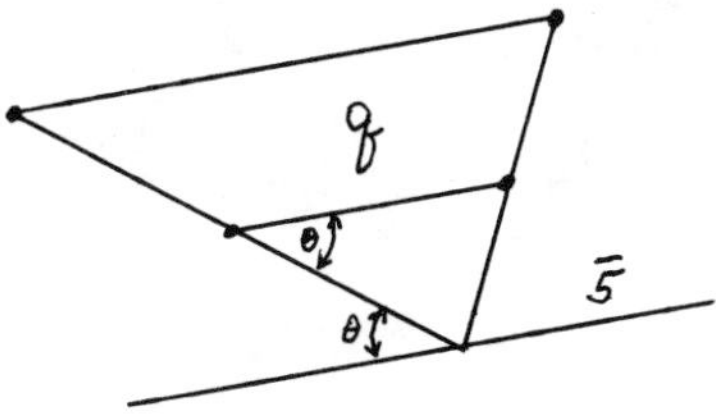

Having chosen $\bar{5}$ in this fashion we observe that

$$W_1 = \frac{\underline{4}}{\underline{4}_1} \text{ on } \bar{1} \quad \text{and} \quad W_1 = \frac{\underline{3}}{\underline{3}_1} \text{ on } \bar{2}\,.$$

The other three wedges are defined in similar fashion:

$$W_2 = \frac{\underline{5}_2}{\underline{1}_2\,\underline{4}_2} \cdot \frac{\underline{1}\,\underline{4}}{\underline{5}}\;;\quad W_3 = \frac{\underline{5}_3}{\underline{1}_3\,\underline{2}_3} \cdot \frac{\underline{1}\,\underline{2}}{\underline{5}}\;;\quad W_4 = \frac{\underline{5}_4}{\underline{2}_4\,\underline{3}_4} \cdot \frac{\underline{2}\,\underline{3}}{\underline{5}}\,. \tag{4}$$

We have yet to establish property P7.

Consider the function

$$g(x,y) = \sum_{i=1}^{4} W_i(x,y) - 1 = \frac{k_1\,\underline{3}\,\underline{4} + k_2\,\underline{1}\,\underline{4} + k_3\,\underline{1}\,\underline{2} + k_4\,\underline{2}\,\underline{3} - \underline{5}}{\underline{5}}\,.$$

Linearity on the sides and normalisation to unity at the vertices establish that $g(x,y) = 0$ on the perimeter of q. The only bilinear which vanishes on all four sides of the quadrilateral is the zero function. Thus the numerator above is identically zero and P7 is verified. A sketch of $W_1(x,y)$ is given in the following figure:

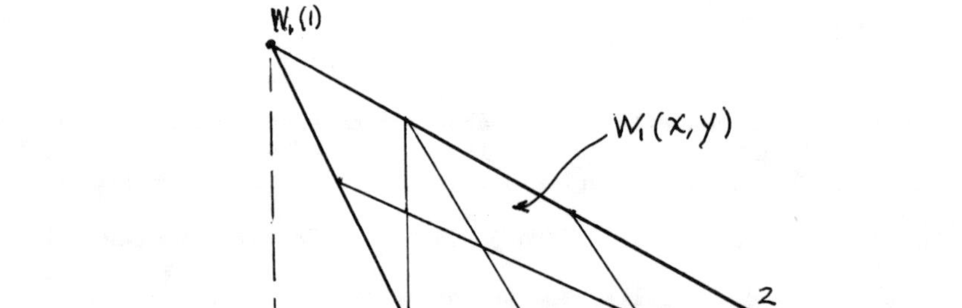

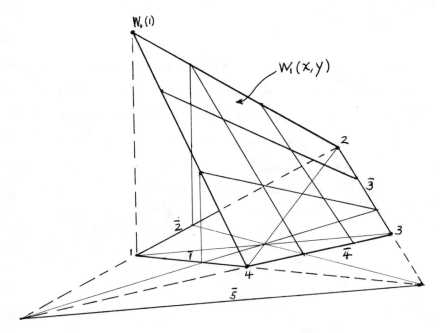

A Quadrilateral Wedge: $W_1 = k_1 \dfrac{\underline{3}\,\underline{4}}{\underline{5}}$.

The wedge construction for the quadrilateral generalises to the n-gon. We seek a wedge of the form:

$$W_i = k_i \frac{\underline{i+2}\ \underline{i+3}\ \underline{i+4}\ \ldots\ldots\ \underline{i-1}}{\underline{x}\ \underline{y}\ \underline{z}\ \ldots\ldots} \tag{5}$$

where the linear forms in the numerator are the (n-2) forms associated with the n-gon sides opposite vertex i and where (n-3) linear forms in the denomonater are chosen so that:

(a) The "right-slant-ratios" $\frac{\underline{i+2}}{\underline{x}}, \frac{\underline{i+3}}{\underline{y}}, \ldots$ are constant on adjacent side $\overline{i}$, and

(b) The "left-slant-ratios" $\frac{\underline{i+3}}{\underline{x}}, \frac{\underline{i+4}}{\underline{y}}, \ldots$ are constant on adjacent side $\overline{i+1}$.

$$\begin{aligned} \text{Then} \quad W_i &= \frac{\underline{i-1}}{(\underline{i-1})_i} \quad \text{on } \overline{i} \\ &= \frac{\underline{i+2}}{(\underline{i+2})_i} \quad \text{on } \overline{i+1} \ . \end{aligned} \tag{6}$$

The "exterior diagonals" of a convex pg are defined as the exterior diagonals of all quadrilaterals formed from the sides of the pg. Each of these quadrilaterals contains the pg. Hence, these exterior diagonals are indeed exterior to the pg. The linear forms in the denominator of W_i are appropriate combinations of the pg exterior diagonals.

It will be shown that wedges determined in this manner for $n > 4$ will not satisfy property P7 except in special cases. We therefore denote these wedges by $W_i'(x,y)$ and normalise the sum over i by defining:

$$W_i(x,y) = \frac{W_i'(x,y)}{\sum_{i=1}^{n} W_i'(x,y)} \tag{7}$$

We illustrate the contruction for a pentagon:

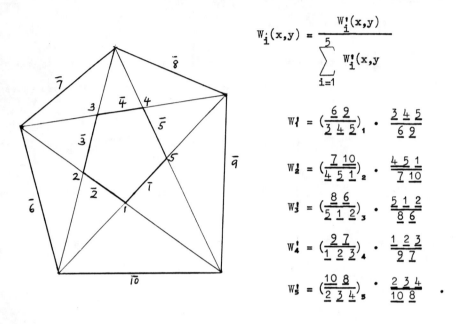

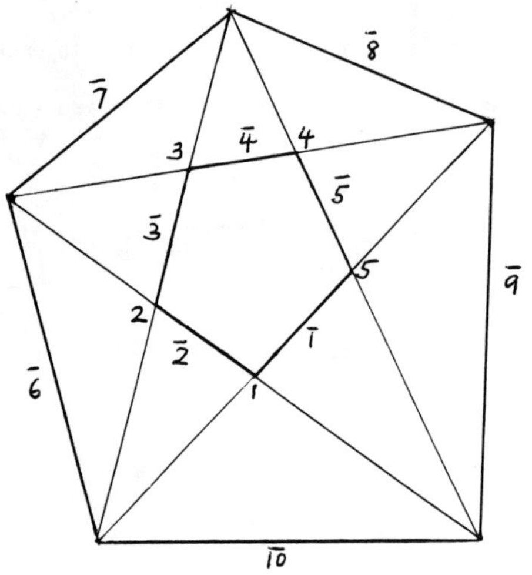

$$W_i(x,y) = \frac{W_i'(x,y)}{\sum_{i=1}^{5} W_i'(x,y}$$

$$W_1' = \left(\frac{\underline{6}\ \underline{9}}{\underline{3}\ \underline{4}\ \underline{5}}\right)_1 \cdot \frac{\underline{3}\ \underline{4}\ \underline{5}}{\underline{6}\ \underline{9}}$$

$$W_2' = \left(\frac{\underline{7}\ \underline{10}}{\underline{4}\ \underline{5}\ \underline{1}}\right)_2 \cdot \frac{\underline{4}\ \underline{5}\ \underline{1}}{\underline{7}\ \underline{10}}$$

$$W_3' = \left(\frac{\underline{8}\ \underline{6}}{\underline{5}\ \underline{1}\ \underline{2}}\right)_3 \cdot \frac{\underline{5}\ \underline{1}\ \underline{2}}{\underline{8}\ \underline{6}}$$

$$W_4' = \left(\frac{\underline{9}\ \underline{7}}{\underline{1}\ \underline{2}\ \underline{3}}\right)_4 \cdot \frac{\underline{1}\ \underline{2}\ \underline{3}}{\underline{9}\ \underline{7}}$$

$$W_5' = \left(\frac{\underline{10}\ \underline{8}}{\underline{2}\ \underline{3}\ \underline{4}}\right)_5 \cdot \frac{\underline{2}\ \underline{3}\ \underline{4}}{\underline{10}\ \underline{8}} .$$

It is easily shown that for the pentagon $\sum_{i=1}^{5} W_i'(x,y) \not\equiv 1$ in general.

We need only consider

$$\sum_{i=1}^{5} W_i'(x,y) - 1 = \frac{\begin{array}{c}(k_1\ \underline{3}\ \underline{4}\ \underline{5}\ \underline{7}\ \underline{8}\ \underline{10} + k_2\ \underline{4}\ \underline{5}\ \underline{1}\ \underline{6}\ \underline{8}\ \underline{9} + k_3\ \underline{5}\ \underline{1}\ \underline{2}\ \underline{7}\ \underline{9}\ \underline{10} \\ + k_4\ \underline{1}\ \underline{2}\ \underline{3}\ \underline{6}\ \underline{8}\ \underline{10} + k_5\ \underline{2}\ \underline{3}\ \underline{4}\ \underline{6}\ \underline{7}\ \underline{9} - \underline{6}\ \underline{7}\ \underline{8}\ \underline{9}\ \underline{10})\end{array}}{\underline{6}\ \underline{7}\ \underline{8}\ \underline{9}\ \underline{10}} .$$

The numerator cannot be identically zero, for at $\bar{6} \cdot \bar{8}$ the numerator reduces to

$$k_3\ \underline{5}\ \underline{7}\ \underline{10}\ \underline{1}\ \underline{2}\ \underline{9} ,$$

none of whose factors vanish.

We note, however, that linearity and normalization of the W_i' does yield $\sum_{i=1}^{p} W_i'(x,y) = 1$ on p, so that $W_i' = W_i$ for (x,y) on p. This ensures P5 after normalization. As an interior angle of an n-gon approaches π, the n wedges for the n-gon degenerate into the n-1 wedges for the resulting (n-1)-gon. This will be demonstrated now for n = 4.

Let vertex 4 of quadrilateral 1234 approach 4' along the line $\overline{44'}$ and let

$\frac{\underline{3}_{4'}}{\underline{3}_1} = \theta$

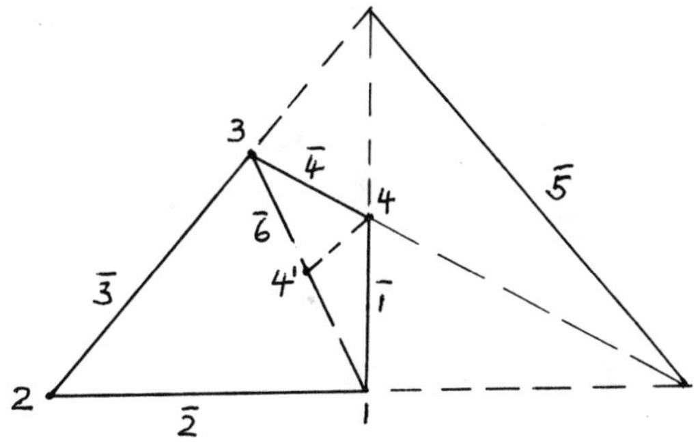

As $4 \to 4'$: $\underline{1}$, $\underline{4}$ and $\underline{5} \to \underline{6}$, where $\overline{6}$ is actually $\overline{1}$ of triangle 123 but has been labelled $\overline{6}$ to distinguish this line from $\overline{1}$ of the quadrilateral.

We have

$$\lim_{4 \to 4'} [W_2(x,y)] = \lim_{4 \to 4'} [(\frac{\underline{1}\,\underline{4}}{\underline{5}}) (\frac{\underline{5}}{\underline{1}\,\underline{4}})_2] = \frac{\underline{6}}{\underline{6}_2} = W_2(x,y) \text{ for triangle 123.}$$

The other wedges are not continuous in the limit.

$$\lim_{4 \to 4'} [W_4(x,y) = (\frac{\underline{2}\,\underline{3}}{\underline{5}}) (\frac{\underline{5}}{\underline{2}\,\underline{3}})_4] \underset{(x,y) \notin \overline{6}}{=\!=} \frac{\underline{2}\,\underline{3}}{\underline{6}} \cdot \lim_{4 \to 4'} (\frac{\underline{5}}{\underline{2}\,\underline{3}})_4 = 0$$

$$\underset{(x,y) \in \overline{6}}{=\!=}$$

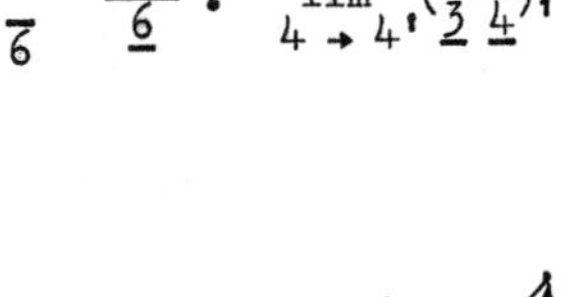

$$\lim_{4 \to 4'} [W_1(x,y) = (\frac{\underline{3}\,\underline{4}}{\underline{5}}) (\frac{\underline{5}}{\underline{3}\,\underline{4}})_1] \underset{(x,y) \notin \overline{6}}{=\!=} \frac{\underline{3}\,\underline{6}}{\underline{6}} \cdot \lim_{4 \to 4'} (\frac{\underline{5}}{\underline{3}\,\underline{4}})_1 = \frac{\underline{3}}{\underline{3}_1}$$

$$\underset{(x,y) \in \overline{6}}{=\!=}$$

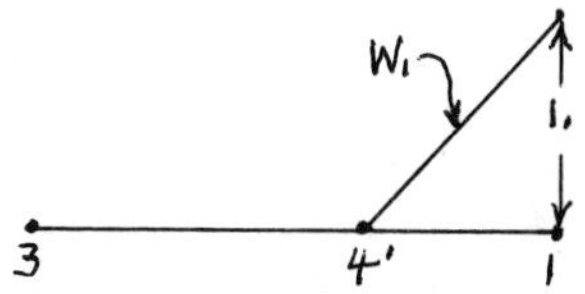

Although W_1 and W_4 are not continuous in the limit,

$$W_1(x,y) + \theta\, W_4(x,y) \xrightarrow[(4 \to 4')]{} \frac{\underline{\underline{3}}}{\underline{3}_1} = W_1(x,y) \text{ for triangle } 123.$$

We also have

$$\lim_{4 \to 4'} [W_3(x,y) = \left(\frac{\underline{\underline{1}}\;\underline{\underline{2}}}{\underline{5}}\right)\left(\frac{\underline{\underline{5}}}{\underline{1}\;\underline{2}}\right) \underset{(x,y) \notin \overline{6}}{=\!=} \frac{\underline{\underline{2}}}{\underline{2}_3}$$

$$\underset{(x,y) \in \overline{6}}{=\!=}$$

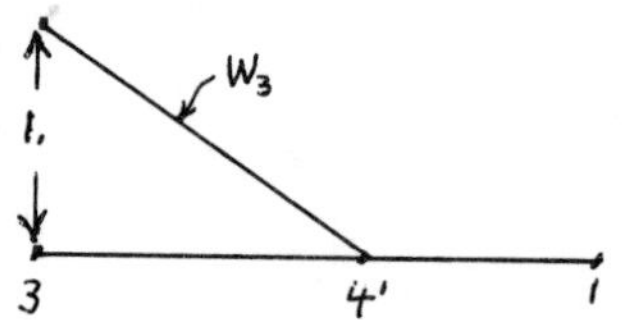

We note that

$$W_3(x,y) + (1-\theta)\, W_4(x,y) \xrightarrow[4 \to 4']{} \frac{\underline{\underline{2}}}{\underline{2}_3} = W_1(x,y) \quad \text{for triangle } 123.$$

It is thus seen that the discontinuous limit functions of the quadrilateral rational wedges may be combined to yield the continuous linear wedge basis functions for the limiting triangle. Areal coordinates are a degenerate case of rational wedges.

Collocation

We now consider the problem of collocation on a poly. Data along side $\overline{i+1}$ may be fit by a polynomial in the wedges W_i and W_{i+1}. These wedges are appropriate for fitting in the "natural coordinate" of side $\overline{i+1}$.

The polynomial fit on each side of the poly is made unique by initial problem specification (interpolation, least square of given degree, Chebyshev fit of given degree, etc.). However, in certain cases the continuation into the gon is not unique. This is true even after we have accepted the rational wedge functions as our continuation basis. A simple example is a triangle where the vertices and one side point are to be fit:

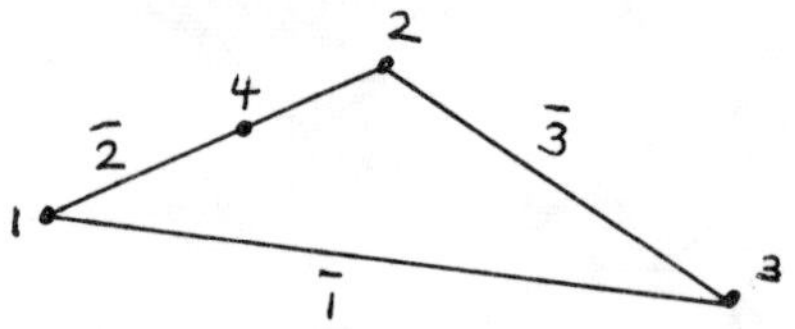

Here the ambiguity is resolved readily. The wedge basis on $\overline{2}$ must not destroy linearity on $\overline{1}$ and $\overline{3}$. The only quadratic on $\overline{2}$ which preserves linearity on $\overline{1}$ and $\overline{3}$ is W_1W_2. Thus, the basis on $\overline{2}$ is $\{W_1, W_2, W_1W_2\}$ and the basis for fitting the four points is $\{W_1, W_2, W_3, W_1W_2\}$.

In general, all basis functions introduced to fit side points on $\overline{i+1}$ must have a factor W_iW_{i+1} in order that there be no interaction between sides.

Suppose we introduce a fifth point on side $\overline{2}$:

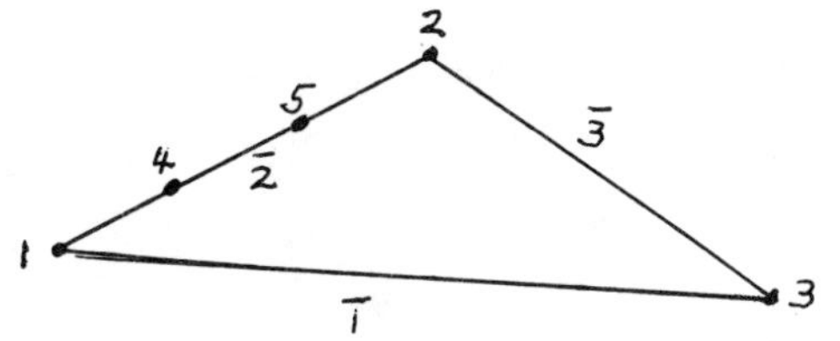

The fit is no longer uniquely determined by invariance and side non-interaction requirements. (By invariance we mean that the basis should not depend upon triangle orientation or vertex labelling.) We resolve this ambiquity by choosing a polynomial basis which is independent of the number and location of side points. When the structure of a specific problem does not indicate a natural set of polynomials, and this is probably the usual case, we choose a set with equally spaced zeros. The following diagram is self explanatory:

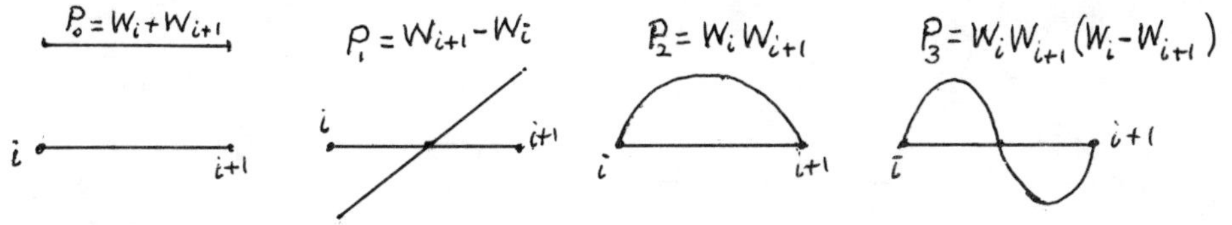

$$\{W_i, W_{i+1}\} \Rightarrow \{P_0, P_1\}$$

$$P_4 = W_i W_{i+1} (2W_i - W_{i+1})(W_i - 2W_{i+1})$$

$$\text{Basis}: \{W_i, W_{i+1}, W_iW_{i+1}, \ldots\}.$$

A procedure for fitting 9 points on a triangle perimeter by a cubic is described by Zlámal in reference [5]. It appears, however, that the polynomial basis concept introduced here has more general application than Zlámal's procedure and has an additional advantage of simplicity.

Some common examples using this basis are:

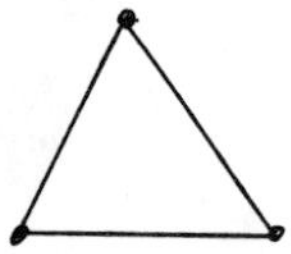

$\{W_1, W_2, W_3\}$

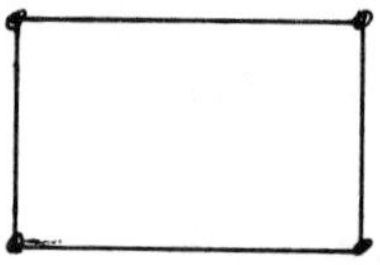

$\{W_1, W_2, W_3, W_4\}$

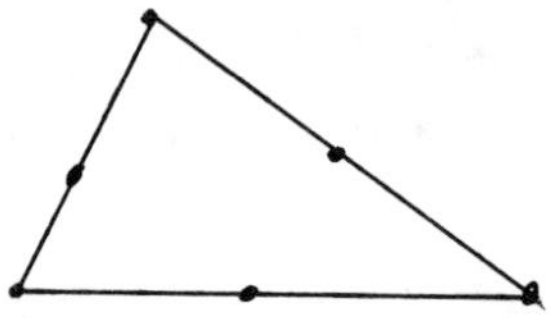

$\{W_1, W_2, W_3, W_1W_2, W_2W_3, W_3W_1\}$.

Some less common examples, heretofore presenting certain mechanical difficulties but now quite straightforward, are:

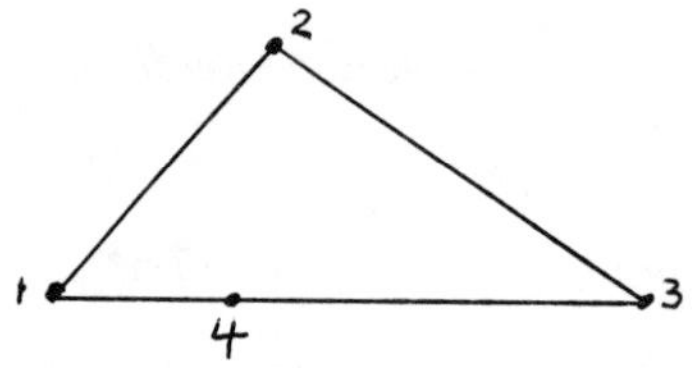

$\{W_1, W_2, W_3, W_3W_1\}$.

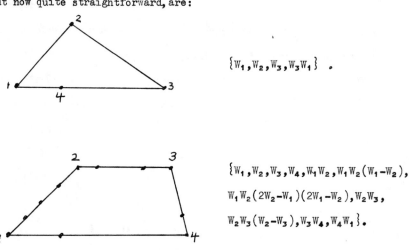

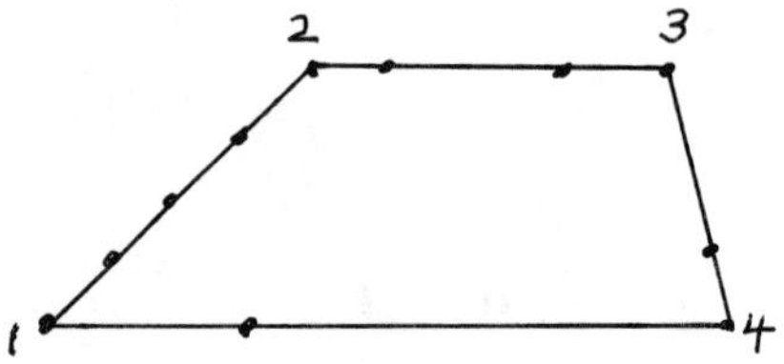

$\{W_1, W_2, W_3, W_4, W_1W_2, W_1W_2(W_1-W_2),$
$W_1W_2(2W_2-W_1)(2W_1-W_2), W_2W_3,$
$W_2W_3(W_2-W_3), W_3W_4, W_4W_1\}$.

When the collocation is performed over a network of polygons, any variety of convex polygons and data points may be handled with reasonable ease. The resulting function is continuous over the entire domain.

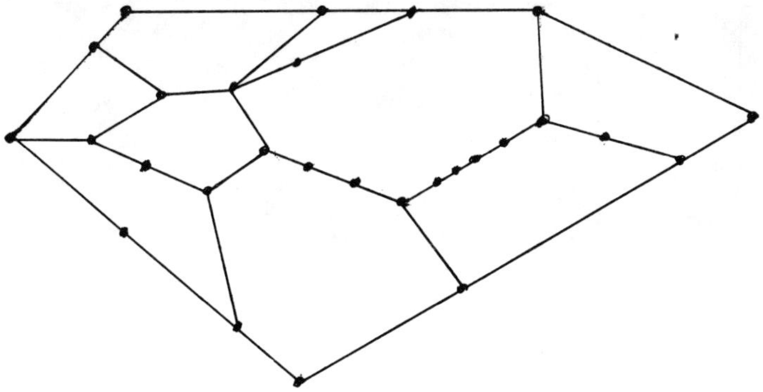

We have thus accomplished the objectives posed initially.

The convenience of the wedge basis will now be illustrated by example.

Example 1

One side point on each side of a triangle.

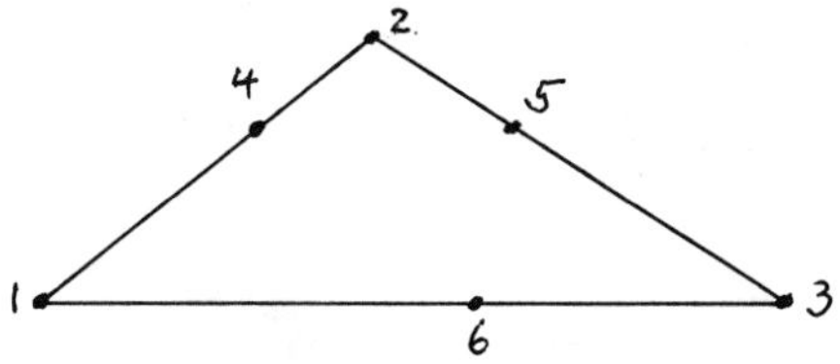

$$\text{Let } u(x,y) = \sum_{i=1}^{3} (c_i W_i + d_i W_i W_{i+1})$$

Then

$$\begin{bmatrix} 1 & 0 & 0 & 0 & 0 & 0 \\ 0 & 1 & 0 & 0 & 0 & 0 \\ 0 & 0 & 1 & 0 & 0 & 0 \\ W_1(4) & W_2(4) & 0 & W_1(4)W_2(4) & 0 & 0 \\ 0 & W_2(5) & W_3(5) & 0 & W_2(5)W_3(5) & 0 \\ W_1(6) & 0 & W_3(6) & 0 & 0 & W_1(6)W_3(6) \end{bmatrix} \equiv B$$

$$B \cdot \begin{bmatrix} c_1 \\ c_2 \\ c_3 \\ d_1 \\ d_2 \\ d_3 \end{bmatrix} = \begin{bmatrix} u_1 \\ u_2 \\ u_3 \\ u_4 \\ u_5 \\ u_6 \end{bmatrix}$$

so that

$$c_1 = u_1 \qquad c_2 = u_2 \qquad c_3 = u_3$$

$$d_1 = \frac{1}{W_1(4)W_2(4)} [u_4 - W_1(4)u_1 - W_2(4)u_2]$$

$$d_2 = \frac{1}{W_2(5)W_3(5)} [u_5 - W_2(5)u_2 - W_3(5)u_3]$$

$$d_3 = \frac{1}{W_1(6)W_3(6)} [u_6 - W_1(6)u_1 - W_3(6)u_3] \quad ,$$

Collecting coefficients of u_i for $i = 1,\ldots,6$, we obtain:

$$u(x,y) = W_1(x,y)\,[1 - \frac{W_3(x,y)}{W_3(6)} - \frac{W_2(x,y)}{W_2(4)}]u_1 + W_2(x,y)\,[1 - \frac{W_1(x,y)}{W_1(4)} - \frac{W_3(x,y)}{W_3(5)}]u_2$$

$$+ W_3(x,y)\,[1 - \frac{W_1(x,y)}{W_1(6)} - \frac{W_2(x,y)}{W_2(5)}]u_3 + \frac{W_1(x,y)\ W_2(x,y)}{W_1(4)\ W_2(4)}\,u_4 \;+$$

$$+ \frac{W_2(x,y)\ W_3(x,y)}{W_2(5)\ W_3(5)}\,u_5 + \frac{W_1(x,y)\ W_3(x,y)}{W_1(6)\ W_3(6)}\,u_6 \quad .$$

Example 2

The case of one point on each side of a quadrilateral is the generalization of the above with four wedges:

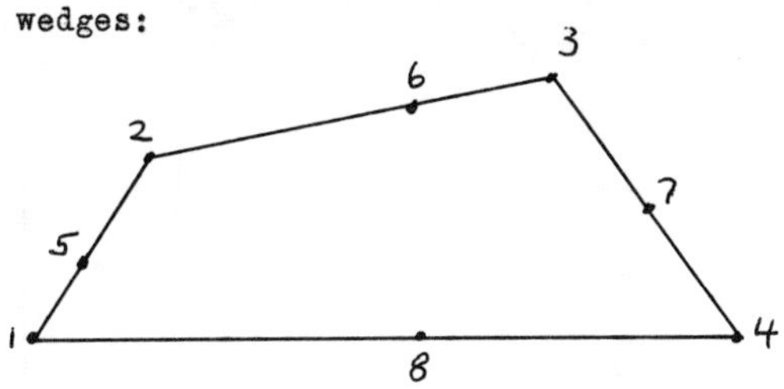

$$u(x,y) = W_1\left(1 - \frac{W_4}{W_4(8)} - \frac{W_2}{W_2(5)}\right)u_1 + \dots + W_4\left(1 - \frac{W_3}{W_3(7)} - \frac{W_1}{W_1(8)}\right)u_4$$

$$+ \frac{W_1\ W_2}{W_1(5)\ W_2(5)}\ u_5 + \dots + \frac{W_4\ W_1}{W_4(8)\ W_1(8)}\ u_8 \quad .$$

Example 3

Collocation on a quadrilateral with side points

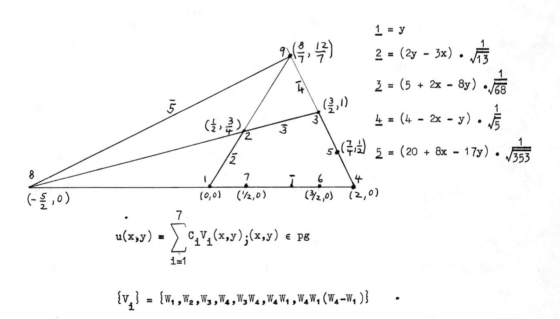

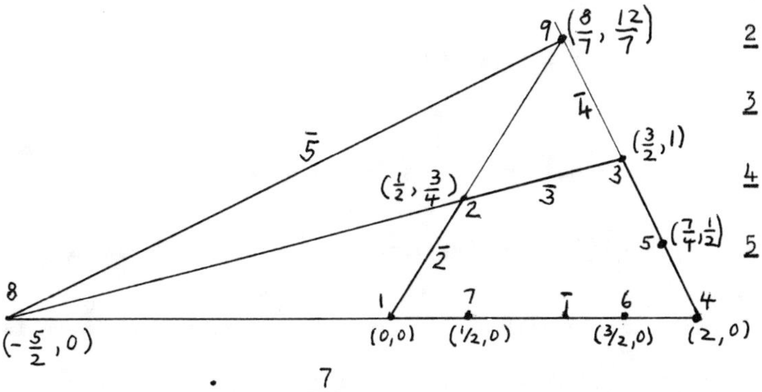

$\underline{1} = y$

$\underline{2} = (2y - 3x) \cdot \frac{1}{\sqrt{13}}$

$\underline{3} = (5 + 2x - 8y) \cdot \frac{1}{\sqrt{68}}$

$\underline{4} = (4 - 2x - y) \cdot \frac{1}{\sqrt{5}}$

$\underline{5} = (20 + 8x - 17y) \cdot \frac{1}{\sqrt{353}}$

$$u(x,y) = \sum_{i=1}^{7} C_i V_i(x,y);\ (x,y) \in pg$$

$$\{V_i\} = \{W_1, W_2, W_3, W_4, W_3W_4, W_4W_1, W_4W_1(W_4-W_1)\} \quad .$$

$$
\begin{bmatrix}
1 & 0 & \multicolumn{4}{c}{\rule{4cm}{0.4pt}} & 0 \\
0 & 1 & 0 & \multicolumn{3}{c}{\rule{3cm}{0.4pt}} & 0 \\
0 & 0 & 1 & 0 & \multicolumn{2}{c}{\rule{2cm}{0.4pt}} & 0 \\
0 & 0 & 0 & 1 & 0 & 0 & 0 \\
0 & 0 & \frac{1}{2} & \frac{1}{2} & \frac{1}{4} & 0 & 0 \\
\frac{1}{4} & 0 & 0 & \frac{3}{4} & 0 & \frac{3}{16} & \frac{3}{32} \\
\frac{3}{4} & 0 & 0 & \frac{1}{4} & 0 & \frac{3}{16} & \frac{3}{32}
\end{bmatrix}
\cdot
\begin{bmatrix} c_1 \\ c_2 \\ c_3 \\ c_4 \\ c_5 \\ c_6 \\ c_7 \end{bmatrix}
=
\begin{bmatrix} u_1 \\ u_2 \\ u_3 \\ u_4 \\ u_5 \\ u_6 \\ u_7 \end{bmatrix}
$$

$$c_i = u_i, \ i = 1,2,3,4 \qquad c_5 = 4u_5 - 2u_3 - 2u_4 \qquad c_6 = \frac{8}{3} \ [(u_6 + u_7) - (u_1 + u_4)]$$

$$c_7 = \frac{16}{3} \ [(u_6 - u_7) + (u_1 - u_4)] \ .$$

$$W_1 = \frac{(5 + 2x - 8y)(4 - 2x - y)}{(20 + 8x - 17y)} \qquad W_2 = \frac{20y(4 - 2x - y)}{3(20 + 8x - 17y)} \ ;$$

$$W_3 = \frac{6y(3x - 2y)}{(20 + 8x - 17y)} \qquad W_4 = \frac{2(3x - 2y)(5 + 2x - 8y)}{3(20 + 8x - 17y)} \ .$$

Thus

$$u(x,y) = \sum_{i=1}^{7} c_i \ V_i(x,y) = W_1 \{1 - \frac{8}{3} W_4[1 - 2(W_4 - W_1)]\}u_1$$

$$+ W_2u_2 + W_3(1 - 2W_4)u_3 + W_4\{1 - 2W_3 - \frac{8}{3} W_1[1 + 2(W_4 - W_1)]\}u_4$$

$$+ 4W_3W_4u_5 + \frac{8}{3} W_1W_4[1 + 2(W_4 - W_1)]u_6 + \frac{8}{3} W_1W_4[1 - 2(W_4 - W_1)]u_7 \ .$$

Having gone through these sample collocation problems, we note that there is a beneficial property of separability in the determination of the expansion coefficients $\{C_i\}$ and the determination of the wedge functions.

We determine the necessary combinations of wedges and their coefficients in terms of the data location. We may then collect terms to get a relationship of the form

$$u(x,y) = \sum_{i=1}^{I} b_i(x,y)u_i$$

for I data points. The $b_i(x,y)$ are expressed in terms of the $W_i(x,y)$ $(i = 1,2,\ldots,n \leq I)$, independent of the functional form of the W_i.

The wedges, on the other hand, are rational functions of x and y which depend only on the polygon and not on the location of data points. Thus, wedges and any integrals of combinations of them over a polygon need only be computed once for each class of similar polygons.

Data on any triangle side may be fit by other than interpolating polynomials. If we demand that the polynomial fit pass through the vertices and be a "best" approximation to data along the sides in some consistent sense, the entire theory applies. For example, ten intermediate side points could be fit by a quadratic in a least square sense. IN EFFECT, DATA ALONG EACH SIDE IS FIT BY A POLYNOMIAL WHICH IS INDEPENDENT OF THE SURROUNDING CONFIGURATION. THE BASIS FUNCTIONS CONTINUE THIS FIT INTO ITS NEIGHBOURING REGIONS.

Part II : Projective Quadrature

1. Projective Coordinates

We recall a few concepts of projective geometry.

"If four points in a plane are joined in pairs by six distinct lines, they are called the vertices of a complete quadrangle, and the lines are its six sides. Two sides are said to be opposite if they have no common vertex. Any point of intersection of two opposite sides is called a diagonal point." [Coxeter(1,P.19)]

A description of projective coordinates may be found on Pp. 234-237 of Coxeter. A key statement therein which indicates the power of projective coordinates for our problem is: "Just as in affine geometry, all triangles are alike, so in projective geometry all quadrangles are alike."

To obtain a system of projective coordinates, we select four points no three of which are collinear. Three of these points are chosen as vertices of the "triangle of reference" with barycentric coordinates (1,0,0) (0,1,0) and (0,0,1). The projective coordinates of these points are the same three triplets. The fourth point has barycentric coordinates $(\lambda_4, \mu_4, \nu_4)$ and projective coordinates (1,1,1).

Thus, the coordinate relationship is:

$$(\lambda_4\zeta,\ \mu_4\xi,\ \nu_4\eta) \quad = \quad (\lambda,\ \mu,\ \nu)$$

where (λ,μ,ν) is the barycentric triplet and (ζ,ξ,η) is the projective triplet. These are homogeneous coordinates: $(\zeta,\xi,\eta) = (a\zeta,\ a\xi,\ a\eta)$ for any $a \neq 0$. By exercise 2 on p,237 of Coxeter: "The four points $(\zeta,\xi,\eta) = (1,\pm1,\pm1)$ form a complete quadrangle whose diagonal triangle is the triangle of reference." We choose this system of projective coordinates:

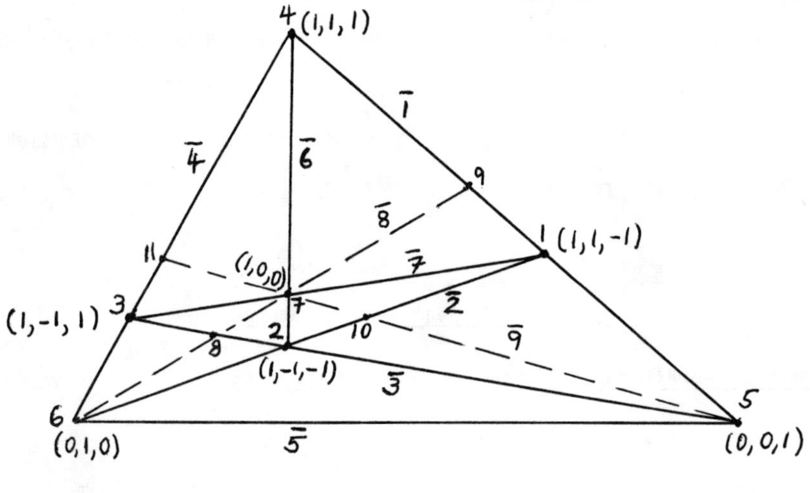

<u>Figure II.1</u> Projective Coordinate System for the Quadrangle $(\zeta,\ \xi,\ \eta)$

In barycentric coordinates the equation of the line through points $(\lambda_1, \mu_1, \nu_1)$ and $(\lambda_2, \mu_2, \nu_2)$ is

$$\det \begin{bmatrix} \lambda & \mu & \nu \\ \lambda_1 & \mu_1 & \nu_1 \\ \lambda_2 & \mu_2 & \nu_2 \end{bmatrix} = 0$$

and the same is true in projective coordinates with (ζ,ξ,η) replacing (λ,μ,ν).

The line $\bar{s} = \{(\zeta,\xi,\eta) | a_s\zeta + b_s\xi + c_s\eta = 0\}$ is denoted by $[a_s,b_s,c_s]$. The nine lines drawn in the quadrangle are:

$$\bar{1} = [1,-1,0] \quad \bar{2} = [1,0,1] \quad \bar{3} = [1,1,0] \quad \bar{4} = [1,0,-1] \quad \bar{5} = [1,0,0]$$

$$\bar{6} = [0,-1,1] \quad \bar{7} = [0,1,1] \quad \bar{8} = [0,0,1] \quad \bar{9} = [0,1,0] \ .$$

The <u>principle of duality</u> [P.231 of Coxeter] asserts that all theorems remain true after a consistent interchange of the words "point" and "line". Thus, the point of intersection of lines $\bar{s}$ and $\bar{p}$ has as its coordinates the coefficients of e_ζ, e_ξ and e_η in

$$k.\det \begin{bmatrix} e_\zeta & e_\xi & e_\eta \\ a_s & b_s & c_s \\ a_p & b_p & c_p \end{bmatrix} .$$

We note that $\zeta = 0$ on $\bar{5}$. Off line $\bar{5}$, we choose k to **yield** $\zeta = 1$. Thus, for point 9 at $\bar{1}.\bar{8}$:

$$\det \begin{bmatrix} e_\zeta & e_\xi & e_\eta \\ 1 & -1 & 0 \\ 0 & 0 & 1 \end{bmatrix} \begin{matrix} \\ \leftarrow \bar{1} \\ \leftarrow \bar{8} \end{matrix} \qquad = (-e_\zeta - e_\xi)$$

and we let k = -1 so that:

$$(\zeta,\xi,\eta)_9 = (1,1,0) \ .$$

Similarly $(\zeta\ \xi\ \eta)_8 = (1,-1,0)$, $(\zeta,\xi,\eta)_{10} = (1,0,-1)$, $(\zeta\ \xi\ \eta)_{11} = (1,0,1)$.

Normalization to $\zeta = 1$ off line $\bar{5}$ provides a coordinate couplet (ξ, η) which may be compared with the "isoparametric" coordinates developed and used by Irons [3], Ergatoudis [2], Zienkiewicz [4] and others in finite element studies.

We shall prove the following assertions:

(1) The transformations between (x,y) and (ξ,η) coordinates are of a simple form in both directions.

(2) The rational wedges are easily expressed as simple functions of ξ and η.

(3) All integrals of products of wedges and their derivatives (of any order) over the quadrilateral may be evaluated in closed form.

Thus, some of the deficiencies of the isoparametric coordinates are eliminated, thereby yielding a formalism for general quadrilaterals of simplicity comparable to that of triangles.

We note first that

$$\iint_q dx\, dy\, f(x,y) = \iint_q d\mu\, d\nu\, f(\mu,\nu).\left|\det J\left(\frac{x,y}{\mu,\nu}\right)\right|$$

$$= \iint_q d\xi\, d\eta\, f(\xi,\eta).\left|\det J\left(\frac{x,y}{\mu,\nu}\right).\ \det J\left(\frac{\mu,\nu}{\xi,\eta}\right)\right|. \tag{9}$$

The barycentric Jacobian is a constant and is twice the area of the reference triangle:

$$J_1 \equiv \left|J\left(\frac{x,y}{\mu,\nu}\right)\right| = \left|\det\begin{bmatrix}1 & 1 & 1\\ x_5 & x_6 & x_7\\ y_5 & y_6 & y_7\end{bmatrix}\right| . \tag{10}$$

The transformation matrix relating (ζ,ξ,η) to (λ,μ,ν) is the matrix M which is determined as follows*:

$$M\begin{bmatrix}\zeta_0 & 0 & 0 & \tau\\ 0 & \xi_0 & 0 & \tau\\ 0 & 0 & \eta_0 & \tau\end{bmatrix} = \begin{bmatrix}1 & 0 & 0 & \lambda_4\\ 0 & 1 & 0 & \mu_4\\ 0 & 0 & 1 & \nu_4\end{bmatrix} \qquad M = \begin{bmatrix}1/\zeta_0 & & \\ & 1/\xi_0 & \\ & & 1/\eta_0\end{bmatrix}$$

$$\begin{aligned}\tau/\zeta_0 &= \lambda_4 \quad \text{or} \quad \frac{1}{\zeta_0} = \frac{\lambda_4}{\tau}\\ \tau/\xi_0 &= \mu_4 \quad \text{or} \quad \frac{1}{\xi_0} = \frac{\mu_4}{\tau}\\ \tau/\eta_0 &= \nu_4 \quad \text{or} \quad \frac{1}{\eta_0} = \frac{\nu_4}{\tau}\end{aligned} \; . \tag{11}$$

---

* I am indebted to Professor Edge of the University of Edinburgh for describing this technique to me.

Hence

$$M = \frac{1}{\tau} \cdot \begin{bmatrix} \lambda_4 & & \\ & \mu_4 & \\ & & \nu_4 \end{bmatrix} . \tag{12}$$

The triangle barycentric coordinates are normalized to $\lambda+\mu+\nu = 1$ everywhere. It follows that

$$\frac{1}{\tau}(\lambda_4\zeta + \mu_4\xi + \nu_4\eta) = 1 \quad \text{or} \quad \frac{1}{\tau} = \frac{1}{\lambda_4\zeta + \mu_4\xi + \nu_4\eta} .$$

For all points where $\zeta \neq 0$ we may normalize the projective coordinates to $\zeta = 1$, thereby obtaining:

$$M \begin{bmatrix} 1 \\ \xi \\ \eta \end{bmatrix} = \begin{bmatrix} \lambda \\ \mu \\ \nu \end{bmatrix} \quad \text{where } M = \frac{1}{\lambda_4 + \mu_4\xi + \nu_4\eta} \begin{bmatrix} \lambda_4 & & \\ & \mu_4 & \\ & & \nu_4 \end{bmatrix} . \tag{13}$$

Let $$Q(\xi,\eta) \equiv \lambda_4 + \mu_4\xi + \nu_4\eta . \tag{14}$$

Then $$\left|\det J\left(\frac{\mu,\nu}{\xi,\eta}\right)\right| = |\det M| = \left|\frac{\lambda_4\mu_4\nu_4}{Q^3}\right| . \tag{15}$$

Our projective coordinates are such that $Q > 1$ in the quadrilateral and $\lambda_4\mu_4\nu_4 > 0$ ($\lambda_4 > 1$, $\mu_4 < 0$ and $\nu_4 < 0$). The absolute signs are not required in q. We have:

$$J_2 = |\det M| = \frac{\lambda_4\mu_4\nu_4}{Q^3(\xi,\eta)} . \tag{16}$$

The wedge functions include factors of $\underline{1}$, $\underline{2}$, $\underline{3}$, $\underline{4}$ and $\underline{5}$. Expressed in (λ,μ,ν) coordinates, these are:

$$\underline{1} = k_1(\mu_4\lambda - \lambda_4\mu) \quad \underline{2} = k_2(\nu_4\lambda + \lambda_4\nu) \quad \underline{3} = k_3(\mu_4\lambda + \lambda_4\mu) \quad \underline{4} = k_4(\nu_4\lambda - \lambda_4\nu)$$

$$\underline{5} = k_5\lambda \text{ (the k's are normalizing factors.)} \tag{17}$$

The transformation to (ξ,η) is

$$\lambda = \frac{\lambda_4}{Q} \quad \mu = \frac{\mu_4\xi}{Q} \quad \nu = \frac{\nu_4\eta}{Q} . \tag{18}$$

Hence, in (ξ,η) coordinates:

$$\underline{1} = \frac{k_1}{Q}(\mu_4\lambda_4 - \lambda_4\mu_4\xi) = k_1\mu_4\lambda_4 \frac{(1-\xi)}{Q} \qquad \underline{2} = \frac{k_2}{Q}(\nu_4\lambda_4 + \lambda_4\nu_4\eta) = k_2\nu_4\lambda_4 \frac{(1+\eta)}{Q}$$

$$\underline{3} = \frac{k_3}{Q}(\mu_4\lambda_4 + \lambda_4\mu_4\xi) = k_3\mu_4\lambda_4 \frac{(1+\xi)}{Q} \qquad \underline{4} = \frac{k_4}{Q}(\nu_4\lambda_4 - \lambda_4\nu_4\eta) = k_4\nu_4\lambda_4 \frac{(1-\eta)}{Q}$$

$$\underline{5} = \frac{k_5\lambda_4}{Q} \quad . \tag{19}$$

We may normalize directly in (ξ,η) coordinates:

$$W_k(\xi,\eta) = \frac{Q(\xi_k,\eta_k)}{4} \; \frac{(1+\eta_k\eta)(1+\xi_k\xi)}{Q(\xi,\eta)} \quad ; \quad k = 1,2,3,4. \tag{20}$$

We observe that any $u(\xi,\eta) = \sum_{k=1}^{4} a_k W_k(\xi,\eta)$ satisfies $\nabla^2(Q(\xi,\eta) . u(\xi,\eta)) = 0$ in the quadrilateral. Therefore, amongst all continuous $\theta(\xi,\eta)$ satisfying the prescribed linear conditions on the quadrilateral boundary, the wedge approximation minimizes the integral

$$I[\theta] = \int_{-1}^{1}\int_{-1}^{1} d\xi \, d\eta \, |\nabla_{\xi,\eta} (Q(\xi,\eta).\theta(\xi,\eta))|^2 \; . \tag{21}$$

In fact, either $|\frac{\partial}{\partial\xi}(Q\theta)|^2$ or $|\frac{\partial}{\partial\eta}(Q\theta)|^2$ may be used as an integrand here instead of $|\nabla(Q\theta)|^2$. This family of Lagrangians may be transformed back into (x,y) coordinates to give "smoothness norms" for quadrilateral approximation.

Alternatively, the existence of such a family of norms having been established, we may determine the Lagrangians directly in (x,y) coordinates. This has been done. For the quadrilateral oriented with $\overline{5}$ as the line $\theta = 0$ and origin at point 6 (see Figure II.1), one such norm in polar coordinates is

$$I[\psi] = \left\{ \iint_q \frac{\psi_r^2 \, dr \, d\theta}{(\sin 2\theta)^{3/2}} \right\}^{\frac{1}{2}} \; , \tag{22}$$

2. Quadrature over Quadrilaterals

In general, products of wedges and their derivatives integrated over the quadrilateral will require evaluation of integrals of the form

$$\int_{-1}^{1}\int_{-1}^{1} d\xi\, d\eta\, \frac{P_1(\xi)\, P_2(\eta)}{[Q(\xi,\eta)]^n} \tag{23}$$

where P_1 and P_2 are polynomials in ξ and η respectively and where n is a positive integer. Such integrals are easily evaluated.

The following recursion formula, obtained from integration by parts, is quite useful:

$$f(0,n-m) \equiv \int_{-1}^{1} d\xi\, \frac{1}{(\lambda+\mu\xi+\nu\eta)^{n-m}} = \frac{1}{\mu(1+m-n)}\left[\frac{1}{(\lambda+\mu+\nu\eta)^{n-m-1}} - \frac{1}{(\lambda-\mu+\nu\eta)^{n-m-1}}\right] \qquad (n-m) > 1$$

$$= \frac{1}{\mu}\ln\frac{\lambda+\mu+\nu\eta}{\lambda-\mu+\nu\eta} \qquad n-m = 1 \tag{24}$$

$$f(m,n) \equiv \int_{-1}^{1} d\xi\, \frac{(1+\xi)^m}{(\lambda+\mu\xi+\nu\eta)^n} = \frac{2^m}{(1-n)\mu(\lambda+\mu+\nu\eta)^{n-1}} + \frac{m}{(n-1)\mu}\, f(m-1,n-1). \qquad n > m > 0. \tag{25}$$

Henceforth, we shall use the symbol (ξ,η) to represent the linear form $(\lambda_4+\mu_4\xi+\nu_4\eta)$. Thus:

$$f(0,n-m) = \frac{1}{\mu(1+m-n)}\left[(1,\eta)^{1+m-n} - (-1,\eta)^{1+m-n}\right]; \quad (n-m) > 1$$

$$= \frac{1}{\mu}\ln\frac{(1,\eta)}{(-1,\eta)}\ ; \quad (n-m) = 1 \tag{26}$$

$$f(m,n) = \frac{2^m}{\mu(1-n)}(1,\eta)^{1-n} + \frac{m}{\mu(n-1)}\, f(m-1,n-1); \quad n > m > 0\ .$$

The integrals of the rational wedges over their quadrilaterals are of the form:

$$I_k = \iint W_k(x,y)\, dx\, dy = C(\xi_k, \eta_k) \int_{-1}^{1}\int_{-1}^{1} d\xi\, d\eta\, \frac{(1+\eta_k\eta)(1+\xi_k\xi)}{(\xi,\eta)^4} \tag{27}$$

evaluated at $(\lambda,\mu,\nu) = (\lambda_4,\mu_4,\nu_4)$.

We need only evaluate this integral for k = 4, in view of the relationship:

$$\int_{-1}^{1}\int_{-1}^{1} d\xi\, d\eta\, \frac{(1+\eta_k\eta)(1+\xi_k\xi)}{(\xi,\eta)^4} = \int_{-1}^{1}\int_{-1}^{1} d\xi\, d\eta\, \frac{(1+\eta)(1+\xi)}{(\xi_k\xi,\eta_k\eta)^4} \quad . \tag{28}$$

At vertex 4, $\eta_k = \xi_k = 1$. We obtain the other integrals by substituting $\xi_k\mu_4$ for μ and $\eta_k\nu_4$ for ν. We must not replace $\lambda_4 + \mu_4 + \nu_4$ by unity until after this substitution.

The recursion formulas yield

$$I_k = C\ \{(\xi_k,\eta_k)\ \ln \frac{(-\xi_k,\eta_k)(\xi_k,-\eta_k)}{(\xi_k,\eta_k)(-\xi_k,-\eta_k)} - 4\xi_k\eta_k\mu_4\nu_4\ [\frac{1}{(-\xi_k,\eta_k)} + \frac{1}{(\xi_k,-\eta_k)} - \frac{1}{(\xi_k,\eta_k)}]\}.$$

Summing over k, we find that the log terms drop out:

$$(1,1) + (-1,-1) - (1,-1) - (-1,1) = 1 + 1 - 2\ (\mu_4+\nu_4) - (1-2\nu_4) - (1-2\mu_4) = 0.$$

We have:

$$K_q = \sum_1^4 I_k = 12\ C\ \mu_4\nu_4\ [\ \frac{1}{(1,1)} + \frac{1}{(-1,-1)} - \frac{1}{(-1,1)} - \frac{1}{(1,-1)}]\ ,$$

and solving for C:

$$C = \frac{K_q}{12\ \mu_4\nu_4\ [\ \frac{1}{(1,1)} + \frac{1}{(-1,-1)} - \frac{1}{(-1,1)} - \frac{1}{(1,-1)}]} \quad .$$

We find that

$$K_q = J_1\ \lambda_4\mu_4\nu_4 \int_{-1}^{1}\int_{-1}^{1} d\xi\, d\eta/Q^3 = \lambda_4 K_{567}[\frac{1}{(1,1)} + \frac{1}{(-1,-1)} - \frac{1}{(-1,1)} - \frac{1}{(1,-1)}]$$

$$\text{at}\ \ (\lambda,\mu,\nu) = (\lambda_4,\mu_4,\nu_4) \quad . \tag{29}$$

Thus:

$$I_k = \frac{\lambda_4 K_{567}}{12\mu_4\nu_4} \{(\xi_k,\eta_k) \ln \frac{(-\xi_k,\eta_k)(\xi_k,-\eta_k)}{(\xi_k,\eta_k)(-\xi_k,-\eta_k)} + 4\xi_k\eta_k\mu_4\nu_4[\frac{1}{(\xi_k,\eta_k)} - \frac{1}{(-\xi_k,\eta_k)} - \frac{1}{(\xi_k,-\eta_k)}]\}$$

$$\text{evaluated at } (\lambda,\mu,\nu) = (\lambda_4,\mu_4,\nu_4). \qquad (30)$$

One scheme for approximation over a quadrilateral is the "triangle-average" procedure [4, P.65]. Here the quadrilateral is subdivided into two triangles, first by splitting along one diagonal and then along the other diagonal. The average of these two piecewise planar triangulations is denoted as the "TA" approximation.

We may compare the TA method with a bilinear (B) over the unit square for a few common integrals. Let

$$\iint_q f\, dx\, dy = \sum_{\substack{i,j=1 \\ j\geq i}}^{4} a_{ij}\, u_i\, u_j + \sum_{i=1}^{4} b_i\, u_i \qquad (31)$$

where the vertex values are the $u_{i,j}$. The following values are obtained quite easily:

f	b_i	$a_{i\,i}$	$a_{i\,i+1}$	$a_{i\,i+2}$	$a_{i\,i+3}$	Method
u	$\frac{1}{4}$	0	0	0	0	TA,B
u^2	0	$\frac{1}{8}$	$\frac{1}{12}$	$\frac{1}{12}$	$\frac{1}{12}$	TA
	0	$\frac{1}{9}$	$\frac{1}{9}$	$\frac{1}{18}$	$\frac{1}{9}$	B
$\lvert\nabla u\rvert^2$	0	1	-1	0	-1	TA
	0	$\frac{2}{3}$	$-\frac{1}{3}$	$-\frac{2}{3}$	$-\frac{1}{3}$	B

It is thus seen that the TA approximation agrees with the bilinear only for the integral of the function. It is not surprising that the "creased" TA approximation differs most from the continuously differentiable bilinear approximation when f involves derivatives.

We note, that the entries for $|\nabla u|^2$ yield the difference approximations to the Laplacian operator:

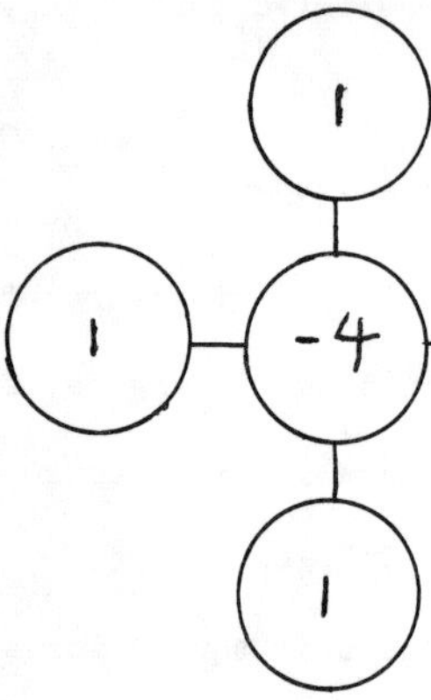

TA

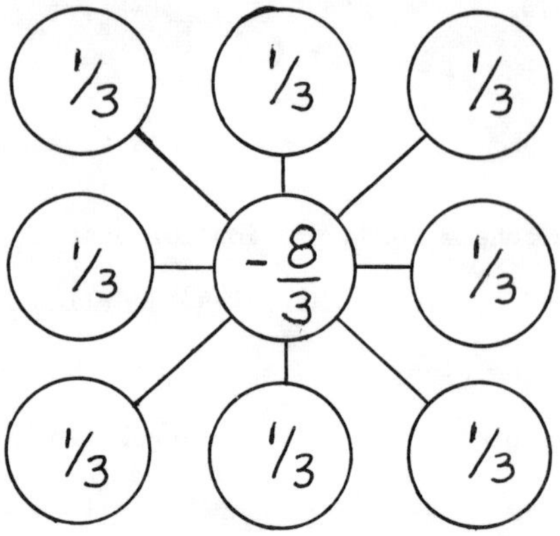

B

We now investigate $\iint_q u \, dx \, dy$ for some more general quadrilaterals to ascertain differences between the TA and wedge quadrature formulas. The table already indicates significant differences for the u^2 and $|\nabla u|^2$ integrals, and we shall not examine these further at this time.

<u>The trapezoid</u> is a special case for which considerable simplification is possible. Referring to the figures, let $K_{124} = K_{134} = B$ and $K_{234} = K_{123} = A$. (triangle areas)

Then $K_q = A + B$.

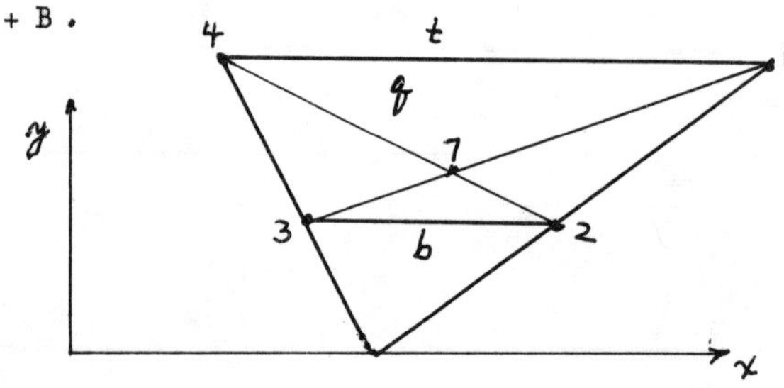

The triangle integration approximation is

$$\iint_T u \, dx \, dy = \frac{K_T}{3} \, (u_a + u_b + u_c)$$

where u_a, u_b and u_c are the vertex values.

Thus, the TA approximation is:

$$\iint_q u \, dx \, dy \underset{(TA)}{=} \frac{K_q}{6} \left[\left(1 + \frac{B}{K_q}\right)(u_1 + u_4) + \left(1 + \frac{A}{K_q}\right)(u_2 + u_3) \right]. \tag{32}$$

We note that

$$\lambda_4 = \frac{y_4}{y_7} \quad . \tag{33}$$

It can be proved that

$$\frac{A}{K_q} = \frac{1}{2\lambda_4} \quad \text{and} \quad \frac{B}{K_q} = 1 - \frac{1}{2\lambda_4} \quad . \tag{34}$$

Hence,

$$\iint_q u \, dx \, dy \underset{(TA)}{=} \frac{K_q}{12} \left[\left(4 - \frac{1}{\lambda_4}\right)(u_1 + u_4) + \left(2 + \frac{1}{\lambda_4}\right)(u_2 + u_3) \right] \quad . \tag{35}$$

We now compute the rational wedge approximation for this integral.

The trapezoid may be treated as a special case by a limiting process. We have:

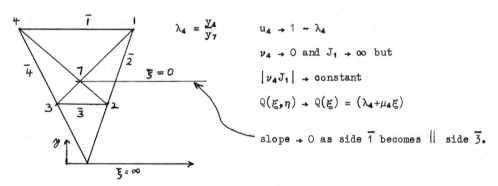

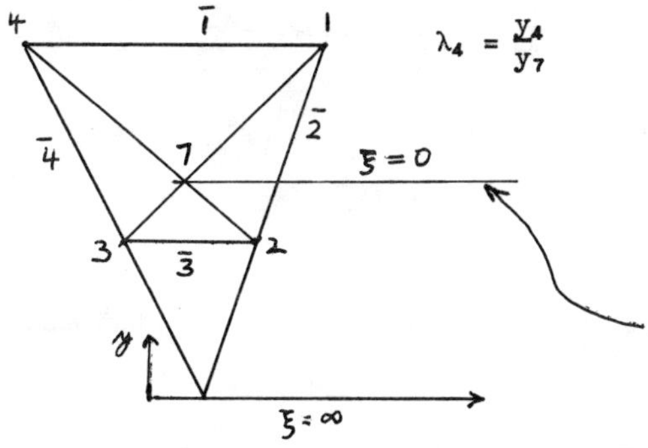

$\lambda_4 = \frac{y_4}{y_7}$

$u_4 \to 1 - \lambda_4$

$\nu_4 \to 0$ and $J_1 \to \infty$ but

$|\nu_4 J_1| \to$ constant

$Q(\xi,\eta) \to Q(\xi) = (\lambda_4 + \mu_4 \xi)$

slope $\to 0$ as side $\bar{1}$ becomes $\|$ side $\bar{3}$.

We define

$$V(\mu) \equiv \frac{3}{2} \int_{-1}^{1} \frac{1+\xi}{(\lambda_4 + \mu\xi)^4} \, d\xi = \frac{(3\lambda_4 - \mu)}{(\lambda_4 - \mu)^2 (\lambda_4 + \mu)^3} \quad . \tag{36}$$

The wedge integrals are:

$$I_k = C \, Q(\xi_k) \, V(\xi_k \mu_4) \, , \quad k = 1,2,3,4 \quad . \tag{37}$$

The constant C is to be determined. We have

$$Q(1) = 1 \quad \text{and} \quad Q(-1) = 2\lambda_4 - 1.$$

Therefore,

$$I_1 = I_4 = C\,\frac{4\lambda_4-1}{(2\lambda_4-1)^2}$$

and

$$I_2 = I_3 = C\,\frac{2\lambda_4+1}{(2\lambda_4-1)^2} \quad .$$

The value of C is determined from $\sum_{k=1}^{4} I_k$ = area of quadrilateral $\equiv K_q$:

$$C = \frac{(2\lambda_4-1)^2 K_q}{12\lambda_4} \quad .$$

Thus,

$$I_4 = I_1 = \frac{K_q}{12\lambda_4}\,(4\lambda_4-1) = \frac{K_q}{12}\,(4-\frac{1}{\lambda_4})$$

and (38)

$$I_3 = I_2 = \frac{K_q}{12}\,(2+\frac{1}{\lambda_4}) \; .$$

This analysis provides a "new" quadrature formula over a trapezoid in terms of vertex values:

$$\iint_q u(x,y)\,dx\,dy = [(u_1 + u_4)(4-\frac{1}{\lambda_4}) + (u_2 + u_3)(2+\frac{1}{\lambda_4})]\,\frac{K_q}{12} \quad . \qquad (39)$$

This is precisely the result obtained previously for the average of the two triangulation formulas!

We now consider a "kite" chosen to facilitate hand computation, but which illustrates the salient features of wedge quadrature. Triangle 456 is equilateral with sides of length $4\sqrt{3}$. Points 1 and 3 are at midpoints of sides $\overline{46}$ and $\overline{45}$, respectively:

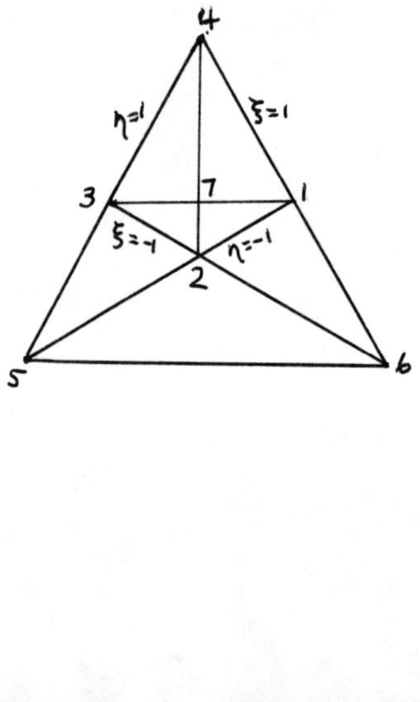

We have:

$$\lambda_4 = \frac{y_4}{y_7} = 2 \quad \mu_4 = \nu_4 = -\tfrac{1}{2} \quad K_q = 4\sqrt{3}$$

$$(\xi,\eta) = [2 - \tfrac{1}{2}(\xi,\eta)] \Rightarrow (1,1) = 1 \quad (-1,-1) = 3 \quad (1,-1) = (-1,1) = 2$$

$$(\lambda\ \mu\ \nu) = (\lambda_4,\mu_4,\nu_4) \ .$$

We define

$$I(\mu,\nu) = (1,1)\ \ln \frac{(1,-1)(-1,1)}{(1,1)(-1,-1)} - 4\mu\nu \left[\frac{1}{(-1,1)} + \frac{1}{(1,-1)} - \frac{1}{(1,1)}\right] .$$

Then

$$I_k = C\ I(\xi_k\mu_4,\ \eta_k\nu_4)$$

and

$$I_4 = C \ln \frac{4}{3} \qquad\qquad I_3 = 2C(\ln \frac{3}{4} + \frac{5}{12})$$

$$I_2 = 3C(\ln \frac{4}{3} - \frac{2}{9}) \qquad\qquad I_1 = 2C(\ln \frac{3}{4} + \frac{5}{12}) \ .$$

The normalization is

$$\sum_{k=1}^{4} I_k = C\ (\frac{5}{6} - \frac{2}{3} + \frac{5}{6}) = C = K_q \ .$$

Hence:

$$I_1 = I_3 = 2K_q(\frac{5}{12} - \ln \frac{4}{3})$$

$$I_2 = 3K_q(\ln \frac{4}{3} - \frac{2}{9})$$

$$I_4 = K_q \ln \frac{4}{3} \ .$$

Substituting 0.287 for $\ln \frac{4}{3}$:

$$\iint_q u\ dx\ dy \underset{\substack{\text{wedge}\\ \text{approx.}}}{=} K_q[.26(u_1 + u_3) + .194u_2 + .286u_4] \ . \qquad (40)$$

The TA approximation for this quadrilateral is

$$\iint_q u\ dx\ dy = K_q[\frac{1}{4}\ (u_1 + u_3) + \frac{5}{24}\ u_2 + \frac{7}{24}\ u_4]$$
$$= K_q[.25(u_1 + u_3) + .208u_2 + .292u_4]. \qquad (41)$$

The proximity of these two formulas is quite remarkable. This analysis provides theoretical support for the TA approximation to $\iint_q u \, dx \, dy$ over a quadrilateral.

A Final Remark

This analysis provides the foundation for practical use of rational wedges over convex quadrilaterals. However, it is not claimed or even implied at any point that quadrilateral elements are superior to rectangles or triangles. One should not deviate from these simple oft-used elements unless specific problem characteristics demand the more general shapes. The following "ϵ-Lyra" is an axiom, not to be taken as lightly as the phraseology might indicate:

It is a fourgon conclusion that the rectangle is a paragon of virtue.

REFERENCES

1. COXETER, H.S.M. Introduction to Geometry (Wiley, (1961)).
2. ERGATOUDIS, J. "Quadrilateral Elements in Plane Analysis" M.Sc. Thesis, University of Wales, Swansea (1966).
3. IRONS, B.M. "Numerical Integration Applied to Finite Element Methods" Conference on use of Digital Computers in Structural Engineering, University of Newcastle, (July 1966).
4. ZIENKIEWICZ, O.C. and CHEUNG, Y.K. Finite Element Methods in Structural and Continuum Mechanics (McGraw Hill, 1967).
5. ZLÁMAL, M. "A Finite Element Procedure of the Second Order of Accuracy" Numer. Math. 14 394-402 (1970).

SOME RESULTS ON BEST POSSIBLE ERROR BOUNDS FOR FINITE ELEMENT METHODS AND APPROXIMATION WITH PIECEWISE POLYNOMIAL FUNCTIONS*

O. B. Widlund

1. Introduction. It is the purpose of this paper to discuss certain questions concerning the best possible error bounds for the finite element method to solve elliptic equations and related problems in the theory of approximation and interpolation with piecewise polynomial functions. Our attention will be restricted to equations with sufficiently smooth periodic coefficients and with the simplest possible boundary condition, namely the one defined by a periodicity assumption. We further restrict the finite element methods to use subspaces of functions generated in a very regular way from one or a few compactly supported basis functions by translation on a uniform rectangular lattice of points and a scaling of the independent variables. The use of such subspaces allows for the construction of arbitrarily accurate methods but many methods of practical interest, many of which allow a quite general triangularization of the region considered, are not included in this class.

The restrictions on the boundary etc. are of course quite severe. They do however allow the use of Fourier techniques to a great advantage. This has been very clearly demonstrated by Fix and Strang [7,18,19]. An entire framework of notations and ideas will in fact be borrowed from these papers and in section 3 we will survey some of their results and compare them with those derived by the more conventional variational approach. It is hoped that a detailed knowledge of these rather special problems will provide insight into problems with more complicated boundaries and boundary conditions; (i.e. these results should play a role somewhat similar to that of the theorems on interior regularity in the development of the theory for elliptic differential equations).

Four main results are presented in this paper. In section 3, we will relate the order of approximation that may be achieved for smooth functions, to the degree of the polynomials which can be represented exactly by elements in the subspace. A best possible bound is also given for Poisson-type problems with sufficiently smooth solutions.

---

* This work was supported by the U. S. Atomic Energy Commission, Contract AT(30-1)-1480, at the Courant Institute of Mathematical Sciences, New York University.

Both these results are due to Strang and Fix. In section 4 we present a result by Ole Hald and the author [23] which gives precise error bounds for the eigenfunctions calculated by the Rayleigh-Ritz procedure. Here we note an interesting difference between problems with constant and variable coefficients. Finally in section 5, we address ourselves to the problem of determining the extent to which the rate of convergence suffers, when we approximate or interpolate functions which are not smooth enough to allow the use of error bounds such as those given in section 3. We state precise upper and lower bounds for the error deriving a result quite similar to those of the classical Jackson-Bernstein-Zygmund theory for best approximation by polynomials and trigonometric polynomials (see e.g. Shapiro [17] or Timan [20]). The proof of this result as well as those on the eigenfunctions will appear elsewhere [22,23].

2. Preliminaries. In this section we will introduce our notation and briefly review some well known results on the Rayleigh-Ritz-Galerkin method for elliptic problems.

We will consider elliptic problems of the form

$$(2.1) \qquad Lu(x) \equiv \sum_{|\alpha|,|\beta|\leq m} (-1)^{|\beta|}\, \partial_x^\beta q_{\alpha\beta}(x)\, \partial_x^\alpha u(x) = f(x) .$$

We use the standard multi-index notation $\alpha = (\alpha_1,\dots,\alpha_n)$, $|\alpha| = \sum \alpha_i$ and $\partial_x^\alpha u = \partial_{x_1}^{\alpha_1}\dots\partial_{x_n}^{\alpha_n} u$. The order of the equation is $2m$, since we require for some α, β with $|\alpha|+|\beta| = 2m$, $q_{\alpha\beta}(x) \neq 0$. For convenience, we restrict ourselves to one dependent variable, assume symmetry, i.e., $q_{\alpha\beta}(x) = \bar{q}_{\beta\alpha}(x)$ and require that all the coefficients $q_{\alpha\beta}(x)$ as well as $f(x)$ are sufficiently smooth functions and periodic with period one in each of the variables x_i. The equation (2.1) is written in so called operational form. In order to present its variational form we introduce the bilinear form associated with L,

$$a(u,v) = \sum_{|\alpha|,|\beta|\leq m} (q_{\alpha\beta}\partial_x^\alpha u, \partial_x^\beta v)_\#$$

where the inner product is defined by

$$(u,v)_\# = \int u(x)\, \bar{v}(x)\, dx$$

with integration over $[0,1]^n$.

The fundamental ellipticity assumption takes the form

$$a(u,u) \geq c||u||^2_{W^m_{2,\#}} .$$

Throughout this paper c denotes a generic positive constant bounded away from zero and C a generic constant bounded from above. We will use the following semi-norms and norms,

$$|u|_{W_q^\ell} = \sup_{|\alpha|=\ell} |\partial_x^\alpha u|_{W_q}, \qquad q = 2 \text{ or } \infty,$$

$$||u||_{W_q^m} = \sum_{\ell=0}^{m} |u|_{W_q^\ell},$$

where W_2 is the space of L_2-functions, with its usual norm, and W_∞ is the space of continuous functions which tend to zero when $|x| \to \infty$. The corresponding spaces of one-periodic functions will be denoted by $W_{2,\#}^m$ and $W_{\infty,\#}^m$.

The variational form of (2.1) is given by

$$\min_{u \in W_{2,\#}^m} \{a(u,u) - 2(f,u)_\#\}$$

which, as is well known, is equivalent to

(2.2) $$a(u,v) = (f,v)_\# \quad \text{for all} \quad v \in W_{2,\#}^m .$$

The corresponding eigenvalue problem has the form

$$L\phi_k = \lambda_k \phi_k .$$

The eigensystem can also be found by considering the stationary points of the Rayleigh quotient,

$$R(u) = a(u,u)/(u,u)_\# ,$$

for $u \in W_{2,\#}^m$.

The Ritz-Galerkin idea amounts to looking for approximate solutions $u^h(x)$ of the form $\sum v_j \Omega_j^h(x)$ in finite dimensional subspaces S^h of $W_{2,\#}^m$. That is, u^h is determined by

$$\min_{u^h \in S^h} \{a(u^h,u^h) - 2(f,u^h)_\#\}$$

or equivalently

(2.3) $$a(u^h,v^h) = (f,v^h)_\# \quad \text{for all} \quad v^h \in S^h .$$

For the eigenvalue problem, we similarly consider the stationary values of $R(u^h)$, $u^h \in S^h$.

The superscript h will soon be given a natural interpretation as a mesh size. The dimension of the subspaces will increase as h^{-n}.

These finite dimensional problems, to determine the coefficients

v_j^h etc., can be formulated as a linear system of equations and a generalized eigenvalue problem respectively.

$$A^h v^h = f^h ,$$

$$A^h \psi_k^h = \lambda_k^h G^h \psi_k^h ,$$

where the elements of the matrices A^h and G^h are defined by $a(\Omega_i^h,\Omega_j^h)$ and $(\Omega_i^h,\Omega_j^h)_\#$, while the components of the vector f^h are defined by $(f,\Omega_j^h)_\#$. Observe that these matrices are symmetric. The vectors v^h and ψ_k^h have components v_j^h and $\psi_{k,j}^h$ respectively. The approximate solutions of the original problems have the form

$$u^h(x) = \sum v_j \, \Omega_j^h(x)$$

and

$$\phi_k^h(x) = \sum \psi_{k,j}^h \, \Omega_j^h(x) .$$

The finite element methods can be characterized as those Ritz-Galerkin methods that have sparse matrices, more precisely those for which the number of nonzero elements in the rows of A^h and G^h is bounded independently of h. The finite element methods thus share an important property of the more conventional finite difference schemes, in fact following Strang and Fix, the class of finite element methods can be characterized as the intersection of the class of finite difference schemes and the class of Ritz-Galerkin methods. What makes the finite element methods different from the conventional finite difference schemes, for a Poisson-type problem, is that the components of v^h do not necessarily have the interpretation of an approximate value of the solution we want to find, that the approximate solution u^h is defined for all x by $\sum v_j \Omega_j^h(x)$ and that the data, represented by the vector f^h, is evaluated in terms of certain mean values of $f(x)$ and not by taking point values.

We will almost exclusively restrict ourselves to subspaces S^h generated by one function $\Omega(x)$. Assume that $\Omega(x) \in W_2^m$ is a function with compact support and define $\Omega_j^h(x)$ by

$$\Omega_j^h(x) = h^{-n/2}\, \Omega((x-jh)/h), \qquad j \in Z^n .$$

It is well known that spline spaces can be generated in this way. In order to discuss Hermite spaces as well as the piecewise polynomial subspaces frequently used in the finite element calculations, we need to consider several basis functions $\Omega^{(1)}(x),\ldots,\Omega^{(N)}(x)$. We refer the reader to the papers of Fix and Strang for a detailed treatment of the general case. Also compare the further discussion in section 3.

This construction will not give us subspaces of one-periodic functions. A natural way to achieve periodicity is to restrict h to take values such that $1/h$ is an integer and define $\Omega^h_{j,\#}(x)$ by

$$\Omega^h_{j,\#}(x) = \sum_{k\in Z^n} \Omega^h_{j+k/h}(x) .$$

It is easily seen that there are only $(1/h)^n$ different functions of this kind. In the periodic case we will therefore restrict our summation to the subset $Z^n_\#$ of Z^n for which the components of j vary between 0 and $(1/h)-1$.

An important feature of subspaces generated in this way is that the Gram matrix G^h is of convolution type, i.e. the value of its elements are functions of i-j alone. The same is true for the matrix A^h if the differential equation has constant coefficients.

We conclude this section by presenting a standard variational argument that establishes the great importance of approximation theory in the study of the finite element methods. (For details cf. Ciarlet, Schultz and Varga [5].)

The so called Galerkin conditions (2.2) and (2.3) yield

(2.4) $a(u^h-u,w^h) = (f,w^h)_\# - (f,w^h)_\# = 0$ for all $w^h \in S^h \subset W^m_{2,\#}$.

By using successively the ellipticity assumption, the relation (2.4), the fact that $w^h-u^h \in S^h$ and the Schwarz inequality we find

$$\begin{aligned} c||u^h-u||^2_{W^m_{2,\#}} &\le a(u^h-u,u^h-u) = a(u^h-u,u^h-u) + a(u^h-u,w^h-u^h) \\ &= a(u^h-u,w^h-u) \le C||u^h-u||_{W^m_{2,\#}} ||w^h-u||_{W^m_{2,\#}} . \end{aligned}$$

Therefore

$$||u^h-u||_{W^m_{2,\#}} \le (C/c) \min_{w^h\in S^h} ||w^h-u||_{W^m_{2,\#}} . \tag{2.5}$$

The same estimate holds for much more general cases e.g. elliptic problems defined on quite general regions, a less restrictive choice of boundary conditions etc.

3. Some Results of Strang and Fix. We begin this section by considering the approximation problem in W_2 and W_∞. We first consider circumstances under which, for a given function $\tilde{\Omega}(x)$ of compact support,

$$\tilde{u}^h(x) = \sum_{j\in Z^n} u(jh)\, \tilde{\Omega}^h_j(x)$$

will approximate any sufficiently smooth function u(x) with an error of order h^{p+1} as h goes to zero. A Taylor series argument shows that the error cannot be of order h^{p+1} unless $\tilde{\Omega}(x)$ satisfies

$$(3.1) \qquad \sum_{j\in Z^n} j^{\alpha}\,\tilde{\Omega}(t-j) = t^{\alpha} \quad \text{for all} \quad |\alpha| \le p\ .$$

It has been shown by Fix and Strang that an estimate of the form

$$\min_{u^h\in S^h} ||u^h-u||_{W_q} \le C\,h^{p+1}|u|_{W_q^{p+1}}\ , \qquad q = 2 \text{ or } \infty,$$

is possible if and only if one can construct a finite linear combination of the basis function $\Omega(x)$ and translates of it (or correspondingly $\Omega^{(1)}(x),\ldots,\Omega^{(N)}(x)$) which satifies (3.1) provided we require that the vector v^h, which defines $u^h(x)$, lies in ℓ_2 for all values of h. In the light of this result the choice of piecewise polynomial functions as basis functions becomes quite natural.

It can be shown that we get as good an error bound for the corresponding interpolation problem in S^h provided that the interpolation matrix B, with elements defined by $\Omega(i-j)$, has a uniformly bounded inverse. By using Fourier series techniques one can show that this inverse is bounded if and only if the trigonometric polynomial $b(\theta) = \sum_{k\in Z^n} \Omega(k)\exp(ik\theta)$ is different from zero.

In order to use the estimate (2.5) we need an estimate for the derivatives of the error. Such an estimate is provided by

$$(3.2) \qquad ||w^h-u||_{W_q^{\ell}} \le C\,h^{p+1-\ell}|u|_{W_q^{p+1}}\ , \quad q = 2 \text{ and } \infty, \quad \ell \le p\ ,$$

where w^h is the best approximation in W_q and by corresponding results for the spaces of periodic functions. The estimate (2.5) therefore implies that for problems with sufficiently smooth solutions we get an error bound of the form

$$||u^h-u||_{W^m_{2,\#}} \le C\,h^{p+1-m}|u|_{W^{p+1}_{2,\#}}\ .$$

By using a sophisticated variational argument, cf. Nitsche [9], Blair [3], Strang and Fix [19] or Schultz [16], one can show that

$$||u^h-u||_{W^{\ell}_{2,\#}} \le C\,h^{r(p,m,\ell)}|u|_{W^{p+1}_{2,\#}}\ , \qquad \ell \le p\ ,$$

where $r(p,m,\ell) = \min(p+1-\ell, 2(p+1-m))$.

Notice that for some combinations of p, m and ℓ the exponent of h is smaller than that in formula (3.2). Two questions now arise. Can the exponent in this estimate be improved and can we get as good a bound in terms of the maximum norm?

The answers to these questions were given by Strang and Fix who showed that the exponent $r(p,m,\ell)$ is indeed the best one and that the corresponding result is true in the maximum norm.

Their main idea is to look upon the finite element methods as finite difference methods. To each choice of an origin one can define a rectangular lattice of meshpoints with mesh size h. One can study the truncation error of the system of difference equations which is defined by the finite element method and the chosen lattice. This is in particular convenient when the differential equation has constant coefficients. The Fourier technique then allows one to calculate the error and to estimate it quite accurately. A second component of the error is due to the evaluation of $u^h(x)$, for all values of x, by what amounts to an interpolation procedure. This error component is of the order h^{p+1}, while the component due to the truncation error is of order $h^{2(p+1-m)}$.

Finally, we mention that Fix and Strang have considered the problem of numerical stability. The condition number of A^h, of importance in the implementation of the finite element method, is determined essentially by the mesh size, the order of the equation and the condition number of the Gram matrix G^h. The condition number of G^h can be studied in a way quite similar to the study of the properties of the interpolation matrix B.

4. The Eigenvalue Problem. Certain error bounds for the eigenvalues and eigenfunctions can be derived by variational methods. For example Birkhoff, de Boor, Swartz and Wendroff [2] have shown that

$$0 \leq \lambda_k^h - \lambda_k \leq C_k h^{2(p+1-m)} ,$$

$$||\phi_k^h - \phi_k||_{W_{2,\#}^m} \leq C_k h^{p+1-m} .$$

We are particularly interested in the possibility of improvement of the bound of the error in the calculation of the eigenfunctions. In their manuscript [19], Strang and Fix used Fourier techniques to show that in the constant coefficient case the error is no worse than the error of best approximation, i.e.

$$||\phi_k^h - \phi_k||_{W_{q,\#}} \leq C_k h^{p+1} , \qquad q = 2 \text{ and } \infty.$$

Fix [6], Pierce and Varga [12] and Schultz [16] have shown, by other methods that the error is of order h^r, $r = \min(p+1, 2(p+1-m))$ for problems with variable coefficients. Most of those results hold only in W_2. The question thus arises: can the estimate for the constant coefficient case be extended to the general case? We will now outline a technique which enables us to determine the exact order of the error.

Consider the generalized eigenvalue problem

$$A^h \psi_k^h = \lambda_k^h G^h \psi_k^h .$$

We can interpret this as a finite difference equation. By expanding the difference operators A^h and G^h in asymptotic series in h, with Hermitian differential operators as coefficients, it can be verified that corresponding expansions exist for λ_k^h and ψ_k^h. The exact order of the error terms, which are due to the truncation error, can now be determined by a study of the first nonvanishing terms in these expansions. The main observation is that at least for some k there will be an error component of the order $h^{2(p+1-m)}$ in the expansion for the eigenfunctions if the leading operator in the expansion of the truncation error does not commute with the differential operator corresponding to $h = 0$. In the case of a problem with constant coefficients all the differential operators in the expansion have constant coefficients and they therefore commute. The error in the approximate eigenfunctions is then due entirely to the second error component we discussed in section 3, namely the interpolation error.

A simple example is provided by the problem

$$\partial_x^4 \phi_k + q(x)\phi_k = \lambda_k \phi_k .$$

We approximate this problem using a subspace of quadratic spline functions. A stable method can be constructed, i.e. we can generate such a space with the aid of a $\Omega(x)$ such that G^h is uniformly well conditioned. We note that in this case $p = 2$, $m = 2$ and that therefore $r(p,m) = 2$. By using the above ideas, one can show that all ϕ_k are apppoximated with an accuracy of order h^3, only if $q(x)$ is a constant.

5. Direct and Inverse Bounds for Approximation and Interpolation Errors

In this section we will discuss the development of a Jackson-Bernstein-Zygmund type theory (cf. Shapiro [17] or Timan [20]) for interpolation and approximation by elements in linear subspaces like those introduced in sections 2 and 3. The technique we use no longer restricts us to spaces with uniform mesh sizes or to spaces generated by just a few basis elements. The arguments which lead to the direct results are quite general and require very few assumptions on the methods considered while for the proof of the inverse results, we will assume, among other things, that the elements of the subspaces considered are piecewise polynomials.

There exists a large literature on error bounds for splines, generalized splines etc. For a survey cf. Birkhoff [1], Schultz [15] and Varga [21]. Many of these error bounds are similar to the estimate (3.2). In view of the results of section 3, this situation is quite natural i.e. the semi-norm on the right-hand side of formula (3.2) vanishes for all polynomials of degree p while some polynomial of degree p+1 cannot be exactly represented by elements of the subspace, if the accuracy is no higher than p+1. We also expect that the error should be represented locally as a polynomial of degree p+1, with coefficients of order h^{p+1} plus higher order terms when we approximate a smooth function $u(x)$. For a further discussion along these lines, the question of best possible constants etc. we refer to Strang [18].

Some work has also been done in order to determine the relation between the rate of convergence and the smoothness of the function $u(x)$, which we seek to approximate, when u does not belong to W_q^{p+1}. Hedstrom and Varga [8] used a technique of interpolation in Banach spaces to derive certain direct results. In an interesting paper, on which we will comment further below, Scherer [14] derived bounds for polynomial spline approximation. The author [23] has recently been able to prove certain results using techniques somewhat similar to those of Hedstrom, Varga and Scherer.

Suppose that for an interpolation or approximation method we know that

$$(5.1) \qquad ||w^h-u||_{W_q} \leq C\, h^{p+1}|u|_{W_q^{p+1}} \quad \text{for all} \quad u \in W_q^{p+1}$$

and also that

$$(5.2) \qquad ||w^h-u||_{W_q} \leq C||u||_{W_q} \quad \text{for all} \quad u \in W_q \ .$$

Then

$$(5.3) \qquad ||w^h-u||_{W_q} \leq C\, h^{\sigma}|u|_{W_q^{\sigma,p+1}} \quad \text{for all} \quad u \in W_q^{\sigma,p+1}, \quad 0 < \sigma \leq p+1.$$

The space $W_q^{\sigma,p+1}$ is the subspace of W_q for which

$$|u|_{W_q^{\sigma,p+1}} = \sup_{h>0,\,|\alpha|=p+1} (h^{-\sigma}||\Delta_{+h}^{\alpha}u||_{W_q})$$

is bounded. The difference operator Δ_{+h} is defined by

$$\Delta_{+h}\phi(x) = \phi(x+h) - \phi(x)$$

in the case of one variable. We define the difference operator Δ_{+h}^{α}, α a multi-index, in a similar way.

We remark that (5.1) is identical to (3.2) for $\ell = 0$ and that the estimate (5.2) amounts to a requirement of stability.

The space $W_q^{\sigma,p+1}$ can be identified as a Besov space for $\sigma < p+1$ and as a Lipschitz space in the case $\sigma = p+1$.

We now turn to the problem of finding inverse, Bernstein-Zygmund type bounds. The only previous complete result in this area known to the author is due to Scherer [14], who used work by Butzer and Scherer [4] to give a complete theory for the best approximation by one dimensional polynomial splines. Certain results have also been found by Nitsche [10,11].

As we will see, the estimate (5.3) is in a certain sense the best possible. It is natural to conjecture that the error is of order h^σ in W_q, only if $u \in W_q^{\sigma,p+1}$, with $\sigma \leq p+1$, and that if the error is $o(h^{p+1})$, u must be a polynomial, which together with the requirement that $u \in W_q$ means that $u(x) \equiv 0$.

This conjecture is however not quite correct without some further assumptions. Thus, by considering the quite small error which can be achieved by using many knots close to a point at which u(x) is not very smooth, cf. Rice [13], we are led to the requirement that the mesh should be quasi-uniform. This means, in one dimension, that the ratio between the largest and the smallest mesh is uniformly bounded. We also have to avoid certain other cases, where a special choice of knots leads to a very good approximation of functions that are not so smooth. Thus consider a roof function of one variable for which the direction of the tangent changes only at the points -1, 0 and 1. Such a function can obviously be represented exactly by elements in the subspaces S^h generated as in sections 2 and 3 by the same roof function provided that we restrict the value of h so that 1/h is an integer.

In order to illustrate the results of Widlund [23] we specialize to subspaces generated as in sections 2 and 3. In fact we will consider a family of subspaces S_γ^h, $0 \leq \gamma < 1$, generated by $\Omega_\gamma(x)=\Omega(x-\gamma)$ where $\Omega(x)$ is a compactly supported function built up by a finite number of picewise polynomials of degree no larger than p. The rate of convergence is measured by

$$E(h,u) = \sup_\gamma ||w_\gamma^h - u||_{W_q} ,$$

where w^h is the best approximation (or the interpolating function) of $u \in W_q$ found by using elements from the space S_γ^h. We can then conclude that E(h,u) is of order h^σ, $0 < \sigma \leq p+1$, only if u $W_q^{\sigma,p+1}$ and that if E(h,u) is $o(h^{p+1})$ then $u(x) \equiv 0$.

For more general results, the periodic case etc. cf. Widlund [23].

References

[1] Birkhoff, G. article in "Approximation with Special Emphasis on Spline Functions," edited by I. J. Schoenberg, Academic Press, New York, 1969.

[2] Birkhoff, G., de Boor, C., Swartz, B. and Wendroff, B. J. SIAM Num. Anal., 3, 188-203 (1966).

[3] Blair, J. J., Ph.D. thesis, Univ. Calif. at Berkeley, 1970.

[4] Butzer, P. L. and Scherer, K. Aequationes Mathematica, 3, 170-185 (1969).

[5] Ciarlet, P. G., Schultz, M. H. and Varga, R. S. Numer. Math., 9, 394-430 (1967).

[6] Fix, G. personal communication.

[7] Fix, G. and Strang, G., Studies in Appl. Math, 48, 265-273 (1969).

[8] Hedstrom, G. W. and Varga, R. S., to appear in J. Approx. Theory.

[9] Nitsche, J., Numer. Math. 11, 346-348 (1968).

[10] Nitsche, J. Math. Zeitschr. 109, 97-106 (1969).

[11] Nitsche, J., Compositio Mathematica 21, 400-416 (1969).

[12] Pierce, J. and Varga, R. S., to appear in J. SIAM Num. Anal.

[13] Rice, J. R., article in same volume as [1].

[14] Scherer, K. J. SIAM Num. Anal., 7, 418-423, (1970).

[15] Schultz, M. H., article in same volume as [1].

[16] Schultz, M. H., Research report, Yale, 1971, to appear.

[17] Shapiro, H. S., "Smoothing and Approximation of Functions," van Nostrand Reinhold, New York, 1969.

[18] Strang, G., article in "Numerical solution of partial Differential Equations - II," edited by B. Hubbard.

[19] Strang, G. and Fix, G., to appear.

[20] Timan, A. F., "Theory of approximation of functions of a real variable," MacMillan, New York, 1963.

[21] Varga, R. S., article in same volume as [1].

[22] Widlund, O. B., to appear.

[23] Widlund, O. B. and Hald, O. H., to appear.

Note

Gilbert Strang has informed the author, after the completion of this manuscript, that he and George Fix are about to complete the manuscript of a book "An Analysis of the Finite Element Method," which is to be published by Prentice-Hall. This manuscript contains a detailed discussion of irregular meshes and boundaries. Much of the material in reference [19] is going to appear in the proceedings of a C.I.M.E. meeting to be held during the summer of 1971.

REMOVAL OF AN INSTABILITY IN A FREE CONVECTION PROBLEM

L.S. Caretto, A.D. Gosman and D.B. Spalding

Mechanical Engineering Department
Imperial College of Science and Technology
London, S.W.7.

ABSTRACT

A new computation scheme, using pressures and velocities as the principal variables, has been developed for problems in steady fluid flow and heat transfer. An instability in this scheme, due to interlinkages between the flow and energy equations, was discovered when the solution procedure was applied to the problem of buoyancy in a stably-stratified fluid. This instability was removed by devising a special underrelaxation procedure for the temperature equation.

Further computations revealed a second instability which arises when too large a perturbation is applied to a zero-flow situation.

1. INTRODUCTION

A fluid heated from the top and cooled below is in stable mechanical equilibrium; any perturbation in the fluid should die away. This physical situation provides a good test for the stability of a numerical analysis of free convection since the ultimate result of any perturbation is known.

Our interest in this problem arose during the development of computational schemes for three-dimensional, confined steady boundary-layer flows. In such flows the stream function-vorticity system used in previous work here (Gosman et al. 1969) for two-dimensional recirculating flows cannot be used. The two methods which are under development for the three-dimensional flows - velocity/vorticity and velocity/pressure - have been described by Gosman and Spalding (1971) and Caretto, Curr, and Spalding (1971). Preliminary computations with the velocity/pressure procedure (called the SIVA procedure since it performs SImultaneous Variable Adjustments) were carried out on a variety of two- and three-dimensional problems including the stably-stratified fluid. These computations showed that there was a maximum value of the Grashof number for which a converged solution could be obtained. The purpose of this paper is to describe the nature of the instability encountered and to give the prescription for its removal.

2. THE PROBLEM CONSIDERED

A fluid is contained in a long, horizontal, square cavity of height L. The temperatures at the top and bottom are maintained at T_H and T_C (hot and cold) respectively. The temperature along the sides varies linearly with height. The walls are impermeable and have no fluid slip. The differential equations describing the problem are written with velocity, pressure, and temperature as independent variables. Thermodynamic and transport properties are taken as constant except for density variations in the buoyancy term. The equations are:

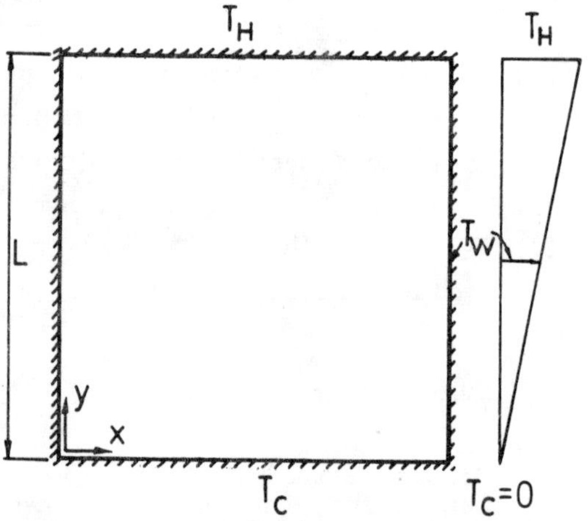

Figure 1. Geometry and Temperature Boundary Conditions.

Continuity

$$\frac{\partial u}{\partial x} + \frac{\partial v}{\partial y} = 0 \qquad (1)$$

x-momentum

$$\frac{\partial uu}{\partial x} + \frac{\partial vu}{\partial y} = -\frac{1}{\rho}\frac{\partial P}{\partial x} + \nu\left(\frac{\partial^2 u}{\partial x^2} + \frac{\partial^2 u}{\partial y^2}\right) \qquad (2)$$

y-momentum

$$\frac{\partial uv}{\partial x} + \frac{\partial vv}{\partial y} = -\frac{1}{\rho}\frac{\partial P}{\partial y} + \nu\left(\frac{\partial^2 v}{\partial x^2} + \frac{\partial^2 v}{\partial y^2}\right) + \beta g T \qquad (3)$$

Energy

$$\frac{\partial uT}{\partial x} + \frac{\partial vT}{\partial y} = \frac{\nu}{Pr}\left(\frac{\partial^2 T}{\partial x^2} + \frac{\partial^2 T}{dy^2}\right) \qquad (4)$$

with the boundary conditions

$u(x,0) = u(x,L) = u(0,y) = u(L,y) = 0$

$v(x,0) = v(x,L) = v(0,y) = v(L,y) = 0$

$T(x,0) = T_C; T(x,L) = T_H > T_C; \; T(0,y) = T(L,y) = (T_H - T_C)(y/L) + T_C$.

All symbols have their usual meaning.

The finite-difference equations for the problem are obtained by integrating the differential equations over a small control volume of side h. A displaced grid system is used so that the physical location of u is midway between P_{ij} and $P_{i+1,j}$ and v_{ij} is midway between P_{ij} and $P_{i,j+1}$ as indicated in figure 2.

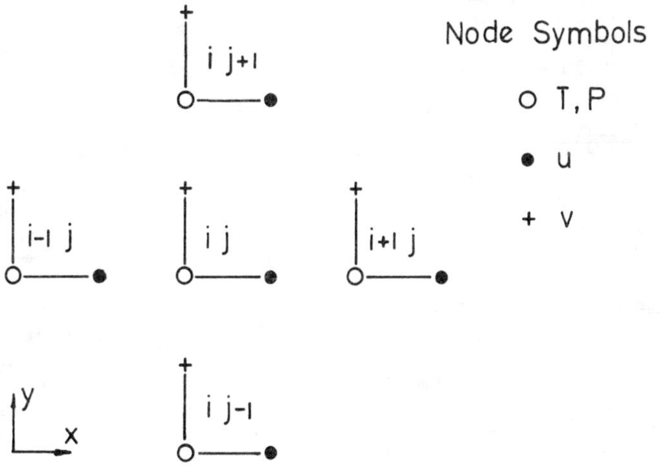

Figure 2. Displaced Grid Notation

The physical location of the temperature grid nodes is the same as the location of the pressure nodes. The finite difference equations are

$$u_{ij}-u_{i-1,j} + v_{ij}-v_{i,j-1} = 0 \qquad (6)$$

$$a_{xN}(u_{i,j+1}-u_{ij}) + a_{xE}(u_{i+1,j}-u_{ij}) + a_{xS}(u_{i,j-1}-u_{ij}) + a_{xW}(u_{i-1,j}-u_{ij}) + (1/\rho)(P_{ij}-P_{i+1,j}) = 0 \qquad (7)$$

$$a_{yN}(v_{i,j+1}-v_{ij}) + a_{yE}(v_{i+1,j}-v_{ij}) + a_{yS}(v_{i,j-1}-v_{ij}) + a_{yW}(v_{i-1,j}-v_{ij}) + (1/\rho)(P_{ij}-P_{i,j+1}) + \beta g h(T_{ij}+T_{i,j+1})/2 = 0 \qquad (8)$$

$$a_{TN}(T_{i,j+1}-T_{ij}) + a_{TE}(T_{i+1,j}-T_{ij}) + a_{TS}(T_{i,j-1}-T_{ij}) + a_{TW}(T_{i-1,j}-T_{ij}) = 0 \; . \qquad (9)$$

The form of the coefficients a_{xN} etc, which link the neighbouring nodes, gives central differences for the diffusion terms and upwind differences for the convection terms. The authors have prepared a report giving the detailed form of these coefficients.

The convection contribution to these coefficients contains the velocity components u and v. In order to linearise the difference equations these coefficients are assumed constant, with velocity values from the previous iteration, during a given iteration on one point.

The SIVA procedure is an iterative one, in which the grid is swept node by node, and the values of the variables are updated by reference to the finite-difference equations. This procedure is repeated until the changes produced are less than some pre-specified value.

The SIVA procedure solves the finite-difference equations simultaneously rather than successively. The simultaneity is obtained by algebraic reduction of the 5 difference equations which relate to the 5-point cluster of P_{ij} and the 4 neighbouring velocities v_{ij}, u_{ij}, $v_{i,j-1}$ and $u_{i-1,j}$. The equations (continuity, and two pairs of (linearised) x- and y-momentum) are combined to yield a single equation for P_{ij}, which does not contain the surrounding velocities. The value of P_{ij} found from this equation is then used to calculate new values for the neighbouring velocities. The effect of these adjustments is to leave satisfied the continuity and linearised momentum equations for the cluster.

Although, as will be shown later, it would be desirable to include the temperature equation in the SIVA procedure, it is not possible to do so; for this equation is effectively decoupled from the others (for a particular cycle of iteration) by the practice of taking the convection velocities from the previous iteration. T_{ij} is therefore excluded from the SIVA procedure, although for reasons of convenience, it is evaluated during the same grid sweep.

3. THE INSTABILITY DISCOVERED

Applications of the SIVA procedure to the staby-stratified fluid problem was carried out by first setting the values of velocity and temperature equal to their known final values. A single temperature was then perturbed. In addition, the pressure field was left uniform instead of at its final equilibrium solution, which is uniform in x and parabolic in y. Although a new T_{ij} was always computed each time P_{ij} was adjusted, three distinct ways of doing so were tried. In the first the new value of T_{ij} was calculated before the calculations of pressure and velocity. Thus the temperature was updated before the buoyancy term was evaluated but old values of the velocities were used in computing the temperature convection. In the second, T_{ij} was evaluated after the hydrodynamic computation; this meant that T_{ij} was evaluated with the latest convection velocities, but an old value of T_{ij} was used in the buoyancy term. The third procedure - found to give best results - was to evaluate the temperature twice at each point, once before the hydrodynamic calculations and again after the new values of the velocities were computed.

With this third method the correct solution to the problem, i.e. the decay of the perturbation and the establishment of the correct velocity,temperature and pressure fields, was obtained for cell Grashof numbers up to 120. (The cell Grashof number is given by

$$Gr_c = \frac{\beta g(\Delta T)_{CELL} h^3}{\nu^2} \tag{10}$$

and is related to the usual Grashof number Gr, based on the overall length and temperature difference by $Gr_c = Gr/N^4$ where $Gr = \beta g \Delta T L^3/\nu^2$ and N is the number of grid nodes in one direction). All computations were made for a Prandtl number of unity. Above this critical value of Gr_c, the perturbation would not die down and values of temperature and pressure would continuously climb. This behaviour was found to be independent of the size of the temperature perturbation. Above the critical value of Gr_c any temperature perturbation, however slight, produced divergence.

In order to determine the nature of the instability, a one-dimensional analysis of the single perturbed temperature point was performed. Details of this analysis are given in the Imperial College Report by the authors. The results of this analysis give the temperature T, and the velocity v (contained in the Peclet number) by the following equations.

Energy:

$$\frac{T - T_{CA}}{T_{UW} - T_{CA}} = \frac{|Pe|}{2 + Pe} \tag{12}$$

Flow:

$$\frac{T - T_{CA}}{T_{UW} - T_{CA}} = \frac{4 + |Re|}{Pr \frac{\beta g h^3}{\nu^2} (T_{UW} - T_{CA})} Pe \qquad (13)$$

In these equations it is assumed that all variables outside the perturbed cell have their equilibrium values. The Peclet number Pe ($\equiv hv\ Pr/\nu$) and Reynolds number Re (hv/ν) are defined in terms of the cell length and the velocity through the cell. T_{CA}, the conduction-average temperature of the surrounding T nodes is the same as the final equilibrium temperature. T_{UW} is the temperature of the node upwind from the temperature node considered. At equilibrium, the difference $T_{UW} - T_{CA}$ is simply the cell temperature difference $(\Delta T)_{cell}$ used in the definition of Gr_c. The Peclet and Reynolds numbers in absolute value signs arise from the computational form used to introduce upwind differencing; these are taken as constant during a given iteration.

The form of the energy and flow equations for this analysis is shown in figure 3.

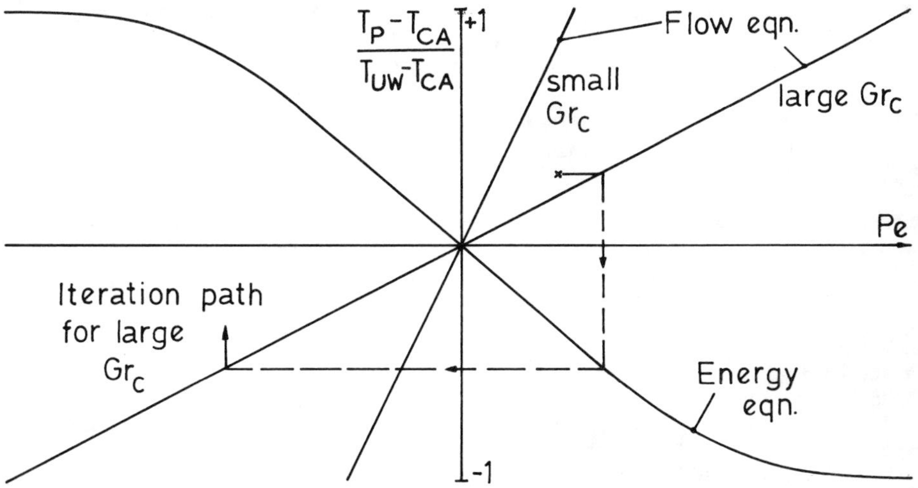

Figure 3: Instability due to interlinkage between the equations

Since the slope of the flow equation is inversely proportional to $(\Delta T)_{cell}$ (i.e. to the cell Grashof number), large values of Gr_c will produce a small slope in the linearised flow equation. Under these conditions a successive-substitution procedure which calculates the hydrodynamic variables and then the temperature will lead to divergence by the path shown (dotted). Conversely, if the value of Gr_c is sufficiently small such that the slope of the flow equation is large, it is easy to show that the successive calculation procedure will lead to the correct solution $T = T_{CA}$ and $Pe = 0$.

This analysis also confirms the observation that divergence will occur above a certain value of Gr_c regardless of the magnitude of the temperature perturbation. The vertical displacement of the "x" at the start of the dotted path represents the size of the temperature perturbation. So long as the flow equation retains its low slope - caused by a high value of Gr_c - any initial vertical displacement above the $T - T_{CA} = 0$ axis will lead to divergence.

It is apparent from figure 3 that the iteration procedure entails too great a correction for the temperature; it is thus necessary to make an appropriate underrelaxation. The underrelaxation factor, α, required to do this is given by

$$\alpha \equiv (T_{CA} - T')/(T - T') \qquad (14)$$

where T' is the initial-guess value of T, and where the 'conduction average' temperature, T_{CA}, is known to be the desired solution. Using the energy equation, we can write T - T' as

$$T - T' = T_{CA} + \frac{T_{UW} - T_{CA}}{2/|Pe| + 1} \qquad (15)$$

where the value of Pe used in this calculation is that found from the hydrodynamic calculation based on a guessed value of Reynolds number, Re', in equation 13. By substituting this value of Pe into the above equation and use of the resulting expression for T - T', the underrelaxation factor, α, is found to be

$$\alpha = [1 + \frac{|T_{UW} - T_{CA}|}{(KG)^{-1} + |T' - T_{CA}|}]^{-1} \qquad (16)$$

where

$$G \equiv \frac{\beta g h^3}{\nu^2}, \text{ and } K \equiv \frac{Pr}{8 + 2|Re'|} \quad . \qquad (17)$$

4. USE OF THE UNDERRELAXATION FACTOR AND A NEW INSTABILITY

The analysis used in deriving the underrelaxation factor was confirmed by a computer experiment. Here all variables were set at their equilibrium values, the temperature of one cell was perturbed, and this perturbed cell was the first cell visited in the grid sweep. In this simple test the correct temperature was obtained in a single iteration, as indeed it should be if the analysis just presented is correct. This result was obtained for arbitrarily large values of Gr_c.

When however tests were performed in which the perturbed cell was not the first visited, and the pressure was initially uniform, the procedure did not give convergence for arbitrarily large values of Gr_c. Although solutions could be obtained for $Gr_c = 1.2 \times 10^4$, a hundredfold increase over the maximum value of cell Grashof number previously solved, there appeared to be no reason for this new ceiling on convergence. It seemed, then, that a new form of instability had been encountered, which was not associated with the interaction between the energy and hydrodynamic equations, but was some other form of instability in the iteration procedure. To check this supposition, a second simple calculation was carried out in which the temperatures were held at their equilibrium values throughout a calculation, providing a fixed buoyancy term. The pressure of a single cell was perturbed and the iteration cycle was performed in the normal manner. In these tests it was found divergence would occur if the pressure perturbation ΔP exceeded a certain value. Indeed, this observation was true even if the buoyancy term was set equal to zero. For a calculation in which all velocities and pressures were set to zero, and the pressure of a single node was perturbed, it was found that the perturbation would not decay, but all pressures would increase indefinitely if the following dimensionless parameter

$$(\frac{\Delta P}{\rho})^{\frac{1}{2}} \frac{h}{\nu} \qquad (18)$$

exceeded a critical value of about 225.

Based on these results, subsequent computations were made for the stably-stratified fluid in which the size of the initial temperature perturbation was varied. With the underrelaxation factor given by equation 16, the correct solution could be obtained for arbitrary values of cell Grashof number provided the initial temperature perturbation was sufficiently small. This was in direct contrast with the earlier calculations; there, with no underrelaxation, the size of the temperature perturbation was unimportant. Subsequent calculations with various values of the physical parameters showed that the factor governing the newly encountered instability for a given temperature perturbation $\Delta T \equiv |T - T_e|$ and pressure perturbation $\Delta P \equiv |P - P_e|$ was:

$$\frac{h^2}{\nu^2} [\beta g h \, \Delta T + \frac{\Delta P}{\rho}] \quad . \qquad (19)$$

When this combined perturbation parameter exceeded a value of about 2×10^4, solutions could not be obtained. Below this critical value a converged solution was obtained regardless of the size of Gr_c.

In summary, the computations reported in this section clearly showed that the second instability encountered is different from the one cured by underrelaxation. Work is currently in progress to find a method for removing the new instability.

5. CONCLUSIONS

In the treatment of a stably-stratified fluid with a perturbation, using pressure, velocity, and temperature as dependent variables, there is an instability caused by interlinkages between the energy and momentum/continuity equations. This instability is removed by an appropriate underrelaxation procedure.

A further instability is present, but this is not limited to the problem of the stably-stratified fluid; rather, it is present in problems involving perturbations on zero-flow situations, using the SIVA solution procedure. This instability is characterised by a dimensionless parameter involving the size of the perturbations.

6. REFERENCES

1. Caretto, L.S., R.M. Curr, and D.B. Spalding, 'Two Numerical Methods for Three-Dimensional Boundary Layers', Submitted for Publication, (1971).

2. Caretto, L.S., A.D. Gosman, and D.B. Spalding, 'Removal of an Instability in a Free Convection Problem', Imperial College Mech. Engr. Dept. Report, EF/TN/A/35 (March, 1971).

3. Gosman, A.D., W.M. Pun., A.K. Runchal, D.B. Spalding, and M. Wolfshtein, Heat and Mass Transfer in Recirculating Flows, Academic Press, (1969).

4. Gosman, A.D., and D.B. Spalding, 'The prediction of confined three-dimensional boundary layers', Symposium on Internal Flows, University of Salford, (20 - 22 April 1971).

BOUNDS FOR THE ERROR IN APPROXIMATE SOLUTIONS OF ORDINARY DIFFERENTIAL EQUATIONS

G.J. Cooper

Summary

The article outlines the derivation of error bounds for numerical solutions of initial value problems for ordinary differential equations. These bounds depend on a bound for the norm of the fundamental matrix of a linear system and on a bound for the norm of the approximation defect. Error bounds are usually expressed as the solution of a linear differential equation but may also be obtained as the solution of a Riccati equation.

1. Introduction

It is convenient to start by considering an initial value problem for an autonomous system of first order differential equations,

$$x^{(1)} = [f]\,x\ ,\quad x(t_0) = x_0,\quad x = (x^1, x^2, \ldots, x^n)^T,\quad x \in R_n\ ,$$

where f is a mapping of R_n into R_n. The norm of a vector w is denoted by $\|w\|$ while for a matrix A, $\|A\| = \sup\{\|Aw\|\ ;\ \|w\| = 1\}$.

For $t_0 < t_1 < \ldots < t_N$ let $y_0, \ldots, y_N$ be given real valued vectors which represent approximations to $x(t_0), \ldots, x(t_N)$. Let $y(t) \in C\,[t_0, t_N)$ with $y(t_j) = y_j$, $j = 0(1)N$, and assume that $y^{(1)}(t) = \lim_{\alpha \to 0+} \alpha^{-1}\{y(t+\alpha) - y(t)\}$ exists for all $t \in [t_0, t_N)$. A posteriori bounds are established for $\|x-y\|$ which are independent of x. Although the bounds are obtained for initial value problems they are applicable to boundary value problems. Thus consider a system of first order differential equations with given boundary conditions. An approximate solution defines an initial value problem whose exact solution satisfies the original system of differential equations with perturbed boundary conditions. The error bound bounds the perturbation.

Let $D \subset R_n$ be a convex domain with $y \in D$ and $x \in D$ for all $t \in [t_0, t_N)$. Assume that the partial derivatives of $f[w]$ exist for all $w \in D$ and assume that the Jacobian matrix of f satisfies a Lipsichiz condition in D,

$$J[w] = (f_1[w], \ldots, f_n[w]),\qquad f_i[w] = \frac{\partial}{\partial w^i} f[w]\ ,$$

$$\|J[w] - J[z]\| \leqslant 2k\,\|w-z\|\ ,\qquad w, z \in D\ .$$

The initial value problem has a unique solution on $[t_0, t_N)$ with $x^{(1)}(t) \in C\,[t_0, t_N)$.

The results given here are mainly due to Dahlquist [2] and Lozinskii [6]. The essential feature is that the error bound may be a decreasing function of t. Additional results have been obtained by Kahan [4] and Strom [8]. This article differs in the use of a Lipschitz condition for the Jacobian matrix which allows an alternative treatment of the non-linear term in the differential equation for the error. Also as a consequence, the norm of the approximation defect, $d(t) = f[y] - y^{(1)}$, may be bounded differently.

It is, perhaps, worth mentioning that $y(t)$ may be obtained as the (approximate) solution of an initial value problem which approximates the given problem. Then an error bound may be used to attempt to validate such a procedure. Similarly, steady state behaviour may be examined.

2. Equations for the error bound

2.1 An integral inequality for the error norm

The initial value problem $\nu^{(1)} = f[\nu+y] - y^{(1)}$, $\nu(t_0) = x_0 - y_0$, has a unique solution, in the sense of Caratheodory, for all $t \in [t_0, t_N)$ [7] and hence $\nu = x-y$. Thus if $r[t;\nu] = f[\nu+y] - f[y] - J[y]\nu$ then

$$\nu^{(1)} = J[y]\nu + r[t;\nu] + d(t), \quad \nu(t_0) = \nu_0 = x_0 - y_0, \tag{2.1}$$

defines the error. Let I be the identity matrix. Since the elements of $J[y]$ are continuous functions of t, $\exists\, C[t;s]$, a fundamental matrix of solutions of the homogeneous system,

$$C[t;s]^{(1)} = J[y]\, C[t;s], \quad C[s;s] = I, \tag{2.2}$$

and $C[t;s]^{(1)} = \lim_{\alpha \to 0+} \alpha^{-1}\{C[t+\alpha;s] - C[t;s]\}$ exists for all $t, s \in [t_0, t_N)$. The variation of constants formula [3] gives an integral equation for the error.

<u>Theorem (1)</u> The error is the unique solution on $[t_0, t_N)$ of

$$\nu = C[t;t_0]\nu_0 + \int_{t_0}^{t} C[t;s]\{r[s;\nu] + d(s)\}\, ds .$$

Thus an integral inequality may be established for $\|\nu\|$ provided bounds can be obtained for $\|r[s;\nu]\|$ and $\|C[t;s]\|$.

<u>Theorem (2)</u> For $s \in [t_0, t_N)$, $\|r[s;\nu]\| \leqslant k\|\nu\|^2$.

<u>Proof</u> This follows from the Lipschitz condition and the identity

$$f[\nu+y] - f[y] - J[y]\nu = \int_0^1 \{J[y+\tau\nu] - J[y]\}\, \nu\, d\tau .$$

The next theorem, which bounds $\|C[t;s]\|$, was established by Dahlquist [2] and Lozinskii [6]. The bound is given in terms of $\mu(J)$ where

$$\mu(J) = \inf_{\alpha>0} \frac{\|I+\alpha J\| - 1}{\alpha} = \lim_{\alpha \to 0+} \frac{\|I+\alpha J\| - 1}{\alpha} .$$

The limit exists, and is the infinmum, since $\alpha^{-1}\{\|I+\alpha J\| - 1\}$ is a bounded and increasing function of α. Also $\mu(J) = \sup\{\mathrm{Re}\lambda;\ \lambda \in V(J)\}$ where $V(J)$ is the numerical range of J [1].

<u>Theorem (3)</u> For $s, t \in [t_0, t_N)$, $\ln \|C[t;s]\| \leqslant \int_s^t \mu(J[t(\tau)])\,d\tau = \ln \beta(t;s)$.

<u>Proof</u> It is first shown that $\|C[\tau;s]\|^{(1)}$ exists for $\tau \in [t_0, t_N)$ and

$$\|C[\tau;s]\|^{(1)} = \lim_{\alpha \to 0+} C_\alpha(\tau), \quad C_\alpha(\tau) = \alpha^{-1}\{\|C[\tau;s] + \alpha C[\tau;s]^{(1)}\| - \|C[\tau;s]\|\}.$$

The limit exists since $C_\alpha(\tau)$ is a bounded increasing function of α.

Now

$$\lim_{\alpha \to 0+} \left| \frac{\| C[\tau+\alpha; s] \| - \| C[\tau ; s] \|}{\alpha} - C_\alpha(\tau) \right| \leq \lim_{\alpha \to 0+} \frac{\| C[\tau+\alpha; s] - C[\tau; s] - \alpha C[\tau; s]^{(1)} \|}{\alpha} = 0$$

by the definition of $C[\tau; s]^{(1)}$. Hence $\| C[\tau; s] \|^{(1)}$ exists and is $\lim_{\alpha \to 0+} C_\alpha(\tau)$. Secondly

$$C_\alpha(\tau) \leq \alpha^{-1} \{ \| I + \alpha J[y(\tau)] \| - 1 \} \| C[\tau ; s] \|$$

follows from (2.2). Hence the result follows from the differential inequality

$$\| C[\tau; s] \|^{(1)} \leq \mu(J[y(\tau)]) \| C[\tau; s] \| .$$

In the p-norms, $p = 1, 2, \infty$, $\mu(J)$ is readily calculated and may be negative [2]. It now follows from theorem (1) that $\| \nu \|$ satisfies the integral inequality

$$\| \nu \| \leq \beta(t; t_0) \| \nu_0 \| + \int_{t_0}^{t} \beta(t; s) \{ k \| \nu(s) \|^2 + \| d(s) \| \} ds \quad . \tag{2.3}$$

2.2 Bounds for the error norm

The integral inequality (2.3) leads to an integral equation for $\varepsilon \geq \| \nu \|$. A proof is given by Lakshmikanthan and Leela [5]. It is possible that the solution of this integral equation cannot be continued up to t_N.

Theorem (4) Let $\varepsilon(t_0) = \varepsilon_0 \geq \| \nu_0 \|$. Then $\exists\, t_* > t_0$, t_* independent of ε_0, such that

$$\varepsilon = \beta(t; t_0) \varepsilon_0 + \int_{t_0}^{t} \beta(t; s) \{ k \varepsilon(s)^2 + \| d(s) \| \} ds$$

has a unique solution $\varepsilon(t) \in C[t_0, t_*)$, and $\varepsilon \geq \| \nu \|$ for $t \in [t_0, t_*)$.

Corollary For some $t_* > t_0$, the Riccati equation

$$\varepsilon^{(1)} = k\varepsilon^2 + \mu(J[y(t)])\varepsilon + \| d(t) \|, \qquad \varepsilon(t_0) = \varepsilon_0 \geq \| \nu_0 \| ,$$

has a unique solution $\varepsilon(t) \in C[t_0, t_*)$, and $\varepsilon \geq \| \nu \|$ for $t \in [t_0, t_*)$.

In these results $\mu(J)$ and $\| d \|$ may be replaced by integrable upper bounds.

The Riccati equation may be replaced by a linear differential equation. Indeed, instead of (2.1), the error may be defined by

$$\nu^{(1)} = f[y+\nu] - f[y] + d(t) = \int_0^1 J[y+\eta\nu] d\eta \; \nu + d(t), \quad \nu(t_0) = \nu_0 .$$

In this system, the norm of the fundamental matrix is bounded by

$$\ln \| C[t; s] \| \leq \int_s^t \mu\left(\int_0^1 J[y(\tau)+\eta\nu(\tau)] d\eta\right) d\tau .$$

This gives the result established by Dahlquist [2] for if M_τ is a compact convex set in R_n, with $x(\tau)$, $y(\tau) \in M_\tau$, then

$$\mu(\int_0^1 J[y+\eta\nu]\, d\eta) \leq \lim_{\alpha\to 0+} \int_0^1 \frac{\| I+\alpha J[y+\eta\nu] \| -1}{\alpha}\, d\eta \leq \max_{w\in M_\tau} \mu(J[w]) .$$

A linear transformation may be applied to the differential equation for the error. Thus if S is a non-singular matrix (2.1) may be transformed to

$$u^{(1)} = \{ SJ[y]S^{-1} + S^{(1)}S^{-1}\} u + Sr[t; S^{-1}u] + Sd(t), \qquad u = S\nu .$$

Let $\lambda(J)$ be the maximum of the real parts of the eigenvalues of J. If J is a constant matrix then, for given $\delta > 0$, $\exists$ constant S such that $\mu(SJS^{-1}) \leq \lambda(J) + \delta$. The error bound also depends on $\|S\|$ and $\|S^{-1}\|$ which may be large but, nevertheless, such a transformation can significantly improve the error bound. If J is not a constant matrix a transformation may still substantially improve the bound.

The results so far obtained are for autonomous systems. A non-autonomous initial value problem is equivalent to an autonomous problem with $f^n[x] = 1$ and $x^n(t_0) = t_0$. In this case $J[y]$ has a zero eigenvalue which may cause $\mu(J)$ to be larger than necessary. However, if it is assumed that $\nu^n(t) = 0$, the analysis may be repeated with $\mu(\mathcal{J})$ replacing $\mu(J)$ and

$$\mathcal{J}[w] = (g_1, [w], \dots, g_{n-1}[w]), \qquad g = (f^1, \dots, f^{n-1})^T .$$

3. Solution of the Riccati equation

3.1 Constant coefficients

It is not sufficient to solve the Riccati equation numerically. However, the equation may be replaced by a sequence of Riccati equations with constant coefficients which are upper bounds for $\mu(J)$ and $\| d \|$ on sub-intervals of $[t_0, t_N)$. For $j = 0(1)N-1$ let

$$\mu_j = \sup\{\mu(J[y(t)]);\ t \in [t_j, t_{j+1})\}, \quad d_j = \sup\{\|d(t)\|;\ t \in [t_j, t_{j+1})\} .$$

Theorem (5) If $\varepsilon(t) \in C[t_0, t_*)$ is the solution of the Riccati equations

$$\varepsilon^{(1)} = k\varepsilon^2 + \mu_j\varepsilon + d_j, \quad t \in [t_j, t_{j+1}), \quad j = 0(1)N-1, \quad \varepsilon(t_0) \geq \|\nu_0\|,$$

on $[t_0, t_*)$, then $\varepsilon \geq \|\nu\|$ on $[t_0, t_*)$.

These Riccati equations may be solved formally and, indeed, ε is monotonic on each sub-interval. Hence only $\varepsilon(t_0)$, $\varepsilon(t_1)$, ... , need (formal) evaluation.

3.2. Bounds for the coefficients

Bounds are easily established for $\mu[J]$ and $\|d\|$ on sub-intervals. The followling theorems are capable of refinement, particularly for special systems. Let $T_j \epsilon [t_j, t_{j+1})$, $j = 0(1)N-1$, be given.

Theorem (6) If $\eta_j = \sup \{ \| y - y(T_j) \| ; \ t \epsilon [t_j, t_{j+1}) \}$, $j = 0(1)N-1$ then

$$\mu_j \leqslant \mu(J[y(T_j)]) + 2k\eta_j , \quad j = 0(1)N-1 .$$

Theorem (7) If $\xi_j = \sup \{ \| f[y(T_j)] - y^{(1)} - J[y(T_j)](y(T_j)-y) \| ; \ t \epsilon [t_j, t_{j+1}) \}$, $j = 0(1)N-1$, then

$$d_j \leqslant \xi_j + k\eta_j^2 , \qquad j = 0(1)N-1 .$$

The first theorem follows from the definition of $\mu[J]$ and the Lipschitz condition on J, while the second result follows from theorem (2).

It is possible to compute $\eta_0, \ldots, \eta_{N-1}$ and $\xi_0, \ldots, \xi_{N-1}$. In particular, this is so if y(t) is piecewise linear. However, for most numerical methods, more elaborate interpolation formulae will give smaller bounds. If y(t) is piecewise cubic with $y(t_j) = y_j$ and $y^{(1)}(t_j) = f[y(t_j)]$, $j = 0(1)N$, then bounds may be obtained for $\eta_0, \ldots, \eta_{N-1}$ and $\xi_0, \ldots, \xi_{N-1}$ which are sometimes satisfactory. Of course, the error bound cannot be expected to be satisfactory if $\mu(J) >> \lambda(J)$ or sign $\mu(J) \neq$ sign $\lambda(J)$. Thus it is more important to search for suitable transformations $u = S\nu$ and to compare $\mu(J)$ for different norms.

Acknowledgements

The author thanks Professor S. Michaelson and D. Kershaw for much encouragement and advice. Various people have kindly pointed out the use of transformations.

References

1. Bonsall, F.F. and J. Duncan;
Numerical Ranges of Operators on Normed Spaces and of Elements of Normed Algebras, London Math. Soc., Lecture Note Series 2, C.U.P. (1971).

2. Dahlquist, G.
Stability and Error Bounds in the Numerical Integration of Ordinary Differential Equations, Trans. Royal Inst. Technology, Stockholm, 130 (1959).

3. Halany, A.,
Differential Equations, Stability, Oscillations, Time Lags, pp. 39-43, Academic Press, New York (1966).

4. Kahan, W.,
An Ellipsoidal Error Bound for Linear Systems of Ordinary Differential Equations, Report of Dept. of Math., Uni. of Toronto, Canada.

5. Lakshmikanthan, V. and S. Leela,
Differential and Integral Inequalities, Vol 1., pp. 315-322. Academic Press, New York (1969).

6. Lozinskii, S.M.,
Error estimate for numerical integration of ordinary differential equations, Soviet Math. Dokl., 163, pp. 1014-1019 (1965).

7. Sansone, G. and R. Conti,
Nonlinear Differential Equations (revised ed.), pp. 10-15, Pergamon Press, Oxford (1964).

8. Strom, J.
On Logarithmic Norms,
Report of Dept. of Computer Science, Royal Inst. Technology, Stockholm.

NUMERICAL SOLUTION OF THE STURM LIOUVILLE PROBLEM WITH PERIODIC BOUNDARY CONDITIONS

D.J. Evans

Abstract

A recursive algorithm for the implicit derivation of the characteristic equation of a symmetric general tridiagonal matrix of order n is derived from a finite difference discretisation of a periodic Sturm Liouville problem. The algorithm yields a Sturmian sequence of polynomials from which the eigenvalues can be obtained by the use of the well known standard bisection process. An extension to Wilkinson's method for deriving the eigenvectors of symmetric tridiagonal matrices yields the required eigenvectors of the periodic Sturm Liouville problem.

1. Introduction

Recent computational techniques for the solution of the algebraic eigenvalue problem involving the Givens, Householder and Lanczos methods all involve determining the eigenvalues of a symmetric or unsymmetric tridiagonal matrix (Wilkinson, 1965). This is an important problem that occurs in its own right since tridiagonal matrices arise naturally in many problems involving ordinary and partial differential equations. The method depends on the technique of determining the characteristic polynomial for numerical values of λ by computing a simple sequence of polynomials derived from the elements of the tridiagonal matrix.

In this paper, we show that similar techniques can be applied to a more general tridiagonal matrix and from which the eigensolutions to a periodic characteristic problem can be obtained in an efficient manner.

2. Formulation of the problem

We consider the periodic characteristic problem

$$\frac{d}{dx}\left(p(x)\frac{dy}{dx}\right) + q(x)y + \lambda r(x)y = 0 \,, \tag{2.1}$$

where we seek numerical values of λ and $y(x)$ which satisfy (2.1) in the range $[a,b]$ subject to the boundary conditions,

$$y(a) = y(b) \tag{2.2}$$

and

$$p(a)\, y'(a) = p(b)\, y'(b) \quad . \tag{2.3}$$

This represents the statement of the Sturm Liouville problem (Froberg, 1965) and for situations in which $p(a) = p(b)$, the boundary condition (2.3) assumes the well-known periodic form.

The direct substitution of the second difference operator in equation (2.1) by the approximation,

$$\frac{d}{dx}\left(p(x)\frac{dy}{dx}\right) \simeq \frac{p_{i+\frac{1}{2}}(y_{i+1}-y_i) - p_{i-\frac{1}{2}}(y_i-y_{i-1})}{h^2} \quad , \tag{2.4}$$

at each of the discrete points x_i, $i = 1, 2, \ldots, n$, in the interval $[a,b]$ where $nh = b-a$ yields a set of homogeneous linear equations of the form,

$$Ay = \lambda\, h^2\, Ry \quad , \tag{2.5}$$

where A is a matrix of the form,

$$\begin{bmatrix} a_{1,1} & a_{1,2} & & & a_{1,n} \\ a_{2,1} & a_{2,2} & a_{2,3} & O & \\ & a_{3,2} & \ddots & \ddots & \\ & O & \ddots & \ddots & a_{n-1,n} \\ a_{n,1} & & & a_{n,n-1} & a_{n,n} \end{bmatrix},$$

with elements

$$a_{i,i} = (p_{i+\frac{1}{2}}+p_{i-\frac{1}{2}}) - q_i h^2 \quad , i = 1, 2, \ldots, n ;$$

$$a_{i,i+1} = -p_{i+\frac{1}{2}} , \; a_{i,i-1} = -p_{i-\frac{1}{2}} , \quad i = 1, 2, \ldots, n;$$

and under the periodic conditions (2.3) which prevail the following relationships $a_{n,n+1} = a_{n,1}$ and $a_{1,0} = a_{1,n}$ are valid.

In addition, when $p(x)$, $q(x)$ and $r(x)$ are periodic functions, then each of the p_i, q_i and r_i are of period n or more generally

$$p_{n+1}/p_1 = q_{n+1}/q_1 = r_{n+1}/r_1 \tag{2.6}$$

and the first n equations correspond to the matrix relation (2.5) where A is a symmetric matrix.

In equation (2.5), R is a diagonal matrix with elements r_i, $(i = 1, 2\ldots,n)$ and for convenience, since each $r_i > 0$, we can make use of the substitution,

$$y = R^{-\frac{1}{2}}u$$

$$C = h^{-2}R^{-\frac{1}{2}}AR^{-\frac{1}{2}}$$

and replace (2.5) by the equivalent system,

$$Cu = \lambda u \tag{2.7}$$

where C is a general symmetric tridiagonal matrix similar to A in structure with diagonal elements c_i,

such that $$c_i = a_{ii}/h^2 r_i \quad (i = 1, 2, \ldots, n) \tag{2.8}$$

and super- and sub-diagonal elements b_{i+1} and b_i such that

$$b_{i+1} = a_{i,i+1}/h^2 (r_i r_{i+1})^{\frac{1}{2}} \; (i = 1, 2, \ldots, n) \tag{2.9}$$

and where the periodic conditions still continue to prevail for i = 1 and n.

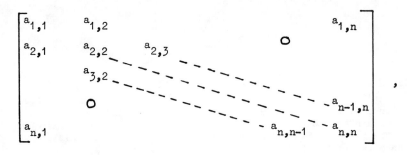

3. Calculation of the eigenvalues of the general symmetric tridiagonal matrix

It is well known that non-trivial solutions to equations (2.7) exist if and only if

$$\det(C - \lambda I) = 0 ,$$

or in full matrix notation,

$$\det\begin{vmatrix} (c_1-\lambda) & b_2 & & & & b_1 \\ b_2 & (c_2-\lambda) & b_3 & & & \\ & b_3 & \ddots & \ddots & & \\ & & \ddots & \ddots & b_{n-1} & \\ & & & b_{n-1} & (c_{n-1}-\lambda) & b_n \\ b_1 & & & & b_n & (c_n-\lambda) \end{vmatrix} = 0. \tag{3.1}$$

A Laplace expansion of $\det(C-\lambda I)$ in terms of the reduced determinants of order $(n-1)$ can be expressed in the form,

$$|C-\lambda I| = (c_n-\lambda)\begin{vmatrix} (c_1-\lambda) & b_2 & & & \\ b_2 & (c_2-\lambda) & b_3 & & \\ & \ddots & \ddots & \ddots & \\ & & & (c_{n-2}-\lambda) & b_{n-1} \\ & & & b_{n-1} & (c_{n-1}-\lambda) \end{vmatrix}$$

$$-b_n\begin{vmatrix} (c_1-\lambda) & b_2 & & & b_1 \\ b_2 & (c_2-\lambda) & b_3 & & \\ & \ddots & \ddots & \ddots & \\ & & b_{n-2} & (c_{n-2}-\lambda) & 0 \\ & & & b_{n-1} & b_n \end{vmatrix}$$

$$+(-1)^{(n-1)}b_1\begin{vmatrix} b_2 & & & & b_1 \\ (c_2-\lambda) & b_3 & & & \\ b_3 & \ddots & \ddots & & \\ & \ddots & \ddots & \ddots & \\ & & b_{n-1} & (c_{n-1}-\lambda) & b_n \end{vmatrix} \tag{3.2}$$

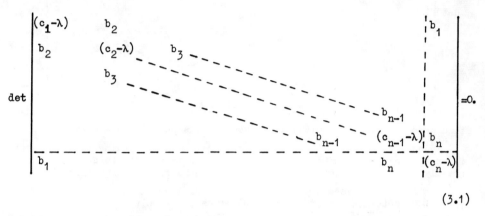

If we now introduce the notation whereby the determinantal expression

$$\det \begin{vmatrix} (c_1-\lambda) & b_2 & & & \\ b_2 & (c_2-\lambda) & b_3 & & \\ & b_3 & \ddots & \ddots & \\ & & \ddots & (c_{n-2}-\lambda) & b_{n-1} \\ & & & b_{n-1} & (c_{n-1}-\lambda) \end{vmatrix} \tag{3.3}$$

is represented by the expression $\det|T^1_{n-1}-\lambda I|$ where T^1_{n-1} denotes a symmetric tridiagonal matrix with diagonal elements c_i, super diagonal elements b_i, with the index i commencing at the superfix value 1 and continuing through until the index i attains the suffix value (n-1). Then we can use this notation to greatly simplify the determinantal expressions derived in equation (3.2).

A further Laplace expansion of the determinantal expressions in equation (3.2) into determinants of order (n-2) gives the result

$$|C-\lambda I| = (c_n-\lambda)|T^1_{n-1}-\lambda I| - \tag{3.4}$$

$$-b_n \left[b_n \begin{vmatrix} (c_1-\lambda) & b_2 & & \\ b_2 & (c_2-\lambda) & b_3 & \\ & \ddots & \ddots & b_{n-2} \\ & & b_{n-2} & (c_{n-2}-\lambda) \end{vmatrix} \right.$$

$$\left. -b_{n-1} \begin{vmatrix} (c_1-\lambda) & b_2 & & & b_1 \\ b_2 & (c_2-\lambda) & b_3 & & \\ & \ddots & \ddots & \ddots & \\ & & b_{n-3} & (c_{n-3}-\lambda) & 0 \\ & & & b_{n-2} & 0 \end{vmatrix} \right]$$

$$+(-1)^{(n-1)} b_1 \left[b_2 \begin{vmatrix} b_3 & & & \\ (c_3-\lambda) & b_4 & & \\ b_4 & \ddots & \ddots & \\ & b_{n-1} & (c_{n-1}-\lambda) & b_n \end{vmatrix} \right.$$

$$+(-1)^{(n-2)} b_1 \begin{vmatrix} (c_2-\lambda) & b_3 & & \\ b_3 & \ddots & \ddots & \\ & \ddots & \ddots & b_{n-1} \\ & & b_{n-1} & (c_{n-1}-\lambda) \end{vmatrix} \Bigg] .$$

Clearing up the terms in equation (3.4) and making full use of the notation previously set up, we have the intermediate result given by a further expansion involving determinants of order (n-3) i.e.,

$$|C-\lambda I| = (c_n-\lambda)\ |T^1_{n-1}-\lambda I| - b_n^2\ |T^1_{n-2}-\lambda I|$$

$$+(-1)^{(n-3)}\, b_n b_{n-1} b_1 \begin{vmatrix} b_2 & (c_2-\lambda) & b_3 & & \\ & b_3 & (c_3-\lambda) & \ddots & \\ & & \ddots & \ddots & \\ & & & \ddots & (c_{n-3}-\lambda) \\ & & & & b_{n-2} \end{vmatrix} \qquad (3.5)$$

$$+(-1)^{(n-1)} \prod_{i=1}^{n} b_i - b_1^2\ |T^2_{n-1}-\lambda I| .$$

Finally, by combining the results given in equations (3.2), (3.4) and (3.5), the determinant of C simplifies to the expression,

$$|C-\lambda I| = |T^1_n-\lambda I| - b_1^2\ |T^2_{n-1}-\lambda I| + (-1)^{n-1}\, 2 \prod_{i=1}^{n} b_i \quad . \qquad (3.6)$$

It can be noticed that the characteristic polynomial of the matrix C has been obtained in terms of the characteristic polynomials of tridiagonal matrices which are sub-matrices of a specified order of the original matrix C. The reason for this is fairly clear when we consider how simple it is to obtain the characteristic polynomial of a tridiagonal matrix.

If we consider the symmetric tridiagonal matrix of order n,

$$T^1_n = \begin{vmatrix} c_1 & b_2 & & & \\ b_2 & c_2 & b_3 & & \\ & \ddots & \ddots & \ddots & \\ & & \ddots & \ddots & b_n \\ & & & b_n & c_n \end{vmatrix}$$

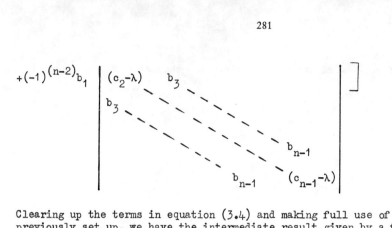

then, the sequence of polynomials which determines the characteristic polynomial det $|T_n^1-\lambda I|$ implicitly for any given value of λ is given by

$$T_{-1} = 0 \quad , \quad T_0 = 1 \quad ,$$

$$T_1(\lambda) = (c_1-\lambda) \quad , \quad T_2(\lambda) = (c_2-\lambda)\, T_1(\lambda) - b_2^2\, T_0 \quad , \tag{3.7}$$

and $T_i(\lambda) = (c_i-\lambda)\, T_{i-1}(\lambda) - b_i^2\, T_{i-2}(\lambda)$ for $i = 3, 4, \ldots, n$.

Similarly we can use identical techniques as in the above to obtain a sequence of polynomials which determines the characteristic polynomials of the matrix C. From equation (3.6) we see that recursion formulae can be set up involving two sequences of polynomials $P(\lambda)$ and $Q(\lambda)$, defined as

$$\begin{array}{lll}
P_{-1} = 0 & ; & \\
P_0 = 1 & ; \; Q_0 = 0 & \\
P_1(\lambda) = (c_1-\lambda) & ; \; Q_1 = 1 & , \; i = 1 , \\
P_2(\lambda) = (c_2-\lambda)\, P_1(\lambda)-b_2^2\, P_0 & ; \; Q_2(\lambda)=(c_2-\lambda) & , \; i = 2 , \\
\text{-----------------} & \text{-----------------} & \\
P_i(\lambda) = (c_i-\lambda)P_{i-1}(\lambda)-b_i^2P_{i-2}(\lambda); & Q_i(\lambda) = (c_i-\lambda)Q_{i-1}(\lambda)-b_i^2\, Q_{i-2}(\lambda) & \\
 & \text{for } i = 3, \; n-1; &
\end{array} \tag{3.8}$$

and finally,

$$P_n(\lambda) = (c_n-\lambda)P_{n-1}(\lambda)-b_n^2\, P_{n-2}(\lambda)-b_1^2Q_{n-1}(\lambda) + (-1)^{(n-1)}\, 2\prod_{i=1}^{n} b_i \quad \text{for } i = n, \tag{3.9}$$

which determines the value of det$(C-\lambda I)$.

Now since the polynomials $P_0, P_1, P_2, \ldots, P_{n-1}, P_n$ form a sequence consisting of the leading principal minors of $|C-\lambda I|$, where C is a symmetric matrix (not necessarily tridiagonal), then they can be shown to form a properly signed interleaved sequence of polynomials (i.e., all $P_k(\lambda) > 0$ for a sufficiently large value of λ either positive or negative and the zeros of $P_k(\lambda)$ strictly separate those of $P_{k+1}(\lambda)$) and thus with the aid of the separation theorem (Wilkinson, 1965), it can be shown that the sequence of polynomials $P_0, P_1, \ldots, P_n$ forms a Sturm sequence of polynomials on the interval $(-\infty, +\infty)$.

We do not in fact compute explicitly the coefficients of $P_i(\lambda)$ but substitute particular values of $\lambda = \lambda_0$ into the equation (3.8) and inspect the consecutive members of the sequence $P_0, P_1(\lambda_0), P_2(\lambda_0), \ldots, P_n(\lambda_0)$ for patterns of plus and minus signs. We know that such polynomials have certain important properties which facilitate the calculation of the roots. This property is such that the number of disagreements in sign $S(\lambda)$ in the sequence $P_i(\lambda)$, $i = 0, 1, 2, \ldots, n$ is equal to the number of roots of $P_n(\lambda)$ smaller than λ_0 in algebraic value. Since each evaluation of a Sturm sequence gives us information about all the eigenvalues, then this result can be used to compute all the roots, or any particular root, the kth root say, which we may require.

The bisection method is briefly described as follows:- Suppose it is known that $S(a) \geq k$ and $S(b) < k$, so that λ_k lies in the interval (a,b). The integer $S(\lambda)$ is then computed for $\lambda = \frac{1}{2}(a+b)$ from the Sturm sequence (3.8). If $S(\lambda) \geq k$ then, λ_k lies between $\frac{1}{2}(a+b)$ and b otherwise it lies between a and $\frac{1}{2}(a+b)$. This halving process can be continued r times always retaining the interval in which λ_k is known to lie. Then we can state that λ_k is located in an interval $< 2^{-r}(b-a)$. Initial values for a and b can be readily obtained by applying the Gersgorin and Collatz eigenvalue bounds to the rows and columns of the matrix C to give the results,

$$\max_{i=1}^{n} \left[(c_i) + |b_i| + |b_{(i+1)}| \right] > b \qquad (3.10)$$

and

$$\min_{i=1}^{n} \left[(c_i) - |b_i| - |b_{(i+1)}| \right] < a \ . \qquad (3.11)$$

In order to carry out the above algorithm in floating point arithmetic without the fear of underflow occurring, we follow a similar procedure to Barth et al. (1967) and replace the sequence of polynomials $P_i(\lambda)$ and $Q_i(\lambda)$ by the sequences $p_i(\lambda)$ and $q_i(\lambda)$ defined by

$$p_i(\lambda) = P_i(\lambda) \,/\, P_{i-1}(\lambda) \quad \text{and} \quad q_i(\lambda) = Q_i(\lambda) \,/\, P_i(\lambda) \ , \quad (\ i = 1,\ 2,\ \ldots,\ n). \qquad (3.12)$$

The polynomials $p_i(\lambda)$ and $q_i(\lambda)$ can be shown with a little analysis to satisfy the relationships,

$$\begin{aligned}
p_0 &= 1 &&; \\
p_1(\lambda) &= (c_1-\lambda) &&; \quad q_1 = 1/p_0 \quad , \\
p_2(\lambda) &= (c_2-\lambda) - b_2^2/p_1(\lambda) &&; \quad q_2(\lambda) = (c_2-\lambda)/p_1(\lambda) \quad , \\
p_i(\lambda) &= (c_i-\lambda)-b_i^2/p_{i-1}(\lambda) &&; \quad q_i(\lambda) = (c_i-\lambda)q_{i-1}(\lambda)/p_{i-1}(\lambda)-b_i^2 q_{i-2}(\lambda)/p_{i-1}(\lambda)p_{i-2}(\lambda),
\end{aligned}$$

for $i = 3,\ 4,\ \ldots,\ n-1$

and finally,

$$p_n(\lambda) = (c_n-\lambda)-b_n^2/p_{n-1}(\lambda)-b_1^2 q_{n-1}(\lambda)+2b_n \prod_{i=1}^{n-1} (-b_i/p_i) \ . \qquad (3.13)$$

with the use of the modified sequence of polynomials $p_i(\lambda)$ in (3.13) we find that the number of negative $p_i(\lambda)$ now gives $S(\lambda)$, the number of eigenvalues smaller than λ .

4. Calculation of the eigenvectors of a general tridiagonal matrix by inverse iteration

Suppose v is taken as the initial vector, then to find the eigenvector corresponding to given values of λ (say λ_i) using two steps of inverse iteration, we must solve

$$(C-\lambda I)x = v \quad (4.1)$$

followed by

$$(C-\lambda I)y = x \quad (4.2)$$

where C is the general tridiagonal matrix expressed by equation (2.9).

Effectively, this requires a knowledge of the inverse of $(C-\lambda I)$. If we write

$$(C-\lambda I) = LU \quad (4.3)$$

where L is a unit lower triangular matrix and U is an upper triangular matrix, then a knowledge of both L and U enables one to solve both (4.1) and (4.2) very easily by forward and backward substitution techniques.

The matrices L and U may be determined by Gaussian elimination, but in order to ensure numerical stability pivoting techniques are essential.

Hence, we eliminate the variables in their natural order but select as pivotal row at the ith stage, the row having the maximum coefficient of x_i. At each stage of the computation there are only three such rows, they are the ith reduced row, the $(i+1)^{th}$ row and the nth row of $(C-\lambda I)x$ with the $(i+1)^{th}$ and n^{th} rows as yet unchanged. We denote the former by

$$u_i x_i + v_i x_{i+1} + w_i x_{i+2} + \ldots + h_i x_{n-1} + f_i x_n \quad (4.4)$$

and the latter two rows by

$$b_{i+1} x_i + (c_{i+1}-\lambda) x_{i+1} + b_{i+2} x_{i+2} \quad (4.5)$$

and

$$g_i x_i + g_{i+1} x_{i+1} + b_n x_{n-1} + (c_n-\lambda) x_n \quad . \quad (4.6)$$

When i=1, equation (4.4) is just the first row i.e.,

$$(c_1-\lambda)x_1 + b_2 x_2 + \ldots\ldots + b_1 x_n \quad , \quad (4.7a)$$

and equation (4.6) is the last row

$$b_1 x_1 + \ldots\ldots + b_n x_{n-1} + (c_n-\lambda) x_n \quad (4.7b)$$

where $g_1=b_1$ and $g_2=0$.

Now, applying pivoting techniques, we have that if $|u_i| > |b_{i+1}|$ and $|g_i|$, then equation (4.4) is chosen, if $|b_{i+1}| > |u_i|$ and $|g_i|$, then equation (4.5) is chosen, otherwise equation (4.6) is taken as the pivotal row. In each case, the pivotal row, in general can be denoted by

$$p_i x_i + q_i x_{i+1} + r_i x_{i+2} + \ldots\ldots + t_i x_{n-1} + s_i x_n \quad . \quad (4.8)$$

Hence, we have the following three cases to consider:-

<u>Case 1</u> For $|u_i| > |b_{i+1}|$ and $|g_i|$, then

$p_i = u_i$, $q_i = v_i$, $r_i = w_i$, $t_i = h_i$, $s_i = f_i$,

$m_{i+1} = b_{i+1}/u_i$,

$u_{i+1} = (c_{i+1}-\lambda)-m_{i+1}v_i$, $v_{i+1}=b_{i+2}-m_{i+1}w_i$, $w_{i+1}=0$, $h_{i+1}=-m_{i+1}h_i$,

$$f_{i+1} = -m_{i+1}f_i, \tag{4.9}$$

$$m_n = g_i/u_i,$$

$$g_i = g_{i+1}-m_nv_i,\ g_{i+1}=0,\ b_n=b_n-h_im_n,\ (c_n-\lambda)=(c_n-\lambda)-f_im_n$$

Case 2 For $|b_{i+1}| > |u_i|$ and $|g_i|$, then

$$p_i = b_{i+1},\ q_i = (c_{i+1}-\lambda),\ r_i = b_{i+2},\ t_i = 0,\ s_i = 0,$$

$$m_{i+1} = u_i/b_{i+1}, \tag{4.10}$$

$$u_{i+1} = v_i-(c_{i+1}-\lambda)m_{i+1},\ v_{i+1}=w_i-m_{i+1}b_{i+2},\ f_{i+1}=f_i,$$

$$h_{i+1} = h_i,\ w_{i+1} = 0,$$

$$m_n = g_i/b_{i+1},$$

$$g_i = g_{i+1}-(c_{i+1}-\lambda)m_n,\ g_{i+1} = -b_{i+2}m_n,\ b_n = b_n,\ (c_n-\lambda) = (c_n-\lambda),$$

Case 3 For $|g_i| > |b_{i+1}|$ and $|u_i|$, then

$$p_i = g_i,\ q_i = g_{i+1},\ r_i = 0,\ t_i = b_n,\ s_i = (c_n-\lambda),$$

$$m_{i+1} = u_i/g_i,$$

$$u_{i+1} = v_i-m_{i+1}g_{i+1},\ v_{i+1}=w_i,\ w_{i+1}=0,\ h_{i+1}=h_i-b_nm_{i+1},\ f_{i+1}=f_i-(c_n-\lambda)m_{i+1},$$

$$m_n = b_{i+1}/g_i, \tag{4.11}$$

$$g_i = (c_{i+1}-\lambda)-g_{i+1}m_n,\ g_{i+1}=b_{i+2},\ b_n=-b_nm_n,\ (c_n-\lambda)=-(c_n-\lambda)m_n.$$

The pivoting at the ith stage is at most a simple interchange of the ith, $i+1^{th}$ or n^{th} rows with fairly trivial modifications when the pivoting has reached the final two rows. Provided the interchanges are noted, the elements m_i, m_n, p_i, q_i, r_i, t_i and s_i provide enough information to solve the equations,

$$(C-\lambda I)x = v \tag{4.12}$$

for any right hand side vector v, by the appropriate forward and backward substitutions. Hence, equation (4.1) can be written in the form

$$LUx = v \tag{4.13}$$

provided we include the interchanges in L.

Now, Wilkinson (1965) has shown that if we take the initial vector v in the form

$$v = Le \tag{4.14}$$

where the vector e is of the form (1, 1, 1, ..., 1), then, substituting equation (4.14) in equation (4.13) gives the result

$$Ux = e \tag{4.15}$$

where we have seen that the upper triangular matrix U has the general form:-

$$U = \begin{bmatrix} p_1 & q_1 & r_1 & & t_1 & s_1 \\ & p_2 & q_2 & r_2 & t_2 & s_2 \\ & & \ddots & \ddots & \ddots & \vdots \\ & & & \ddots & \ddots & r_{n-2} \\ & & & & p_{n-1} & q_{n-1} \\ & & & & & p_n \end{bmatrix} . \qquad (4.16)$$

With this choice of v, x is determined by a back substitution only, and we have no need to determine Le explicitly. Once the vector x is obtained, we can then find the second iterated vector y by a forward and back substitution process.

5. Numerical Results

The algorithmic process given by equation (3.13) was checked for validity by computing the solution to the periodic characteristic value problem,

$$\delta^2 y_k + \lambda y_k = 0, \; y_{2N+2+k} = y_k, \; (k = 1, 2, \ldots, 2N+2). \qquad (5.1)$$

(Hildrebrand, 1968).

Choosing N = 4 and unit interval h, the problem reduces to determining the eigensolutions of the (10×10) matrix C given by

$$C = \begin{bmatrix} 2 & -1 & & & -1 \\ -1 & 2 & -1 & & \\ & \ddots & \ddots & \ddots & \\ & & \ddots & \ddots & -1 \\ -1 & & & -1 & 2 \end{bmatrix} . \qquad (5.2)$$

An ALGOL procedure to compute the Sturm sequence of polynomials $p_i(\lambda)$, $i = 1,2,\ldots 10$ was inserted in the ALGOL program BISECT given by Barth et al (1967) and the numerical results presented in Table 1 confirm the existence of N(=4) coincident eigenvalues $\lambda_j = 4 \sin^2\{j\pi/2(N+1)\}$, $j = 1,2,\ldots N$ in addition to the 2 eigenvalues of unit multiplicity $\lambda_0 = 0$ and $\lambda_{N+1} = 4$.

Similarly, the complete eigensolution to the general three term matrix

$$\begin{bmatrix} 3 & -1 & & & -1 \\ -1 & 4 & -1 & & \\ & -1 & 5 & -1 & \\ & & -1 & 6 & -1 \\ -1 & & & -1 & 7 \end{bmatrix} \qquad (5.3)$$

can be obtained by using the procedures outlined in this paper. The matrix can be shown to possess the eigenvalues and eigenvectors as given in Table 2.

6. Applications

Further generalisations of the algorithm developed in Section 3 can be carried out by setting $n = 3$ and performing further analysis on bordering it with further rows and columns. This yields a similar recursive algorithm for the characteristic equation of a symmetric quindiagonal matrix (Evans, 1971) and indicates a suitable approach for treating sparse symmetric matrices of wider bandwidth.

7. Acknowledgements

The author is indebted to Mrs. L.A. Chester for programming assistance.

8. References

Froberg, C.E. 'Introduction to numerical analysis'
Addison Wesley Pub. p 258. (1965)

Wilkinson, J.H. 'The algebraic eigenvalue problem'
Oxford Univ. Press. (1965)

Wilkinson, J.H. Num. Math. 4, pp 368-376. (1962)

Barth, W., R.S. Martin & J.H. Wilkinson Num.Math. 9, pp 386-393.(1967)

Martin, R.S. & J.H. Wilkinson Num. Math. 9, pp 279-301. (1967)

Hildrebrand, F.B. 'Finite Difference Equations and Simulations'
Prentice Hall Inc. p 53. (1968)

Evans, D.J. to be published. (1971)

p \ λ	λ_0	λ_1	λ_2	λ_3	λ_4	λ_5
	0•0000000	0•3819660	1•3819660	2•6180340	3•6180340	4•0000000
p_1	2•00000,0	1•61803,0	6•18034,-1	-6•18034,-1	-1•61803,0	-1•99999,0
p_2	1•50000,0	1•00000,0	-9•99999,-1	1•00000,0	-9•99999,-1	-1•49999,0
p_3	1•33333,0	6•18034,-1	1•61803,0	-1•61803,0	-6•18034,-1	-1•33333,0
p_4	1•25000,0	2•18279,-10	1•67347,-10	4•80213,-10	1•23691,-9	-1•24999,0
p_5	1•20000,0	-4•58130,9	-5•97561,9	-2•08241,9	-8•08464,8	-1•19999,0
p_6	1•16667,0	1•61803,0	6•18034,-1	-6•18034,-1	-1•61803,0	-1•16667,0
p_7	1•14286,0	1•00000,0	-9•99999,-1	1•00000,0	-9•99999,-1	-1•14286,0
p_8	1•12500,0	6•18034,-1	1•61803,0	-1•61803,0	-6•18034,-1	-1•12499,0
p_9	1•11111,0	4•07454,-10	3•34694,-10	9•60426,-10	2•50293,-9	-1•11111,0
p_{10}	-7•45786,-11	-1•25000,-1	6•25000,-2	-3•12500,-2	-1•56250,-2	2•91402,-9

Table 1.

$p_9 = 0$ for λ_j, $j = 1,2,3,4$ and $p_{10} = 0$ for λ_j, $j = 0,5$ confirm the existence of 4 coincident eigenvalues and 2 single eigenvalues together with the correct number of negative p's.

λ_1	λ_2	λ_3	λ_4	λ_5
7•8652523,0	6•3817653,0	4•8549683,0	3•7995583,0	2•0984556,0
v_1	v_2	v_3	v_4	v_5
-1•6471501,-1	2•3293634,-1	1•2145695,-1	-5•2675806,-1	7•9144409,-1
-6•1600990,-3	-3•5809782,-1	-5•7502016,-1	5•0894206,-1	5•3109112,-1
1•8852487,-1	6•1996895,-1	3•7016712,-1	6•2877105,-1	2•1844925,-1
-5•3401105,-1	-4•9855384,-1	6•2870611,-1	2•4586101,-1	1•0274909,-1
8•0754046,-1	-4•2963843,-1	3•4972127,-1	-8•7768225,-2	1•8243090,-1

Table 2.

Algol Program

The Sturm sequence ALGOL procedure which was used is given below.

```
begin
    comment sturm sequence;
    array p[1:n],q[1:n];
    real prod;
    z:=z+1;
    a:=0; i:=1;
    p[i]:=c[i]-x1;
    q[i]:=(if p[i]≠0 then 1/p[i] else 1/relfeh);
    prod:=(if p[i]≠0 then -b[i]/p[i] else -b[i]/relfeh);
    if p[i]<0 then a:=a+1;
    i:=2;
    p[i]:=c[i]-x1-(if p[i-1]≠0 then beta[i]/p[i-1] else
                beta[i]/relfeh);
    q[i]:=(c[i]-x1)×(if p[i]≠0 then q[i-1]/p[i] else
                q[i-1]/relfeh);
    prod:=-prod×(if p[i]≠0 then b[i]/p[i] else b[i]/relfeh);
    if p[i]<0 then a:=a+1;
    for i:=3 step 1 until n-1 do
    begin
        p[i]:=c[i]-x1-(if p[i-1]≠0 then beta[i]/p[i-1] else
                beta[i]/relfeh);
        q[i]:=(c[i]-x1)×(if p[i]≠0 then q[i-1]/p[i] else
                q[i-1]/relfeh);
        q[i]:=q[i]-beta[i]×(if (p[i]×p[i-1])≠0 then q[i-2]
              /(p[i]×p[i-1]) else q[i-2]/relfeh);
        prod:=-prod×(if p[i]≠0 then b[i]/p[i] else b[i]/relfeh);
        if p[i]<0 then a:=a+1;
    end i;
    i:=n;
    p[i]:=c[i]-x1-beta[i]×q[i-1]+2×b[i]×prod;
    p[i]:=p[i]-(if p[i-1]≠0 then beta[i]/p[i-1] else
                beta[i]/relfeh);
    if p[i]<0 then a:=a+1;
    if a<k then
    begin if a<m1 then    xu:=wu[m1]:=x1
    else
    begin xu:=wu[a+1]:=x1;
    if x[a]>x1 then x[a]:=x1;
    end
    end
    else xo:=x1;
end x1;
```

ONE DIMENSIONAL METHODS FOR THE NUMERICAL SOLUTION OF NONLINEAR HYPERBOLIC SYSTEMS

A. R. Gourlay, G. McGuire and J. Ll. Morris

1. Introduction

We will introduce novel first order and second order accurate methods for the numerical solution of the system of conservation laws

$$\frac{\partial \underset{\sim}{u}}{\partial t} + \sum_{j=1}^{M} \frac{\partial \underset{\sim}{f}_j}{\partial x_j} = \underset{\sim}{0} \qquad (1.1)$$

in M space dimensions where $\underset{\sim}{u}$ is an N-vector $(u_1, u_2, \ldots, u_N)^T$ and the $\underset{\sim}{f}_j$ are nonlinear functions of the unknown $\underset{\sim}{u}$. The methods considered have been developed in order that they satisfy the following criteria:

(a) The introduction of dissipation (in the sense of Kreiss - see Richtmyer and Morton [10]) with the appropriate sign of the coefficient of the dissipative term should be such that the linearized stability of the schemes should not be affected other than, possibly, if the coefficient is zero the scheme may be unstable.

(b) The method in one space variable with dissipation added either explicitly or implicitly in a one dimensional manner forms the basis for schemes in higher space dimensions which preserve condition (a) and so that at any one time , dissipation in one space variable only is added and the stability criterion for one space variable is preserved in the higher space dimensions.

(c) The one space variable method should be optimally stable, i.e. satisfy the C.F.L. condition exactly.

To our knowledge no present method exists which satisfies these three criteria and in our sense is therefore not good. Consequently the methods we introduce will satisfy these three criteria and in this sense are better than existing methods.

The aim of such methods is the good resolution of shock like phenomena which, as is well known, often occur in nonlinear hyperbolic systems. We believe the above criteria are necessary qualities of any scheme which is to produce good shock resolution.

2. Criterion b

At first, criterion (b) might appear to be the most formidable to satisfy. However, with the introduction of the multistep formulations of difference schemes first introduced by Strang [11],[12] and subsequently studied by Gourlay and Morris [4],[5], McGuire and Morris [7] it is now clear that multidimensional problems can be solved in a simple manner by resorting to one dimensional methods with optimal characteristics as the basis of factorizations which produce the methods for equations of the form (1.1). Briefly, if $L_{x_1}^{(1)}$ is a first order approximation to the one space dimensional equation

$$\frac{\partial \underset{\sim}{u}}{\partial t} + \frac{\partial \underset{\sim}{f}_1}{\partial x_1} = \underset{\sim}{0} \tag{2.1}$$

and $L_{x_2}^{(1)}$ is a first order approximation to the similar equation

$$\frac{\partial \underset{\sim}{u}}{\partial t} + \frac{\partial \underset{\sim}{f}_2}{\partial x_2} = \underset{\sim}{0} \tag{2.2}$$

then the approximation

$$\underset{\sim}{u}_{m+1} = L_{x_1}^{(1)} L_{x_2}^{(1)} \underset{\sim}{u}_m$$

with $\underset{\sim}{u}_{m+1} \equiv u(ih, jh, (m+1)k)$ etc., h,k the mesh spacings of a regular rectangular grid in the space and time directions respectively, is a first order approximation to equation (1.1) with M=2 which retains the stability characteristics of $L_{x_1}^{(1)}$ and $L_{x_2}^{(1)}$ and which at any stage of the calculation involves a one dimensional solution (see Gourlay and Morris [4] for the details). In a similar manner the first order scheme can be written down with general M in equation (1.1), namely

$$\underset{\sim}{u}_{m+1} = \prod_{j=1}^{M} L_{x_j}^{(1)} \underset{\sim}{u}_m \tag{2.3}$$

where the $L^{(1)}_{x_j}$ are first order approximations to equations similar to (2.1) and (2.2).

Similarly if $L^{(2)}_{x_j}$ is a second order approximation to an equation like (2.1), and (2.2) then

$$\underset{\sim}{u}_{m+1} = \prod_{j=1}^{M-1} L^{(2)}_{\frac{x}{2}j} \cdot L^{(2)}_{x_M} \cdot \prod_{j=M-1}^{1} L^{(2)}_{\frac{x}{2}j} \quad \underset{\sim}{u}_m \qquad (2.4)$$

is a second order approximation to (1.1) which again, for operators $L^{(2)}_{x_j}$ with optimal characteristics (satisfy the C.F.L. condition exactly for example) can be proved to be optimally stable. Other combinations of the operators are possible which produce difference methods which satisfy criterion (b) namely

$$\underset{\sim}{u}_{m+1} = \tfrac{1}{2}\left\{\prod_{j=1}^{M} L^{(2)}_{x_j} + \prod_{j=M}^{1} L^{(2)}_{x_j}\right\} \quad \underset{\sim}{u}_m$$

and

$$\underset{\sim}{u}_{m+1} = L^{(2)}_{\frac{x_1}{2}} \left\{\prod_{j=2}^{M} L^{(2)}_{x_j} + \prod_{j=M}^{2} L^{(2)}_{x_j}\right\} L^{(2)}_{\frac{x_1}{2}} \quad \underset{\sim}{u}_m$$

where $L^{(2)}_{\frac{x_1}{2}}$ is a second order approximation to the solution of (2.1) at $(m+\frac{1}{2})k$.

These latter two methods however are less efficient computationally than (2.4) on account of the possible combinations of "half-operators" at the beginning of one step and end of the previous one (see Gourlay and Morris [4]).

Consequently using the procedure outlined above criterion (b) can be satisfied in a straight forward manner.

3. A first order method satisfying criteria a-c

By a combination of the well known Lax method [6] with explicit pseudo viscosity term $\sigma\delta_x{}^2 u_i^m$ and the backward difference scheme with a pseudo viscosity term $\sigma\delta_x{}^2 \underset{\sim}{u}_i^{m+1}$, the first order accurate Hopscotch-Lax method is proposed for the solution of the one dimensional equation (2.1); namely

$$\underset{\sim}{u}_i^{m+1} + \theta_i^{m+1} [p/2\ H_x\ \underset{\sim}{f}_i^{m+1} - \sigma\delta_x{}^2 \underset{\sim}{u}_i^{m+1}] = \underset{\sim}{u}_i^m - \theta_i^m [p/2\ H_x\ \underset{\sim}{f}_i^m - \sigma\delta_x{}^2 \underset{\sim}{u}_i^m] \qquad (3.1)$$

where

$$\theta_i^m \equiv \begin{cases} 1 & m+i \text{ odd} \\ 0 & m+i \text{ even}, \end{cases}$$

and $H_x \underset{\sim}{u}_i^m \equiv \underset{\sim}{u}_{i+1}^m - \underset{\sim}{u}_{i-1}^m$, $\delta_x \underset{\sim}{u}_i^m \equiv \underset{\sim}{u}_{i+\frac{1}{2}}^m - \underset{\sim}{u}_{i-\frac{1}{2}}^m$ and $\mu_x \underset{\sim}{u}_i^m \equiv \dfrac{\underset{\sim}{u}_{i+1}^m + \underset{\sim}{u}_{i-1}^m}{2}$ are the usual difference operators. The way in which (3.1) is applied is similar to the Hopscotch scheme described by Gourlay [2] and will hence not be repeated. The stability of this method can be investigated by considering the constituent schemes

$$(1+2\sigma)\underset{\sim}{u}_i^m = \underset{\sim}{u}_i^{m-1} - [p/2\, H_x \underset{\sim}{f}_i^m - 2\sigma\mu_x \underset{\sim}{u}_i^m] \tag{3.2}$$

$$\underset{\sim}{u}_i^{m+1} = (1-2\sigma)\underset{\sim}{u}_i^m - [p/2\, H_x \underset{\sim}{f}_i^m - 2\sigma\mu_x \underset{\sim}{u}_i^m]. \tag{3.3}$$

Investigating the sums of the indices of the functions in (3.2) and (3.3) it can be seen that all the functions, except $\underset{\sim}{u}_i^m$, have the similar property of having the sum either odd or even. Consequently $\underset{\sim}{u}_i^m$ is that value calculated by the complement scheme to (3.2) and (3.3) which together make up (3.1). Thus eliminating $\underset{\sim}{u}_i^m$ from (3.2) and (3.3) we obtain the equation

$$(1+2\sigma)\underset{\sim}{u}_i^{m+1} = (1-2\sigma)\underset{\sim}{u}_i^{m-1} - 2[p/2\, H_x \underset{\sim}{f}_i^m - 2\sigma\mu_x \underset{\sim}{u}_i^m] \tag{3.4}$$

upon which to investigate stability. This can be done in the usual von Neumann manner; linearizing equation (3.4) and assuming a Fourier decomposition of the errors. Doing this we obtain the equation for the eigenvalues of the amplification matrix, namely

$$(1+2\sigma)\rho^2 + 2[p\lambda i \sin\beta h - 2\sigma\cos\beta h]\rho - (1-2\sigma) = 0 \tag{3.5}$$

where ρ is the eigenvalue of the amplification matrix, λ is an eigenvalue of the Jacobian matrix of $\underset{\sim}{f}$ with respect to $\underset{\sim}{u}$, $i = \sqrt{-1}$ and β is a constant. For stability we require the roots of equation (3.5) to be less than or equal to 1 in modulus. To investigate the location of these roots we employ the results given by Miller in the paper contained in this volume [8]. With the notation of that paper we form

$$F(\rho) = (1+2\sigma)\rho^2 + 2[p\lambda i \sin\beta h - 2\sigma\cos\beta h]\rho - (1-2\sigma)$$

and

$$F^*(\rho) = -(1-2\sigma)\rho^2 + 2[-p\lambda i \sin\beta h - 2\sigma\cos\beta h]\rho + (1+2\sigma)$$

from which

$$F^*(0) = (1+2\sigma), \; F(0) = -(1-2\sigma)$$

so that

$$|F^*(0)| > |F(0)| \text{ if } \sigma > 0 .$$

To ensure the roots of (3.5) lie on the unit disk, we employ Theorem 6.1 of Miller's paper so that we need to show

$$F_1 \equiv \frac{F^*(0)\, F(\rho) - F(0)\, F^*(\rho)}{\rho}$$

is von Neumann. Evaluating F_1 we find after some simplification

$$F_1 = 8\sigma[\rho + p\lambda i \sin \beta h - \cos \beta h].$$

To show that F_1 is von Neumann we merely require to restrict the zero of this equation to be less than or equal to 1 in modulus. This is easily seen to be the case if

$$p\,|\lambda max| \leqslant 1 .$$

Thus the Hopscotch-Lax method is optimally stable for all $\sigma > 0$.

Consequently the Hopscotch-Lax method satisfies criteria a-c.

4. A second order method satisfying a-c

A generalization of the two step version of the Lax-Wendroff method due to Richtmyer [9] produces the second order accurate scheme given by

$$\begin{aligned} \underset{\sim}{u}_i^{*m+1} &= \tfrac{1}{2}\,(\underset{\sim}{u}^m_{i+\frac{1}{2}} + \underset{\sim}{u}^m_{i-\frac{1}{2}}) - 2a\ \delta_x\ \underset{\sim}{f}_i^m \\ \underset{\sim}{u}_i^{m+1} &= \underset{\sim}{u}_i^m - \frac{p}{2}\Big[(1-\frac{1}{4a})\ H_x\ \underset{\sim}{f}_i^m + \frac{1}{2a}\ \delta_x\ \underset{\sim}{f}_i^{*m+1}\Big] \end{aligned} \qquad (4.1)$$

where $\underset{\sim}{f}_i^{*m+1} = \underset{\sim}{f}(\underset{\sim}{u}_i^{*m+1})$, where a is a parameter controlling the amount of pseudo viscosity introduced implicitly into the scheme.

The stability of (4.1) can be investigated in the usual linearized manner. We find that on eliminating the * level of (4.1), the resulting linearized equation has the form

$$\underset{\sim}{u}_i^{m+1} = \underset{\sim}{u}_i^m - \tfrac{1}{2}pA\ [H_x\ \underset{\sim}{u}_i^m - pA\ \delta_x^{\,2}\ \underset{\sim}{u}_i^m]$$

where A is the assumed locally constant Jacobian. This equation is independent of the parameter a so that the linearized stability is independent of a. A straight forward analysis (it is the same analysis given by Burstein and Rubin [1] whose scheme is given by substituting $a=\frac{1}{2}$ in (4.1)) produces the result that (4.1) is optimally stable. Hence the method (4.1) satisfies criteria a-c.

5. Numerical results

Because of the considerations of space, the account of the numerical experiments carried out must be brief - a full account of the results presented in this paper can be found in [3].

Problems with smooth solutions were first investigated to study the validity of the linearized analyses carried out in sections 3 and 4. The errors obtained were of the correct magnitude and the linearized analyses appeared to represent well the dependence of the methods on the parameters σ and a.

Discontinuous problems were also investigated and the methods have proved successful in eliminating the common feature in existing methods of oscillation build up behind the shock front.

These methods are currently being investigated upon multidimensional physical problems of hydrodynamic flows. These results should be published in the future.

REFERENCES

[1] S.Z. Burstein and E. Rubin: Difference methods for the inviscid and viscous equations of a compressible gas. J. Comp. Phys., 2, (1967).

[2] A.R. Gourlay: Hopscotch: a fast second order partial differential equation solver. J.I.M.A., 6, (1970).

[3] A.R. Gourlay, G. McGuire and J.Ll. Morris: One dimensional methods for the numerical solution of nonlinear hyperbolic systems. To appear as I.B.M., Peterlee Scientific Centre, technical report No.1, 1971.

[4] A.R. Gourlay and J.Ll. Morris: A multistep formulation of the optimized Lax-Wendroff method for nonlinear hyperbolic systems in two space variables. Maths of Comp., 22, (1968).

[5] A.R. Gourlay and J.Ll. Morris: On the comparison of multistep formulations of the optimized Lax-Wendroff method for nonlinear hyperbolic systems in two space variables. J. Comp. Phys., 5, (1970).

[6] P.D. Lax: Weak solutions of nonlinear hyperbolic equations and their numerical computation. Comm. Pure. Appl. Math., 7, (1954).

[7] G. McGuire and J.Ll. Morris: Boundary techniques for the multistep formulations of the optimized Lax-Wendroff method for nonlinear hyperbolic systems in two space dimensions. (Submitted to J.I.M.A., 1971.)

[8] J.J.H. Miller: On weak stability, stability, and the type of a polynomial. (This volume page 316.)

[9] R.D. Richtymer: A survey of difference methods for nonsteady fluid dynamics. NCAR Tech. Notes, 63-2.

[10] R.D. Richtymer and K.W. Morton: Difference methods for initial value problems. Wiley, Interscience, (1967.)

[11] W.G. Strang: Accurate partial difference methods II. Nonlinear problems, Numer. Math. 13, (1964).

[12] W.G. Strang: On the construction and comparison of difference schemes. S.I.A.M. J. Numer. Anal., 5, (1968).

THE DEVELOPMENT AND APPLICATION OF SIMULTANEOUS ITERATION FOR EIGENVALUE PROBLEMS

A. Jennings

Introduction

The development of the power method for determining the eigenvalues of a matrix has been closely connected with the vibration analysis of mechanical and structural systems. The Stodola-Vianello iterative method for vibration dates from 1904 [1]. When applied to systems having only concentrated masses it can be shown that each iteration involves pre-multiplying a trial vector by the 'dynamical matrix' which is the product of the flexibility and mass matrices of the system. This was generalised into the power method for obtaining the dominant eigenvalue of a matrix by von Mises and Geiringer in 1929. The work of Duncan and Collar [2] in extending the power method to subdominant eigenvalues and eigenvectors was associated with the application to vibration and flutter analysis of aircraft.

The extension of the power method to involve simultaneous iteration with two or more trial vectors has been developed at Belfast primarily with the purpose of facilitating vibration analysis. This paper is a summary of the development indicating the connection with, and potential regarding, some possible applications.

Initial Development of Simultaneous Iteration for Symmetric Matrices

To obtain good idealisations of complicated structures such as aircraft, ships, dams etc. it is desirable to be able to operate with dynamic equations of order many thousand but normally no more than about 20 frequencies of vibration will be required. When translated into an eigenvalue problem this means that a partial eigensolution is needed in which the required eigenvalues and eigenvectors are the dominant ones. On account of this it would appear that the power method or any improvement of it is likely to be of use for large order vibration analyses. Poor convergence of the power method arises when eigenvalues are close together and is due to the difficulty in eliminating components of eigenvectors subdominant to the one being determined. The main advantage of simultaneous iteration over the power method is that by iterating for all the required eigenvectors simultaneously the only elimination that is required is concerned with eigenvector components subdominant to the whole set of required eigenvectors. This results in much more satisfactory convergence characteristics.

A satisfactory simultaneous iteration method for symmetric matrices was first evolved at Belfast by using a set of mutually orthogonal trial vectors which are assumed to be close to the required eigenvectors. The resulting perturbation solution was then modified to a form which would give convergence for any set of mutually orthogonal vectors [3]. Let A be an n x n symmetric matrix with eigenvalues λ_1, λ_2 --- λ_n in descending order of modulus and corresponding eigenvectors q_1, q_2 --- q_n which may be compounded by colums to give an n x n matrix Q. The trial vectors u_1, u_2 --- u_m may also be compounded by columns forming an n x m matrix U. From the initial perturbation assumption and the orthogonal condition it is possible to show that if

$$U = QC \tag{1}$$

then the coefficients of C have the properties

$$\left.\begin{aligned} c_{ii} &\approx 1 \\ c_{ij} &\approx -c_{ji} << 1 \end{aligned}\right\} \tag{2}$$

Simultaneous premultiplication of all the trial vectors gives

$$V = AU = Q\Lambda C \tag{3}$$

where Λ is a diagonal matrix of the eigenvalues. The matrix of inner products

$$B = U^T V = U^T AU \tag{4}$$

can be used to estimate the coefficients in the first m rows of C and has been called the interaction matrix. The linear predictions for off-diagonal elements of C are according to

$$c_{ij} = \frac{b_{ij}}{b_{ii}-b_{jj}} \tag{5}$$

and can be used to eliminate the coupling between the vectors V after premultiplication. The diagonal elements of B give Rayleigh quotient predictions of the dominant eigenvalues.

Two extra facilities have been included to convert the procedure into a viable iterative algorithm which is suitable when any orthogonal set of trial vectors is used. The first is to include a method of curtailing the production of large c_{ij} values due to cancellation in the denominator of equation (5) while permitting small c_{ij} values to be virtually unchanged. Whereas a smooth curve was determined to perform this filtering process (figure 1), it appears that the convergence is fairly insensitive to the form of curve and a simple cut off for large c_{ij} values may be adequate. The second facility involves a re-orthogonalisation of the vectors before starting the next iteration performed by orthogonalising the lower vectors with respect to the higher ones. As it was unlikely that there would be a theoretical proof of convergence from any starting vectors the algorithm was tested numerically over a range of problems starting with any trial vectors and demanding high accuracy. It is significant that this basic method has proved reliable even when a large number of trial vectors are used. The largest number of vectors simultaneously processed at Belfast was 65 for a principal component analysis involving a dense matrix of order 130.

The highest eigenvectors tend to gain accuracy most rapidly as illustrated in figure 2 and iteration with these vectors may be curtailed when they have passed the tolerance test. Slow convergence of the lower eigenvectors can be avoided by carrying extra 'guard' vectors. In the

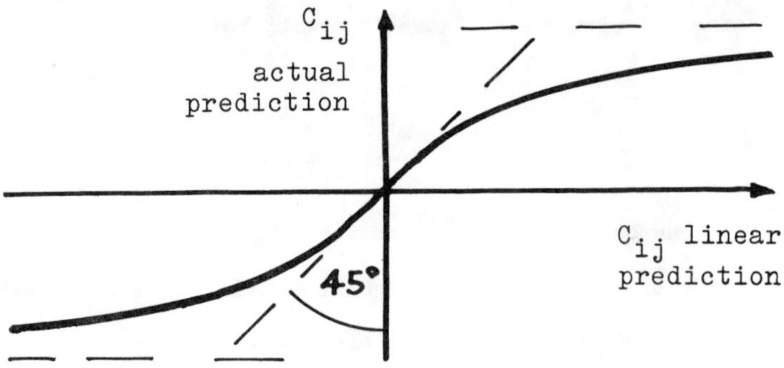

Figure 1. Nonlinear Filter for prediction of coupling coefficients.

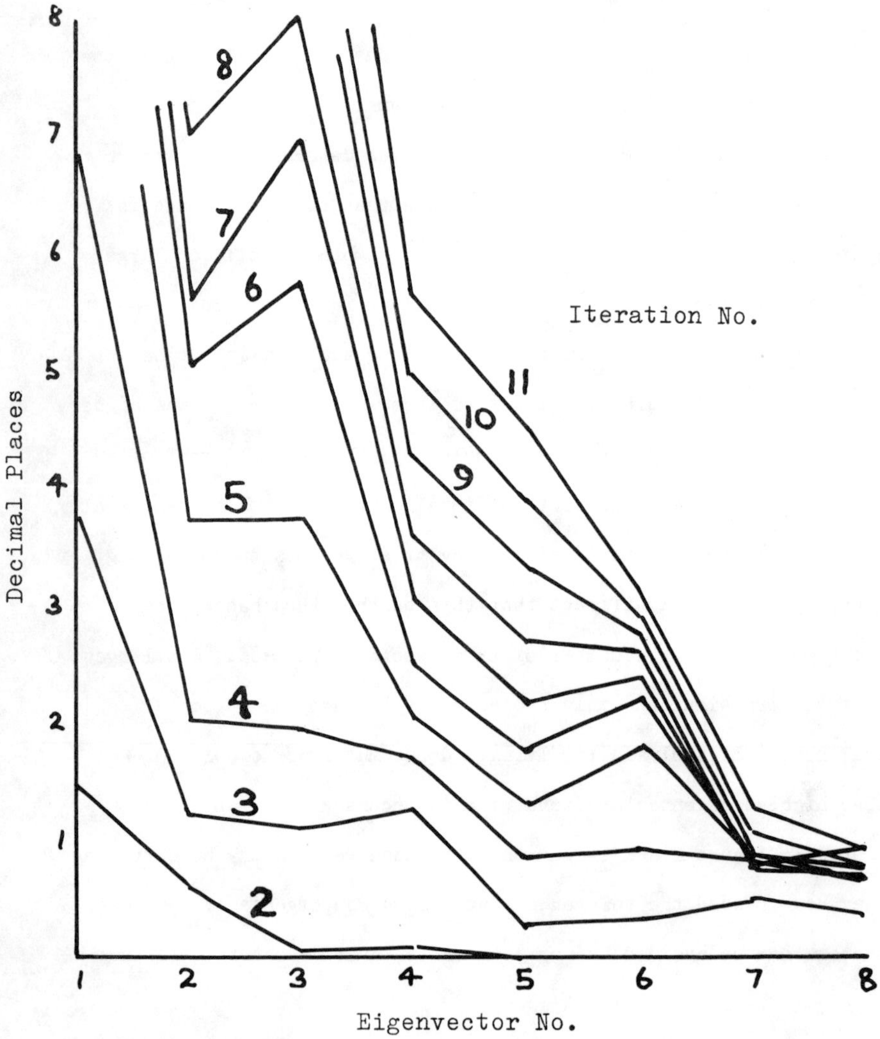

Figure 2. Accuracy of eigenvector prediction after different numbers of iterations.

example of figure 2 if only 5 eigenvectors are required then the extra 3 vectors would be guard vectors and all 5 required eigenvectors would have passed a tolerance of 10^{-3} after 9 iterations.

Alternative Procedures for Symmetric Eigenvalue Problems

Although the given method is possibly the simplest simultaneous iteration procedure many variations and additional features are possible. It was recognised that the interaction analysis was simply an approximate eigensolution of the interaction matrix $B = U^T AU$ and numerical tests involving an eigensolution of B within the iteration loop showed that the speed of convergence could be improved in cases where washout of the lower eigenvector error components is fast [4]. A similar procedure was developed by Rutishauser [5] as an extension of Bauer's bi-iteration method in which an eigensolution of the matrix

$$B_R = V^T V = U^T A^2 U$$

was used within the iteration cycle. This seems to be more satisfactory because the processed vectors are still theoretically mutually orthogonal. However Rutishauser still needs an orthogonalising sequence and special steps have to be taken in the case where eigenvalues may have negative signs [6]. Two extra facilities adopted by Rutishauser are the use of a Chebychev polynomial procedure for accelerating the washing out of unwanted eigenvectors and the insertion of a random vector for the last guard vector at each iteration. Simultaneous iteration has much less risk than the power method of missing out an eigenvector, and this last facility makes the risk even less. Both devices have been found to work satisfactorily with the non-linear prediction method of interaction analysis (despite the footnote in Rutishauser's paper).

Simultaneous iteration methods tend to be more efficient and reliable than the power method and have the same advantages. Not only are they of advantage when a partial eigensolution is required but also they can take advantage of known approximate solutions (when a similar case has been analysed previously) and they can take full advantage of any sparseness in the original matrix. Because the original matrix is only involved in a matrix multiplication operation there is no need for it to be stored explicitly. Computing time is saved if only low accuracy predictions for the eigenvectors are required. It is therefore necessary to review carefully any possible application to determine whether the properties of the problem make simultaneous iteration particularly suitable.

Undamped Vibration

For the analysis of undamped vibration not only is a partial eigensolution required but also most of the sparseness of the vibration equations may be retained. The basic equations may be written in the form

$$\omega^2 Mx = Kx$$

where M and K are sparse symmetric mass and stiffness matrices which are normally positive definite. This may be reduced to a symmetric eigenvalue problem by using auxiliary variables

$$y = L^T x \tag{6}$$

where L is the Choleski lower triangular matrix formed from the decomposition of K according to

$$K = LL^T, \tag{7}$$

the resulting equation being

$$L^{-1}ML^{-T}y = \lambda y \tag{8}$$

where $\lambda = 1/\omega^2$. It is then possible to apply simultaneous iteration methods storing the sparse matrix M and the matrix L which can be in band form instead of the dense matrix $L^{-1}ML^{-T}$ [7, 8].

Simultaneous iteration starts to become significantly better than transformation methods when $n > 100$. For very large n only methods which retain the band form of the equations are feasible from the computer storage standpoint. For large problems the M and K or L matrices will be held in backing store and hence any operation with those matrices will involve calling peripherals. One particular advantage of simultaneous iteration over the power method, Lanczos's method [9] and the sturm sequence method of Peters, Wilkinson and Gupta [10, 11] is that it will require the least number of transfer operations to effect a solution. If in the core store there is sufficient space for two sets of trial vectors and a little extra working space then simultaneous iteration may be programmed with only the operations involving M and L out of core.

Principal Component Analysis

In principal component analysis the dominant eigenvectors of a covariance (or correlation) matrix are determined. The matrix will be dense and symmetric positive definite. Normally only the first two or three eigenvectors will be required and these only to a low accuracy. Simultaneous iteration is suitable for this purpose, but if many more eigenvectors are required the closeness of the lower eigenvalues might make the use of Chebychev acceleration of the simultaneous iteration advisable. Where the population to be analysed is large enough, it is advisable to take several samples and repeat the analysis for each to determine which eigenvalues and eigenvectors show any degree of consistency. If this is done simultaneous iteration will be able to take advantage of the previous solutions for all but the first analysis.

Simultaneous Iteration for Unsymmetric Matrices

Simultaneous iteration may be extended to unsymmetric matrices by using two sets of trial vectors which iterate onto the dominant subsets of left-hand and right-hand eigenvectors [12]. In the simplest procedure it is necessary for the trial vectors to enter the complex plane if any of the required eigenvectors are complex. The trial vectors must be bi-orthogonal, i.e. if $\bar{U}$ and U are sets of m LH and RH trial vectors respectively then

$$\bar{U}^H U = I \qquad (9)$$

when H is the Hermitian transpose. If $\bar{U}$ and U are related to the full set of LH and RH eigenvalues according to

$$\left.\begin{aligned} \bar{U} &= \bar{Q}\bar{C} \\ \text{and}\quad U &= QC \end{aligned}\right\} \qquad (10)$$

a perturbation solution based on the interaction matrix

$$B = \bar{U}^H A U \qquad (11)$$

gives estimates of the off-diagonal elements of the first m rows of $\bar{C}$ and C as follows

$$c_{ij} = -\bar{c}^*_{ji} = \frac{b_{ij}}{b_{ii}-b_{jj}} \quad . \qquad (12)$$

It was found that curtailing of large predictions due to cancellation of the denominator has to be based on the size of the product $c_{ij}c_{ji}$ rather than c_{ij} itself. The full iteration cycle consists of premultiplication of both sets of trial vectors according to

$$\left.\begin{aligned} \bar{V} &= A^H\bar{U} \\ \text{and}\quad V &= AU \end{aligned}\right\} \qquad (13)$$

followed by an interaction analysis to eliminate the coupling arising from the predicted coefficients c_{ij}, and then a bi-orthogonalising process and tolerence test. In the case of a real matrix which can have complex pairs of eigenvalues then an alternative procedure has been developed in which only real trial vectors are used [13]. The presence of complex pairs is predicted by examination of the interaction matrix and the interaction analysis is modified so that convergence is to the real and imaginary parts of the complex eigenvector (i.e. q_r and q_i instead of $q_r + iq_i$ and $q_r - iq_i$). The unsymmetric algorithms have not yet been so effectively tested as the symmetric algorithms and no allowance has been made for the occurence of non-linear elementary divisors. However the possible application of simultaneous iteration in two fields has been considered.

Flutter Analysis

The classical British method of flutter analysis involves finding the characteristic solution of the dynamic response equation

$$A\ddot{z} + B\dot{z} + Cz = 0. \tag{14}$$

Here A is a mass matrix while B and C are damping and stiffness matrices which may both have a structural and an aerodynamic component. Introducing new variables **x** where

$$\mathbf{z} = \mathbf{x} e^{\lambda t} \tag{15}$$

gives

$$(\lambda^2 A + \lambda B + C)x = 0, \tag{16}$$

which may be converted into a conventional matrix eigenvalue problem to give

$$\begin{bmatrix} -A^{-1}B & -A^{-1}C \\ I & 0 \end{bmatrix} \begin{bmatrix} \lambda x \\ x \end{bmatrix} = \lambda \begin{bmatrix} \lambda x \\ x \end{bmatrix} . \tag{17}$$

As the undamped vibration modes are normally used as generalised co-ordinates these equations tend not to be of large order. For dynamic stability it is necessary that the solutions for λ should all lie in the negative half of the Argand diagram. If simultaneous iteration is applied directly to the 2n x 2n flutter matrix then a full eigensolution will be required, but as it is necessary to analyse a range of flight speeds the eigenvectors computed for one frequency will be useful as trial vectors for the next flight speed. The number of iterations per flight speed will depend on the accuracy demanded and the length of the flight speed interval. Tests indicate that it may be possible to keep this to about two iterations/flight speed in which case the computing times are likely to be comparable with those using the QR method.

There is a possibility of increasing the efficiency of simultaneous iteration by transforming the flutter matrix so that only a partial eigensolution is required. For instance Strachey's transformation [14] has the effect of making unstable flutter modes correspond to the eigenvalues of largest modulus of the transformed matrix.

The derivation of the aerodynamic matrices for the dynamic response equation (14) is based on a structure which is oscillating steadily at a given frequency parameter, and this restricts the validity and usefulness of the solutions. However Richardson [15] has formulated a new approach to the representation of the aerodynamic forces which gives a more satisfactory formulation of the dynamic response. This method gives rise to larger order flutter equations, but which only require a partial eigensolution. In a recent survey of methods of flutter analysis [16] the need for efficient computational algorithms for use with Richarson's method is recognised. From these considerations it seems that there is scope for further development of the applications of simultaneous iteration for flutter analysis.

Markov Chains

A Markov chain may be defined according to the equation

$$x^{(k+1)} = Px^{(k)} \qquad (18)$$

where P is an unsymmetric stochastic matrix and $x^{(k)}$ is a column vector of state probabilities for iteration k in which the probability of being in state i is $x_i^{(k)}$. The eigenvector(s) corresponding to the eigenvalue(s) of 1 represent the ultimate state probabilities to which the Markov process tends. No eigenvalue can have modulus greater than 1. If simultaneous iteration is used to determine the dominant eigenvalues and corresponding eigenvectors, then the solution will determine whether more than one recurrent chain exists; it will also determine the probabilities associated with these and the rate of convergence of the Markov process (given by the magnitude of the subdominant eigenvalues). For large sparse stochastic matrices, simultaneous iteration would seem to be an effective method of analysis.

General Comment

Although simultaneous iteration has been developed for use in the analysis of structural vibration, other eigenvalue problems may have special characteristics which make simultaneous iteration an attractive method of analysis.

References

1. S. H. Crandall. "Engineering Analysis, A Survey of Numerical Procedures". McGraw-Hill. New York and London, (1956).

2. W. J. Duncan and A. R. Collar. Matrices applied to the motions of damped systems. Philos. Mag. Ser. 7, 19 (1935), 197-219.

3. A. Jennings. A direct iteration method of obtaining latent roots and vectors of a symmetric matrix. Proc. Camb. Phil. Soc., 63 (1967), 755-765.

4. M. Clint and A. Jennings. The evaluation of eigenvalues and eigenvectors of real symmetric matrices by simultaneous iteration. Comp. J., 13 (1970), 76-80.

5. H. Rutishauser. Computational aspects of F. L. Bauer's simultaneous iteration method. Num. Math., 13 (1969), 4-13.

6. H. Rutishauser. Simultaneous iteration method for symmetric matrices. Num. Math., 16 (1970), 205 - 223.

7. A. Jennings and D. R. L. Orr. Application of the simultaneous iteration method to undamped vibration problems. Int. J. for Num. Meth. in Engng., 3 (1971). 13 - 24.

8. O. E. Brönlund. "Eigenvalues of Large Matrices". Symp. Finite Elem. Techn., Institut für Statik und Dynamik der Luft-und Raumfahrtkonstruktionen, University of Stuttgart (1969).

9. I. U. Ojalvo and M. Newman. Vibration modes of large structures by an automatic matrix-reduction method. AIAA Journal, 8 (1970), 1234-1239.

10. G. Peters and J. H. Wilkinson. Eigenvalues of $Ax = \lambda Bx$ with band symmetric A and B. Comp. J.,12 (1969), 398-404.

11. K. K. Gupta. Vibration of frames and other structures with banded stiffness matrix. Int. J. for Num. Meth. in Engng., 2 (1970), 221-228.

12. M. Clint and A. Jennings. A simultaneous iteration method for the unsymmetric eigenvalue problem. J.I.M.A. to be published.

13. M. Clint. The Eigensolution of Unsymmetric Matrices by Simultaneous Iteration. Doctoral Dissertation, Queen's University, Belfast (1970).

14. G. Strachey. Some matrix problems arising from the consideration of flutter in aircraft, ARC Report Comp. 107a 0.1296a (1956).

15. J. R. Richardson. A more realistic method for routine flutter calculations. AIAA Symposium of Structural Dynamics and Aeroelasticity (1965).

16. A. J. Lawrence and P. Jackson. Comparison of Different Methods of Assessing the Free Oscillitary Characteristics of Aeroelastic Systems. ARC Report CP. 1084 (1970).

AN ITERATIVE PROCEDURE FOR THE SOLUTION OF LINEAR AND NONLINEAR EQUATIONS

J. Le Foll

Abstract

The proposed procedure applies to the general linear equation

$$X \circ A = B$$

where B denotes a given L_2 function (or vector) and $X \circ A$ the product of the unknown function (or vector) X by a given linear operator (or matrix) A which satisfies the inequality condition

$$\|U \circ A\| \leqslant M\|U\| , \qquad M \text{ finite} \tag{1}$$

for any L_2 function (or vector) U ($\|U\|$ is defined by $\|U\| = \sqrt{U^2}$, with

$$UV = \int_a^b U(x)\, V(x)\, dx \qquad \text{or} \quad \sum U_i V_i \ ) .$$

The procedure is iterative and defined by the following formulae, where $\hat{X} = X/\|X\|^2$,

$$X_0 = U_0 = 0 , \quad V_1 = B$$

$$U_p = U_{p-1} + \hat{V}_p \circ A^*$$

$$X_p = X_{p-1} + \hat{U}_p$$

$$V_{p+1} = B - X_p \circ A \qquad .$$

Provided the equation has at least one L_2 solution X_p converges in the quadratic mean towards the solution with the smallest norm. If the equation is Fredholm's second kind with a square summable kernel, the rate of convergence is factorial. The calculation requires only three main subprograms:

- NORM (α) to calculate $\overline{X^2}$ for any X
- AR (X) to calculate product X_oA (right)
- AL (X) to calculate product X_oA^* (left)

where A* is defined by Green's identity.

$$\overline{U(V_oA)} = \overline{(U_oA^*)V} \quad \text{for any } L_2 \text{ functions U and V .} \tag{2}$$

1. "True error" minimization

Given any L_2 function on U, a first approximation X_1 to a solution X is of course λU, where λ is selected to minimize the "transformed error" $\overline{(B-AX_1)^2}$, but specially if the operator is illconditioned, it is much more attractive to minimize the "true error" $\overline{(X-X_1)^2}$ with $\lambda = \overline{XU}/\overline{U^2}$. Then, according to Enskog's remark, provided U is constructed from another L_2 function Y by means of $U = Y_oA^*$, scalar product $\overline{XU}$, and therefore λ can be calculated although X is of course unknown, since then Green's identity (2) gives $\overline{XU} = \overline{BY}$. For instance, with $Y_1 = \hat{B}$, $\overline{XU_1} = 1$ and $X_1 = \hat{U}_1$.

Suppose now that an iteratively derived sequence Y_n satisfies the conditions:

$$\overline{BY}_n = 1, \quad U_n = Y_{n o}A^*, \quad \overline{\hat{U}_p U_n} = 0 \quad \text{if } p < n \quad . \tag{3}$$

Then X_p is expanded as $\sum \lambda_p U_n$, and the "true error" is minimized for $\lambda_p = 1/\|U_p\|^2$ so that

$$X_p = \sum_{q=1}^{p} U_q \quad . \tag{4}$$

The problem therefore reduces to the derivation of sequence U (or Y) by means of simple recursion formulae. Introducing a sequence V in order to give Y a form similar to that of X, or

$$Y_p = \sum_{q=1}^{p} \hat{V}_p , \quad , \tag{5}$$

it will be shown that successive terms can be derived alternatively in U and V sequencies by means of initial values $U_0 = 0$, $V_1 = B$ and the recursion formulae

$$U_p = U_{p-1} + \hat{V}_p \circ A^* \qquad (6)$$

$$\text{for} \quad p \geq 1 \quad .$$

$$V_{p+1} = V_p - \hat{U}_p \circ A \qquad (7)$$

2. Orthogonality of U and V sequencies

First, this recursion defines further L_2 functions except if $U_n = 0$ or $V_{n+1} = 0$. Let us assume that the recursion does not stop before U_{n+2} and that

$$\overline{U_p U_q} = 0 \qquad \qquad p \leq n$$
$$\text{if} \quad p \neq q \quad \text{and if} \qquad \qquad .$$
$$\overline{V_p V_q} = 0 \qquad \qquad q \leq n$$

With $p = n$, recursion formula (7) can be written as

$$V_{n+1} = V_n - \hat{U}_n \circ A \; ,$$

so that after scalar multiplication by $\hat{V}_p$, $p \leq n$

$$\overline{\hat{V}_p V_{n+1}} = \overline{\hat{V}_p V_n} - \overline{(\hat{V}_p \circ A)\hat{U}_n}$$

according to Green's formula, and since

$$V_p \circ A = U_p - U_{p-1}$$

according to recursion formula (6), then

$$\overline{\hat{V}_p V_{n+1}} = \overline{\hat{V}_p V_n} - \overline{U_p \hat{U}_n} + \overline{U_{p-1} \hat{U}_n} \quad .$$

Sequences U and V being orthogonal up to n, all the terms on the R.H.S vanish for $p < n$, and for $p = n$ they reduce to $1-1+0 = 0$. Therefore the orthogonality of V now holds up to n+1.

A similar treatment of recursion formula (6) gives

$$\overline{\hat{U}_p U_{n+1}} = \overline{\hat{U}_p U_n} + \overline{V_p \hat{U}_{n+1}} - \overline{V_{p+1} \hat{U}_{n+1}}$$

where the terms in the R.H.S all vanish if $p < n$ and reduce to $1+0-1=0$ if $p = n$.

Since the assumption obviously holds for $n = 1$, sequencies U and V are separately orthogonal as long as the recursion can be continued, and Y can be defined by relation (5). Then, summing both sides in recursion formula (6) from $p = 1$ to $p = n+1$ gives formula (3)

$$U_{n+1} = Y_{n+1} \circ A^* \quad . \qquad (3)$$

3. Stop on U or V terms

As a consequence of formula (3), the recursion stops on U_p only if there is an L_2 function Q such that $Q \circ A^* = 0$ and this is possible only if the operator is singular. But even then, if the equation has at least one L_2 solution X, then

$$\overline{XU}_p = \overline{X(Y_p \circ A^*)} = \overline{(X \circ A)Y}_p = \overline{BY}_p = \sum_{q=1}^{n} \overline{B\hat{V}}_p = \overline{B\hat{V}}_1 = 1 \quad .$$

Therefore $U_p = 0$ is impossible, and the recursion can only stop on a V term. Now, summing both sides of recursion formula (7) from $p = 1$ to $p = n$ gives

$$V_{n+1} = B - X_n \circ A \quad . \qquad (8)$$

Therefore, if $V_{n+1} = 0$, X_n is a solution.

Since the operator may be singular, there may be L_2 proper functions satisfying $P \circ A = 0$. Then,

$$\overline{PU}_p = \overline{P(Y_p \circ A^*)} = \overline{(P \circ A)Y}_p = 0 \quad .$$

Therefore U_p and X_p are orthogonal to any such function.

4. Field of convergence

If the equation has at least one L_2 solution, then Fischer-Riecz's theorem implies the convergence of the resolving series $\hat{U}$. Since the V are orthogonal and

$$B - X_p \circ A = V_{p+1} = V_{p+2} + \hat{U}_{p+1} \circ A \,, \qquad \text{then}$$

$$\|B - X_p \circ A\|^2 = \|V_{p+1}\|^2 \leqslant \|U_{p+1} \circ A\|^2 \quad .$$

If the resolving series converges, $\|\hat{U}_{p+1}\|^2 < \varepsilon^2$ for p large enough, and therefore limit X_1 is a solution provided operator A satisfies inequality condition (1) since then

$$\|B - X_p \circ A\|^2 < M^2 \varepsilon^2 \qquad \text{for p large enough.}$$

Therefore, if the equation has at least one L_2 solution, the comparatively weak condition (1) is sufficient to ensure convergence in the mean towards the solution with the smallest norm.

5. Factorial rate of convergence for Fredholm's equations of the second kind

Here

$$X_0 A = X + X_0 K \tag{9}$$

with $$(X_0 K)_x = \int_0^1 K(x,y)\, X(y)\, dy.$$

If the kernel is square summable, i.e. if $\|K\|^2 = \int_0^1 K^2(x,y)\, dx\, dy$ is finite, a specially efficient mechanism of convergence takes place, because the two following lemmas can be applied:

Lemma 1 If $Z_p(x)$ denotes any infinite sequence of L_2 orthogonal functions for $0 \leqslant x \leqslant 1$, then;

$$\sum \|H_p\|^2 \leqslant \|K\|^2 ,$$

where

$$H_p(x,y) = Z_p(y) \int_0^1 K(x,y)\, \hat{Z}_p(y) dy ,$$

so that

$$\|H_p\|^2 \leqslant O\left(\frac{1}{p}\right) . \tag{10}$$

Lemma 2 Given any pair of infinite orthogonal sequencies U_p and V_p, if the acute angle w_p between U_p and its projection in the (V_p, V_{p+1}) plane vanishes when $p \to +\infty$, then

$$|\sin (U_p, V_p) \cos(U_p, V_{p+1})| \leqslant O(w_p) .$$

Lemma 1 applies to U_p and K, and since according to (9),

$$X_o K = X_o A - X,$$

then

$$\hat{U}_p{}_o K = V_p - V_{p+1} - \hat{U}_p,$$

so that

$$\|H_p\|^2 = \|U_p\|^2 \, \|V_p - V_{p+1} - \hat{U}_p\|^2 \leqslant O(\tfrac{1}{p}) \quad . \qquad (11)$$

Denoting normed functions by small letters, u_p can be projected in the (v_p, v_{p+1}) plane and on a unit vector t orthorgonal to this plane, so that

$$u_p = \cos w_p (\cos \alpha_p v_p + \sin \alpha_p \, v_{p+1}) + \sin w_p \, t_p \, .$$

The square $\|H_p\|^2$ is then split into 3 squares:

$$a_p{}^2 = | \, \|U_p\| \, \|V_p\| - \cos w_p \cos \alpha_p|^2 \leqslant O(\tfrac{1}{p})$$

$$b_p{}^2 = | \, \|U_p\| \, \|V_{p+1}\| - \cos w_p \sin \alpha_p|^2 \leqslant O(\tfrac{1}{p})$$

$$c_p{}^2 = \sin^2 w_p \leqslant O(\tfrac{1}{p}) \quad .$$

The last inequality shows that Hilbert's vector u_p is progressively squeezed inot the (v_p, v_{p+1}) plane, so that lemma 2 holds, and if the equation has a solution, the only possibility is $|\sin^2 \alpha_p| \leqslant O(\frac{1}{p})$, $|\cos^2 \alpha_p| \to 1$. Then, from the other two inequalities

$$\|\hat{U}_p\| \sim \|V_p\| \, ,$$

$$\|V_{p+1}\|^2 / \|V_p\|^2 \sim \|\hat{U}_{p+1}\|^2 / \|U_p\|^2 \sim O(\tfrac{1}{p}) \, ,$$

or

$$\|\hat{U}_p\|^2 < \frac{C}{p!} \quad .$$

ON WEAK STABILITY, STABILITY, AND THE TYPE OF A POLYNOMIAL

John J. H. Miller

We are concerned here with the qualitative theory of the zeros of polynomials and not with the quantitative problem of finding numerical approximations to zeros. We say that a polynomial is of type (p_1, p_2, p_3) relative to the unit circle if it has p_1 zeros interior to, p_2 on, and p_3 exterior to the unit circle. The unit circle is appropriate for difference approximations, while for differential equations we would define the type relative to the imaginary axis. Below we indicate the connection between the type of polynomials and the stability of difference approximations, and we summarize our results to date.

Most stability problems for difference approximations can be reduced to the determination of the type of one or a family of polynomials. The former occurs in numerical integration and in multistep schemes for ordinary differential equations, the latter in difference schemes for partial differential equations. We are interested mainly in the latter, where the von Neumann condition for weak stability is equivalent to the condition that each member of the family of characteristic polynomials of the scheme's symbol is of type $(p_1, p_2, 0)$. Such polynomials are called von Neumann polynomials. This condition is also a necessary (but generally insufficient) condition for stability. Moreover, for the important dissipative schemes for initial value problems due to Kreiss [3], the characteristic polynomials of the symbol must be of type $(p_1, 0, 0)$ for all values of the dual variables ξ satisfying $0 < |\xi| \leq \pi$ and of type $(p_1, p_2, 0)$ for $\xi = 0$. Indeed for schemes with a given order of dissipativity, stability is guaranteed by an appropriate order of accuracy.

Even more important for physical applications are the recent results of Gustavsson, Kreiss and Sundström [2] for mixed initial-boundary value problems for hyperbolic systems. They show that the stability of difference schemes for such systems is governed by qualitative properties of two families of polynomials, one arising from the interior approximation, the other from the boundary approximation. In particular, if $f(\kappa, z, \xi)$ and $g(\kappa, z, \xi)$ denote the polynomials in $\kappa = \kappa(z, \xi)$ belonging respectively to each of these families, where z is the point at which the resolvent is evaluated and ξ are the dual variables, then f is of type $(p_1, 0, p_3)$ in κ for all $|z| > 1$ and all ξ. For stability then it is necessary and sufficient that the p_1 zeros of f which are interior to the unit circle for $|z| > 1$, are not also zeros of g for any value of z such that $|z| \geqslant 1$. Indeed in the examples considered in §5 I of [2] the polynomials g are all of type $(0, q_2, q_3)$ in κ for all $|z| \geqslant 1$, so that it is only for $|z| = 1$ that common zeros can possibly occur.

To determine the type of a polynomial the well known Schur-Cohn method may always be used, see Marden [4] chapter X. Our methods are essentially the same, but the treatment is more compact and the results are in a form suitable for the applications discussed above.

For each $z \in \mathbb{C}$ and each polynomial in z of degree n we define their inversions in the unit circle respectively by $z^* = \frac{1}{\bar{z}}$ and $f^*(z) = z^n \overline{f(z^*)}$. We assume henceforth, without loss of generality, that $f(0) \neq 0$, $f^*(0) \neq 0$. We define the reduced polynomial corresponding to f as the Bezout resultant $\check{f}(z) = (f^*(0)\, f(z) - f(0)\, f^*(z))/z$, and we call f self-inversive if f and f^* have the same set of zeros and the multiplicity of each distinct zero is the same in both f and f^*.

LEMMA 1. The following conditions are equivalent:

(a) f is self-inversive.

(b) The zeros of f and their multiplicities are symmetric with respect to inversion in the unit circle.

(c) $f^*(0)\, f(z) = f(0)\, f^*(z) \quad \forall z \in \mathbb{C}$.

(d) $|f^*(z)| = |f(z)| \quad \forall z \in \mathbb{C}$.

(e) $\check{f}(z) = 0 \quad \forall z \in \mathbb{C}$.

Furthermore, if f is self-inversive and of degree n then it is of type $(p, n-2p, p)$ for some integer p, $0 \leqslant p \leqslant [n/2]$.

LEMMA 2. Suppose f is of degree n and is not self-inversive. Let $f = \psi g$, where ψ is the maximal self-inversive factor of f and is of degree m. Then

(a) ψ is a factor of $\check{f}$.

(b) $|f^*(0)| - |f(0)|$ and $|g^*(0)| - |g(0)|$ are either both zero or have the same sign.

(c) $\check{f}$ is of degree n-1 iff $|f^*(0)| \neq |f(0)|$

(d) ψ is the maximal self-inversive factor of $\check{f}$ if $|f^*(0)| \neq |f(0)|$

(e) f is of type (p_1, p_2, p_3) iff g is of type $(p_1-q, 0, p_3-q)$, where $q = (m - p_2)/2$.

From Lemmas 1 and 2 we obtain

THEOREM 1. Suppose f is a polynomial such that $|f^*(0)| \neq |f(0)|$. Then f is of type (p_1, p_2, p_3) iff $\check{f}$ is of type (p_1-1, p_2, p_3) if $|f^*(0)| > |f(0)|$ and of type (p_3-1, p_2, p_1) if $|f^*(0)| < |f(0)|$.

In order to obtain a non-trivial reduced polynomial corresponding to a self-inversive polynomial f we introduce $\tilde{f}(z) = f(z) + \xi z f'(z)$, where ξ is small and positive. We have then

LEMMA 3. Suppose f is a self-inversive polynomial of degree n with k distinct zeros on the unit circle. Then for all sufficiently small $\xi \neq 0$

(a) f is of type (p, n-2p, p) iff $\tilde{f}$ is of type (p+k, n-2p-k, p).

(b) $|(\tilde{f})^*(0)| > |\tilde{f}(0)|$.

(c) $\tilde{f}$ is not self-inversive.

(d) $(\tilde{f})^{\vee}$ is of degree n-1.

(e) $(\tilde{f})^{\vee}$ differs from f' by a non-zero constant factor.

From Lemma 3 we obtain

THEOREM 2.

Suppose f is a self-inversive polynomial of degree n with k distinct zeros on the unit circle. Then f is of type (p, n-2p, p) iff f' is of type (p+k-1, n-2p-k, p).

From Theorems 1 and 2 we obtain the following characterization of von Neumann polynomials.

THEOREM 3.

f is a von Neumann polynomial iff either $|f^*(0)| > |f(0)|$ and $\check{f}$ is a von Neumann polynomial or $\check{f} \equiv 0$ and f' is a von Neumann polynomial. The latter possibility arises iff all the zeros of f are on the unit circle.

The proof of this theorem and its application to several well known difference schemes may be found in [5]. An interesting and less trivial application (an outcome of this conference) may be found in Gourlay, McGuire and Morris [1]. The point of Theorem 3 is that it reduces the problem of testing f to that of testing either $\check{f}$ or the derivative f', each of which is of degree one lower than f. Repeated application of the theorem provides a method for finding the type of any von Neumann polynomial. Analogous results for the imaginary axis are given in [6].

We have seen above however that for mixed initial-boundary value problems the polynomials in question are not necessarily of von Neumann type. This means that the intermediate case can arise where $\check{f}$ is neither of degree n-1 nor identically zero. In [4] chapter X Marden presents a device for reducing this case to that in which $\check{f}$ is of degree n-1. However the challenge still remains of discovering how much information about the type of f may be extracted from that of $\check{f}$. Some partial results in this direction are stated below, details of which we hope to publish later. Even these have been found useful in simplifying the testing of the specific examples considered in [2], since the polynomials there are of low degree.

THEOREM 4. Suppose f is of degree n, $|f^*(0)| = |f(0)|$ and $\check{f}$ is not identically zero. Then

(a) f has zeros both interior to and exterior to the unit circle.

(b) The maximal self-inversive factor ψ of f is a factor of $\check{f}$.

(c) $\check{f}(z) = z^q h(z)$ for some integer q, $0 \leq q \leq [n/2] - 1$.

(d) h is self-inversive and is of degree $m = n-2q-2$.

(e) If $m=0$ then $q = (n-2)/2$, f is of type $(n/2, 0, n/2)$ and has no self-inversive factor.

(f) If h is a factor of f and is of type $(r, m-2r, r)$ then $q = (n-m-2)/2$, f is of type $((n-m+2r)/2, m-2r, (n-m+2r)/2)$ and h is the maximal self-inversive factor of f.

(g) If h and f have no common factor and h is of type $(m/2, 0, m/2)$ then f is of type $(n/2, 0, n/2)$ and has no self-inversive factor.

(h) If h and f have no common factor then f has no self-inversive factor and is of type $(p, 0, n-p)$ for some p, $0 < p < n$.

OPEN PROBLEM.

In Theorem 4 (h) can anything more be said about p if it is known that h is of type $(r, m-2r, r)$ for some r, $0 \leq r \leq [m/2]$?

ACKNOWLEDGEMENT.

The author is grateful to Dr. B. Gustavsson for an interesting discussion about mixed initial-boundary value problems.

REFERENCES

[1] A. R. Gourlay, G. McGuire, John Ll. Morris. "One dimensional methods for the numerical solution of nonlinear hyperbolic systems". These proceedings (1971).

[2] B. Gustavsson, H. -O. Kreiss, A. Sundström. "Stability theory of difference approximations for mixed initial boundary value problems, II." Preprint, Dept. of Computer Sciences, Uppsala University (1970).

[3] H. -O. Kreiss. "On difference approximations of the dissipative type for hyperbolic differential equations" Comm. Pure Appl. Math. 17 (1964) 335-353.

[4] M. Marden. "Geometry of Polynomials" Math. Surveys No. 3, Amer. Math. Soc. (1966) 2nd edition.

[5] John J. H. Miller. "On the location of zeros of certain classes of polynomials with applications to numerical analysis" Preprint (1971, To appear in J.Inst.Maths.Applic.)

[6] John J. H. Miller. "On the stability of differential equations" Preprint (1971).

ERROR ESTIMATES FOR CERTAIN INTEGRATION RULES ON THE TRIANGLE

G. M. Phillips

1. The interpolating polynomial

For numerical integration over a triangular region Δ in the x-y plane, Lauffer [3] has obtained a sequence of integration rules which will be denoted here by $\{I_n\}$. The rule I_n $(n = 1,2,3,\ldots)$ is exact for polynomials of degree $\leq n$ in x and y, and requires $\frac{1}{2}(n+1)(n+2)$ function evaluations at certain symmetrically distributed points in the triangle. These points are

$$\left((\beta_1 x_1 + \beta_2 x_2 + \beta_3 x_3)/n,\ (\beta_1 y_1 + \beta_2 y_2 + \beta_3 y_3)/n\right), \tag{1}$$

where $0 \leq \beta_1,\beta_2,\beta_3 \leq n$, $\beta_1 + \beta_2 + \beta_3 = n$ and the points (x_r,y_r), $1 \leq r \leq 3$, denote the vertices of the triangle Δ. We use S_n to denote the above set of $\frac{1}{2}(n+1)(n+2)$ points. The rule I_n uses the approximation

$$\iint_\Delta f(x,y)\,dx\,dy \simeq \iint_\Delta p_n(x,y)\,dx\,dy,$$

where the integrals are taken over the triangular region Δ and p_n is the interpolating polynomial for f constructed on the point set S_n. To obtain a simple expression for $p_n(x,y)$, let us first define

$$u_i(x,y) = \frac{(\tau_i + \eta_i x - \xi_i y)}{(\tau_i + \eta_i x_i - \xi_i y_i)}, \quad 1 \leq i \leq 3, \tag{2}$$

where

$$\tau_1 = x_2 y_3 - x_3 y_2, \qquad \xi_1 = x_2 - x_3, \quad \eta_1 = y_2 - y_3$$

and the other τ_i,ξ_i,η_i are defined cyclically. In (2), for any $i = 1,2,3$, the denominator

$$\tau_i + \eta_i x_i - \xi_i y_i = \begin{vmatrix} 1 & x_1 & y_1 \\ 1 & x_2 & y_2 \\ 1 & x_3 & y_3 \end{vmatrix} = \pm 2A \neq 0,$$

where A denotes the area of Δ. The linear function $u_i(x,y)$ has the value 1 at (x_i,y_i) and is zero at the other two vertices. Thus u_1,u_2 and u_3 are the barycentric co-ordinates for the point (x,y). We write f_β to denote the value of $f(x,y)$ at the point given in (1), and define

$$\pi_\beta(x,y) = \prod_{i=1}^{3}\left[\frac{1}{\beta_i!}\prod_{j=0}^{\beta_i - 1}(nu_i - j)\right], \tag{3}$$

where, if any $\beta_i = 0$, the empty product is taken to have the value 1 and $0! = 1$. It may be verified that $\pi_\beta(x,y)$ has the value 1 at the point with co-ordinates (1) and is zero at every other point of S_n. It follows that the polynomial

$$p_n(x,y) = \sum_\beta f_\beta \, \pi_\beta(x,y), \tag{4}$$

where the summation is over all $\frac{1}{2}(n+1)(n+2)$ terms, one associated with each point of S_n, *interpolates* $f(x,y)$ at each point of S_n. Thus (4) is a generalisation of the Lagrange form of the interpolating polynomial from one to two dimensions. It is easily verified that (4) is the unique polynomial of degree at most n in x and y which interpolates f at the points S_n. Also, if p_n and p_n^* interpolate f on two adjacent triangles Δ and Δ^* with a common edge, the fact that each polynomial interpolates f at n+1 points on the common edge ensures continuity of the approximating polynomials across that edge.

The foregoing results are readily extended to a simplex of any dimension k. For instance, with k = 3, we have the four barycentric co-ordinates u_1, u_2, u_3, u_4 and for the polynomials π_β we simply use the right side of (3) with the outer product taken over $1 \leq i \leq 4$. This allows us to write down interpolating polynomials, analogous to (4), for an arbitrary tetrahedron. Note that, for interpolation on the triangle, each factor of (3) corresponds geometrically to a straight line. In the analogous expression for the tetrahedron, each factor corresponds to a plane.

The above approach to interpolation on simplexes is pursued also by Argyris, Fried and Scharpf [1], Silvester [5] and Nicolaides [4]. In Silvester [5], the integration rules of Lauffer [3] are independently derived, as described briefly here, but without the error estimates given below.

2. Integration rules

Integrating (4) over Δ, we obtain

$$\iint_\Delta p_n(x,y)\, dx\, dy = \sum_\beta w_\beta f_\beta, \tag{5}$$

where the weight w_β is given by

$$w_\beta = \iint_\Delta \pi_\beta(x,y)\, dx\, dy. \tag{6}$$

The right side of (5) is the Lauffer quadrature rule I_n for the triangle. Lauffer [3] does not, in fact, give the weights w_β explicitly, as in (6) here.

It follows from the uniqueness of the interpolating polynomial that the rule I_n integrates exactly any polynomial in x and y of degree $\leq$ n. To derive the rule I_n, it is easier to make a change of variable. From (2) we have

$$\begin{aligned} x &= x_1u_1 + x_2u_2 + x_3u_3 \\ y &= y_1u_1 + y_2u_2 + y_3u_3 \end{aligned} \tag{7}$$

and $u_1 + u_2 + u_3 = 1$. Thus x and y are linear in, say, u_1 and u_2. This entails that any polynomial in x and y of degree $\leq$ n may be expressed in the form

$$\sum_{i+j\leq n} b_{ij}\, u_1^i u_2^j ,$$

a polynomial in u_1 and u_2 of degree $\leq$ n. So instead of finding the w_β from (6), we can find the weights so that the functions $u_1^i u_2^j$ are integrated exactly, for $0 \leq i + j \leq n$. This will entail solving a system of linear equations which must be non-singular, due to the uniqueness of the interpolating polynomial. We have

$$\iint_\Delta u_1^i u_2^j \, dx\, dy = 2A \int_0^1 \int_0^{1-u_2} u_1^i u_2^j \, du_1\, du_2 , \tag{8}$$

where the first integral sign on the right side refers to integration with respect to u_2 and, as before, A denotes the area of Δ. It is easily verified that the right side of (8) has the value

$$2A \frac{i!\, j!}{(i+j+2)!} . \tag{9}$$

In (6) the weight w_β is associated with the triple $(\beta_1, \beta_2, \beta_3)$, where $0 \leq \beta_1, \beta_2, \beta_3 \leq n$ and $\beta_1 + \beta_2 + \beta_3 = n$. By symmetry, the order of the numbers β_1, β_2 and β_3 is irrelevant in determining the weight w_β. Hence the number of distinct weights does not exceed the number of partitions of n into at most three parts. This cuts down the number of linear equations (see note below) which needs be set up to determine the w_β. The weights required for the rules I_n, $1 \leq n \leq 4$, are listed in Fig.1. For a given value of n, $1 \leq n \leq 4$, the w_β are obtained by multiplying the numbers shown in Fig.1 by the factor α_n displayed below the relevant triangle.

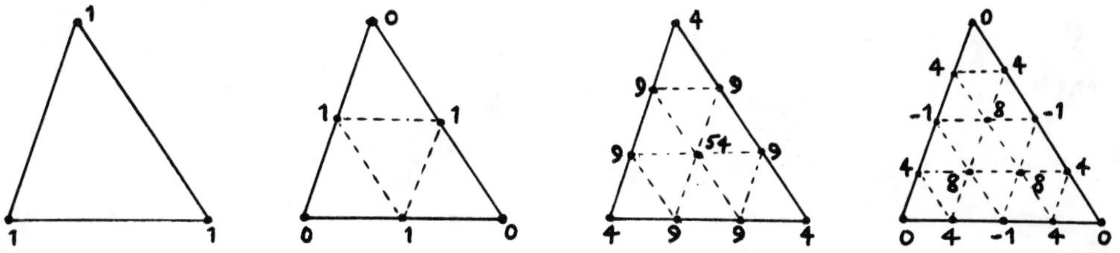

α_1 = A/3 α_2 = A/3 α_3 = A/120 α_4 = A/45

Fig.1 Integration rules for the triangle.

Silvester [5] lists 'open' and 'closed' formulae for $k = 2$ and $1 \leq n \leq 8$ and also for $k = 3$ and $1 \leq n \leq 6$. For the tetrahedron ($k = 3$) of volume V, the formula for $n = 3$ is extremely simple with weights $V/40$ at the vertices and weights $9V/40$ at the centroids of the faces. Lauffer [3] obtains formulae for all k and $1 \leq n \leq 5$. These generalise the Newton-Cotes formulae, which correspond to $k = 1$. It may be observed that the quadratic rule corresponding to general values of k and n has $\binom{n+k}{k}$ weights, which is asymptotically $n^k/k!$ for large n. However, the number of *distinct* weights is not greater than the number of partitions of n into at most k+1 parts, which is (see Wright [8]) asymptotically $n^k/\{k!(k+1)!\}$ for large n.

See Stroud [7], for rules which integrate polynomials over the simplex exactly, using a smaller number of function evaluations than the Lauffer rules.

3. Error estimate

In (7), to simplify the notation for what follows, let us replace u_1 by u, u_2 by v and therefore u_3 by $1 - u - v$ to give

$$\begin{aligned} x &= x_1 u + x_2 v + x_3(1-u-v) \\ y &= y_1 u + y_2 v + y_3(1-u-v) \end{aligned} \qquad (10)$$

The triangle with vertices at (x_r, y_r), $1 \leq r \leq 3$, corresponds to the region $u \geq 0$, $v \geq 0$, $u + v \leq 1$ of the u - v plane. The interpolating polynomial $p_n(x,y)$ defined by (4) is transformed into a polynomial, say $q_n(u,v)$. Let $f(x,y)$, which is interpolated by $p_n(x,y)$ on the point set S_n, be transformed by (10) into the function $g(u,v)$. Then $q_n(u,v)$ is the interpolating polynomial for $g(u,v)$ on the set of points $u = i/n$, $v = j/n$, with $i \geq 0$, $j \geq 0$ and $i + j \leq n$. The polynomial $q_n(u,v)$ is a special case of the interpolation formula due to Biermann [2], which is quoted by Stancu [6]. We write

$$R_n(g) = g(u,v) - q_n(u,v)$$

and write

$$\left[\frac{\partial^{p+q}}{\partial u^p \, \partial v^q} g(u,v)\right]_{u=\alpha, v=\beta} = g_{p,q}(\alpha,\beta),$$

for the sake of brevity. Biermann has shown that if the partial derivatives of g of order $n + 1$ are continuous for $u \geq 0$, $v \geq 0$, $u+v \leq 1$, then there exists a point $u = \alpha$, $v = \beta$ such that

$$R_n(g) = \frac{1}{(n+1)!} \sum_{i=0}^{n+1} \binom{n+1}{i} \rho_i(u)\, \rho_{n+1-i}(v)\, g_{i,n+1-i}(\alpha,\beta), \qquad (11)$$

where

$$\rho_i(u) = \begin{cases} u(u-\frac{1}{n}) \ldots (u-\frac{i-1}{n}), & i > 0 \\ 1 & , \quad i = 0. \end{cases}$$

To estimate the partial derivatives occurring in (11), we write

$$\frac{\partial g}{\partial u} = \frac{\partial f}{\partial x} \cdot \frac{\partial x}{\partial u} + \frac{\partial f}{\partial y} \cdot \frac{\partial y}{\partial u}$$

and, using (10), we obtain

$$\frac{\partial g}{\partial u} = (x_1 - x_3)\frac{\partial f}{\partial x} + (y_1 - y_3)\frac{\partial f}{\partial y} .$$

More generally, we have

$$\frac{\partial^{p+q} g}{\partial u^p \, \partial v^q} = \left((x_1 - x_3)\frac{\partial}{\partial x} + (y_1 - y_3)\frac{\partial}{\partial y}\right)^p \left((x_2 - x_3)\frac{\partial}{\partial x} + (y_2 - y_3)\frac{\partial}{\partial y}\right)^q . f(x,y). \quad (12)$$

Suppose now that each side of the triangle Δ with vertices (x_r, y_r), $1 \leq r \leq 3$, is not greater than h. Also, suppose that each (n+1)*th* partial derivative of f is in modulus not greater than M_{n+1}. Then from (12),

$$\left| \frac{\partial^{n+1} g}{\partial u^i \, \partial v^{n+1-i}} \right| \leq (2h)^{n+1} . M_{n+1}. \quad (13)$$

Applying this to (11), noting that $|\rho_i(u)| \leq 1$ on Δ and $\sum_{i=0}^{n+1} \binom{n+1}{i} = 2^{n+1}$, we obtain

$$|R_n(g)| \leq \frac{1}{(n+1)!} (4h)^{n+1} . M_{n+1}. \quad (14)$$

Suppose that a polygonal region of area A* is dissected into triangles of maximum side $\leq h$ and that the integration rule I_n is used on each triangle. Then, applying (14), the error in integration is bounded by

$$A^* . M_{n+1} . (4h)^{n+1} / (n+1)! \qquad .$$

Acknowledgment. This paper has arisen out of work done jointly with Professor A. R. Mitchell, University of Dundee, to whom I give my grateful thanks.

References

1. Argyris, J.H., Fried, I., and Scharpf, D.W. The TET 20 and TEA 8 Elements for the Matrix Displacement Method, The Aeronautical Journal of the Royal Aeronautical Society 72, 618-623 (1968).

2. Biermann, O. Über näherungsweise Kubaturen, Monatsh. Math. Phys. 14, 211-225 (1903).

3. Lauffer, R. Interpolation mehrfacher Integrale, Arch. Math. 6, 159-164 (1955).

4. Nicolaides, R.A. On Lagrange Interpolation in n Variables, Technical Note I.C.S.I. 274, Institute of Computer Science, University of London (1970).

5. Silvester, P. Symmetric Quadrature Formulae for Simplexes, Math. of Comp. 24, 95-100 (1970).

6. Stancu, D.D. The Remainder of Certain Linear Approximation Formulas in Two Variables, Journal of S.I.A.M. Ser B. Numerical Analysis 1, 137-163 (1964).

7. Stroud, A.H. Some approximate integration formulas of degree 3 for an n-dimensional simplex, Num. Math.9, 38-45 (1967).

8. Wright, E.M. Partitions into k parts, Math. Annalen 142, 311-316 (1961).

LINEAR MULTISTEP METHODS WITH VARIABLE MATRIX COEFFICIENTS

S. Sigurdsson

Summary

A class of linear multistep methods with variable matrix coefficients for integration of systems of ordinary differential equations (I.V.P.) is introduced. The asymptotic behaviour of the numerical solution as we use the class for integration of the time dependent linear system $y' = A(t)y$ is established for a third order subclass. This shows how instability effects arising from the "spurious" roots of the multistep method may be effectively suppressed if certain stabilizing conditions are satisfied.

The following class of multistep methods with variable matrix coefficients:

$$\sum_{j=0}^{k} \left(\alpha_j + \sum_{s=1}^{S} a_j^{(s)} h^s Q_n^s\right) y_{n+j} = h \sum_{j=0}^{k} \left(\beta_j + \sum_{s=1}^{S-1} b_j^{(s)} h^s Q_n^s\right) f_{n+j} \qquad (1)$$

has been considered for the integration of the initial value problem

$$y' = f(t,y) \qquad y(t_0) = y_0 \quad . \qquad (2)$$

Here, y, f are N vectors, Q_n is an N×N matrix which may vary with n; α_j, β_j, $a_j^{(s)}$, $b_j^{(s)}$ $j = 0,1,\ldots,k$; $s = 1,\ldots,S$ are scalar coefficients; h is the stepsize. The class is an extension of the class of constant coefficient linear multistep methods and a generalization of variable coefficient methods introduced by Lambert [2], for integration of scalar equations.

The most important properties of class (1) are best illustrated by restricting our attention to a subclass of (1) given by the following choice of coefficients ($k = 3$, $S = 2$):

$$\alpha_3 = 1 \qquad \alpha_2 = -1-A \qquad \beta_3 = 0 \qquad \beta_2 = \frac{1}{12}(23-5A-B)$$

$$\alpha_1 = A+B \qquad \alpha_0 = -B \qquad \beta_1 = -\frac{2}{3}(2+A-B) \qquad \beta_0 = \frac{1}{12}(5+A+5B)$$

$$\begin{aligned}
a_3^{(1)} &= \frac{1}{3} + 2a & b_3^{(1)} &= 0 \\
a_2^{(1)} &= \frac{1}{12}[-15 + A + B - 24a(1+A)] & b_2^{(1)} &= -\frac{1}{6} + a(4-A) \\
a_1^{(1)} &= \frac{1}{3}[4 - B + 6a(A+B)] & b_1^{(1)} &= \frac{1}{6}A - a(3+A-B) \qquad (3) \\
a_0^{(1)} &= \frac{1}{12}[-5 - A + 3B - 24aB] & b_0^{(1)} &= -\frac{1}{6}B + a(1+B)
\end{aligned}$$

$$a_3^{(2)} = a \qquad a_2^{(2)} = -3a$$

$$a_1^{(2)} = 3a \qquad a_0^{(2)} = -a \quad .$$

Here, $A = \zeta_2+\zeta_3$, $B = \zeta_1\zeta_2$ where ζ_1,ζ_2,ζ_3 are the roots of the characteristic polynomial

$$\sum_{j=0}^{3} \alpha_j \zeta^j = 0 \quad ; \quad \zeta_1 = 1 \ .$$

The subclass (3) is third order in the sense that the truncation error (defined similarly as for a constant coefficient method) is:

$$h^4[\frac{1}{24}(9+A+B)y^{(4)}(t_n)] + [\frac{1}{36}(14+A+2B) + \frac{1}{6}a(17+A-B)]Q_n y^{(3)}(t_n) + O(h^5) \ .$$

The coefficient satisfy the following conditions, referred to as stabilizing conditions (c.f. Lambert [2]):

$$\sum_{j=0}^{3} (a_j^{(1)} + \beta_j)\zeta_p^j = 0 = \sum_{j=0}^{3} (a_j^{(2)} + b_j^{(1)})\zeta_p^j \qquad p = 2,3$$

and in the special case, $\zeta_2 = \zeta_3$, also:

$$\sum_{j=1}^{3} j(a_j^{(1)} + \beta_j)\zeta_2^{j-1} = 0 = \sum_{j=1}^{3} j(a_j^{(2)} + b_j^{(1)})\zeta_2^{j-1} \quad .$$

If, furthermore, we insist on the condition:

$$|\zeta_p| < 1, \quad p = 2,3 \ ,$$

then the subclass (3) is said to be <u>stabilized.</u>

The reason for these definitions and the main motivation for considering the class (1) can now be seen in the light of the following result.

Apply a general method from the subclass (3) to the time-dependent linear system:

$$y' = A(t)y \qquad y(t_0) = y_0 \quad . \tag{4}$$

Let:

(i) $R(-hQ) = [I+(\frac{1}{3}+2a)hQ + a(hQ)^2]^{-1}\ [I-(\frac{2}{3}-2a)hQ + (\frac{1}{6}-a)(hQ)^2] = e^{-hQ} + O(h^4)$,

$B_j(hQ) = [I+(\frac{1}{3}+2a)hQ) + a(hQ)^2]^{-1}\ [\beta_j I + b_j^{(1)} hQ] \quad j = 0,1,2,$

K, w, L_i, $i = 1,2,\ldots,n$ be positive constant such that:

$$\Big\|\prod_{i=1}^{m} R(-hQ_{j_i})\Big\|_\infty \leqslant Kw^m \qquad 0 \leqslant j_1 < j_2 < \ldots < j_m \leqslant n\ , \tag{5}$$

$$\sum_{j=0}^{2} \|B_j(hQ_i)\|_\infty \cdot \|Q_i + A(t_{i+j})\|_\infty = L_i\ ,$$

(ii) $\mu_1 = [(1-\zeta_2)(1-\zeta_3)]^{-1}$;

$\mu_2 = [(\zeta_2-1)(\zeta_2-\zeta_3)]^{-1}$; $\mu_3 = [(\zeta_3-1)(\zeta_3-\zeta_2)]^{-1}$ $\quad \zeta_2 \neq \zeta_3$

$\mu_2 = -(\zeta_2-1)^{-2}$, $\quad \mu_3 = (\zeta_2-1)^{-1} \quad \zeta_2 = \zeta_3$

$\tilde{u}$, u_r, ε_r $\ r = 2,3$ be positive constants such that:

$$u_r = \frac{1}{\varepsilon_r} \left|\frac{\mu_r}{\mu_1}\right| (Kw+1) \geqslant 1 \qquad \tilde{u} = \max(u_2,u_3)\ ,$$

(iii) α be a constant such that:

$$e^{-\alpha} = \begin{cases} \max(w, |\zeta_2| + \varepsilon_2, |\zeta_3| + \varepsilon_3) & \zeta_2 \neq \zeta_3 \\ \max(w, |\zeta_2| + \varepsilon_2, |\zeta_2| + \frac{\varepsilon_2}{\varepsilon_3}|1-\zeta_2| + \varepsilon_3) & \zeta_2 = \zeta_3 \end{cases} \tag{6}$$

(iv) $V = \begin{bmatrix} 1 & \zeta_2^2 & \zeta_3^2 \\ 1 & \zeta_2 & \zeta_3 \\ 1 & 1 & 1 \end{bmatrix}$ if $\zeta_2 \neq \zeta_3$ $\qquad V = \begin{bmatrix} 1 & \zeta_2^2 & 2\zeta_2 \\ 1 & \zeta_2 & 1 \\ 1 & 1 & 0 \end{bmatrix}$ if $\zeta_2 = \zeta_3$,

$$K' = K\tilde{u}\, \|V\|_\infty \cdot \|V^{-1}\|_\infty \qquad K'' = K'e^\alpha\ ,$$

(v) $Y_0 = \max_j (\|y_j\|_\infty)$ $j = 0,1,2, \quad ,$

then

$$\|y_{n+k}\|_\infty \leq K'Y_0 \, e^{-\alpha(n+1)+hK''\sum_{i=1}^{n} L_i} . \tag{7}$$

The proof of this result is omitted, but we make the following observations:

(a) If we choose $Q_n \simeq -A(t_*)$ $t_* \in [t_n, t_{n+2}]$, we can in general expect L_i to remain small, even for a reasonable stepsize h. We advocate such a choice. When integrating the nonlinear system (2) this corresponds to choosing

$$Q_n \simeq \left. \frac{\partial \underset{\sim}{f}}{\partial \underset{\sim}{y}} \right| t = t_0 \in [t_n, t_{n+2}] \quad . \tag{8}$$

(b) Denote by $\mu[A]$ the logarithmic norm of A, ($\mu[A] = \lim_{h\to 0+} \frac{1}{h}(\|I+hA\| - 1)$ and $\lambda[A]$ the maximum real part of an eigenvalue of A. It is known that if $\mu[A(t)] < 0$ $\forall$ $t \geq t_0$ then the solution to (4) will tend to $\underset{\sim}{0}$ as $t \to \infty$ (see Coppel [1]).

Also $\mu[-hQ] < 0 \Rightarrow \|e^{-hQ}\| < 1$. This property, unfortunately, does not fully carry over to $R(-hQ)$. But we can show that if we choose $\mu[A] = \lambda[\frac{1}{2}(A+A^*)]$ i.e. interpret $\mu[A]$ in the sense of the spectral norm (see Coppel [1]), then:

$$\mu[-Q] < 0, \ \mu[-Q^2] < 0, \quad a \geq \frac{1}{12} \Rightarrow \|R(-hQ)\|_2 < 1 \qquad \forall \quad h > 0 .$$

Hence if this is true for all Q_i we are assured of the existence of an $w < 1$ in (5), noting that by equivalence of norms in finite dimensions, our particular choice of norm affects K but not w.

(c) If the subclass (3) is stabilized and $w < 1$ it follows that in (6), $\alpha > 0$ for sufficiently small ε_r. In this sense any possible asymptotic instabilities arising from the "spurious" roots ζ_2, ζ_3 can be seen to be suppressed, for any fixed steplength h. This is the important effect of stabilization. (Note however that small ε_r leads to large $\tilde{u}$).

The conclusion is that we have been able to indicate a favourable asymptotic stability behaviour of the numerical solution obtained, as we apply stabilized methods from class (1), with the convention (8), to time dependent linear systems (4), using a fixed ("reasonable") steplength h.

In particular, we have not had to restrict the analysis to the special case:

$$y' = Ay \qquad y(t_0) = y_0 \qquad \text{A a constant matrix} \tag{9}$$

(c.f. the concept of A-stability). It is readily seen from (7) that if we do apply (3) to (9) and choose $Q_n = -A$, $a \geq \frac{1}{12}$ then for all $h > 0$, the numerical solution will tend to $\underset{\sim}{0}$ as $n \to \infty$ whenever $\lambda[A] < 0$, in accordance with the theoretical solution.

Finally we observe that application of (3) to a general nonlinear system (2) involves at each time-step the solving of the following linear system:

$$[I + (\tfrac{1}{3} + 2a)hQ + a(hQ)^2]\underset{\sim}{y}_{n+3} = \underset{\sim}{g}_n \qquad \text{(known vector)}$$

rather than a nonlinear system. This is brought about by our choice $\beta_3 = b_1^{(3)} = 0$ in (3). We thus refer to the subclass (3) as a <u>linearly implicit</u> class.

Numerical results obtained, when linearily implicit methods from class (1) have been used for integration of small nonlinear stiff systems over large time intervals reflect the favourable asymptotic stability behaviour indicated by our analysis.

This paper has dealt only with the subclass (3) of the general class (1). A full treatment of the general class will be found in Lambert and Sigurdsson [3].

Acknowledgement

This work was carried out whilst the author was in receipt of grants from University of St. Andrews and the Icelandic Science Foundation.

References

[1] Coppel, W.A.: Stability and asymptotic behaviour of differential equations. D.C. Heath and Company, Boston, 1965.

[2] Lambert, J.D.: Linear multistep methods with mildy varying coefficients. Math. Comp. 24 (1970), pp. 81-97.

[3] Lambert, J.D. and Sigurdsson, S.T.: Multistep methods with variable matrix coefficients. To appear.

PARODE: A NEW REPRESENTATIONAL METHOD FOR THE NUMERICAL SOLUTION OF PARTIAL DIFFERENTIAL EQUATIONS

J.C. Taylor and J.V. Taylor

ABSTRACT

We describe a new approach to the numerical solution of partial differential equations of evolution-type. The basic idea is to treat the various space derivatives as independent unknowns and to obtain evolution equations for these. The method is applied to four prototype problems with reasonable results. The specially attractive feature of this approach is that the matrix of the problem is a universal constant matrix so that, together with its inverse, it can be read in as data of any computation using the method.

In the following we consider a method of integrating partial differential equations (linear or non-linear) different from the usual one of direct finite differencing of all differentials. To exemplify the approach we consider a prototype equation fairly typical of those which occur in fluid mechanic, namely

$$\frac{\partial u}{\partial t} = u\frac{\partial u}{\partial x} + \sigma\frac{\partial^2 u}{\partial x^2} \qquad \sigma(x,t) > 0 \text{ given} \tag{1}$$

$$= Lu \text{ say.}$$

One of the first questions to arise is: given the considerable difficulties that occur when (1) is tackled in a "normal" manner by differencing in both x and t can another approach be found? One possibility is to take $\partial u/\partial x$ and $\partial^2 u/\partial x^2$ as new dependent variables writing $u_1 \equiv \partial u/\partial x$, $u_2 \equiv \partial^2 u/\partial x^2$ so that (1) gives, with $\bullet = \partial/\partial t$

$$\dot{u} = uu_1 + \sigma u_2 \quad . \tag{2}$$

We now require "evolution equations" for u_1 and u_2. These are obtained from (1) by differentiating it w.r. to x once (for u_1) twice (for u_2) etc., getting for σ constant,

$$\dot{u}_1 = uu_2 + u_1^2 + \sigma u_3 \tag{2.1}$$

$$\dot{u}_2 = uu_3 + 3u_1 u_2 + \sigma u_4 \tag{2.2}$$

$$\dot{u}_3 = uu_4 + 4u_1 u_3 + 3u_2^2 + \sigma u_5 \tag{2.3}$$

and clearly the evolution equation for any u_n introduces two further unknowns u_{n+1}, u_{n+2} and the system of ordinary differential equations for the u_i as functions of t is not closed, if truncated at any finite n. However, and this is a central point of the method, it will turn out in the later stages of our analysis that this point can be taken care of in a very "natural" way.

For the moment let us consider how we should use these quantities $u(x,t)$, $u_1(x,t) = \partial u(x,t)/\partial x$, $u_2(x,t) = \partial^2 u(x,t)/\partial x^2$, etc. For this we return to the type of complete specification of the problems associated with equation (1). Typically this would be:

Find a function $u(x,t)$ satisfying (1) throughout the (x,t) space R (for example R: $x \in (a,b)$ and $t \geq 0$) and such that

$$u(x,o) = f(x) \qquad x \in (a,b) \qquad \text{initial condition} \tag{3.1}$$

$$u(a,t) = u_a \,, \quad u(b,t) = u_b \qquad t \geq 0 \qquad \text{boundary conditions} \tag{3.2}$$

u_a, u_b being given numbers and $f(x)$ a given function.

As the integration of system (2) provides us with the x-derivatives of u at any given time and for any chosen x we can think of representing $u(x,t)$ via its Taylor expansion <u>in x</u>. As we can only work with a finite number of x-derivatives say to u_3 we can only use a truncated expansion thus, (we drop the t for simplicity of notation and also now write $u1 = u_1 \ldots\ldots un = u_n \ldots\ldots$)

$$u(x_0+\lambda) = u_0 + \lambda . u1_0 + \lambda_2 . u2_0 + \lambda_3 . u3_0 + \lambda_4 . u4_0 + \lambda_5 . u5_0 + O(\lambda^6) \tag{4.1}$$

Similarly, using a prime to denote x-differentiation,

$$u'(x_0+\lambda) = u1_0 + \lambda . u2_0 + \lambda_2 . u3_0 + \lambda_3 . u4_0 + \lambda_4 . u5_0 + O(\lambda^5) \;. \tag{4.2}$$

where $\lambda_n = \lambda^n/n!$ and $un_0 = \partial^n u/\partial x^n$ at (x_0, t) .

The idea now is to use a representation (4.1) for $u(x,t)$ at any t, truncated at a level (order) determined by the level of accuracy required, to apply uniformly within a sub-interval of length say 2δ, chosen by the user. That is, given the numerical accuracy required then (provided $u(x)$ is regular in a certain sub-interval 2δ about x_0) a chosen number of terms in (4.1), (4.2) will be required to guarantee the chosen accuracy throughout 2δ. The smaller δ and the fewer terms will be necessary and vice versa. Different representations will apply in the various sub-intervals. Of course it is quite possible only to require one expansion point for the entire interval.

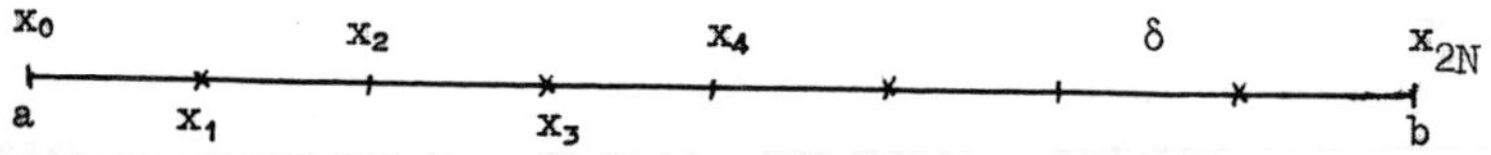

Specifically, applying equations (4) we can write

$$\begin{aligned}
u_0 &= u_1 - \delta.u1_1 + \delta_2.u2_1 - \delta_3.u3_1 + \delta_4.u4_1 - \delta_5.u5_1 + O(\delta^6) \\
u_0' &= u1_1 - \delta.u2_1 + \delta_2.u3_1 - \delta_3.u4_1 + \delta_4.u5_1 + O(\delta^5) \\
u_2 &= u_1 + \delta.u1_1 + \delta_2.u2_1 + \delta_3.u3_1 + \delta_4.u4_1 + \delta_5.u5_1 + O(\delta^6) \\
u_2' &= u1_1 + \delta.u2_1 + \delta_2.u3_1 + \delta_3.u4_1 + \delta_4.u5_1 + O(\delta^5)
\end{aligned} \quad . \qquad (5.1)$$

These are expansions centred on x_1. Similarly, we can express u_2, u_2' via expansions about x_3 to get

$$\begin{aligned}
u_2 &= u_3 - \delta.u1_3 + \delta_2.u2_3 - \delta_3.u3_3 + \delta_4.u4_3 - \delta_5.u5_3 + O(\delta^6) \\
u_2' &= u1_3 - \delta.u2_3 + \delta_2.u3_3 - \delta_4.u4_3 + \delta_4.u5_3 + O(\delta^5)
\end{aligned}$$

and also (5.2)

$$\begin{aligned}
u_4 &= u_3 + \delta.u1_3 + \delta_2.u2_3 + \delta_3.u3_3 + \delta_4.u4_3 + \delta_5.u5_3 + O(\delta^6) \\
u_4' &= u1_3 + \delta.u2_3 + \delta_2.u3_3 + \delta_3.u4_3 + \delta_4.u5_3 + O(\delta^5)
\end{aligned}$$

and so on. In the above equations (5) all terms on right hand side and to left of vertical line are known since the quantities u, u1, u2, u3 can be obtained from equations (2). The terms u4, u5 however, are not known so far. We now show how these can be found, for each interval, by requiring that u and $u' \equiv \partial u/\partial x \equiv u1$ be continuous at the "Matching points" x_2, x_4, x_6, etc. We note that this is a natural requirement, as mentioned above, because if $u(x,t)$, $\partial u/\partial x$ are known at any x for all t, so that u, $\partial u/\partial x$ are also available, then using (1) the higher x-derivatives can be found thus allowing us to generate $u(x,t)$ for all $x \in (0,1)$ via its Taylor expansion.

The simplest way to see this is to consider the situation where only one expansion point, x_1 say, at the mid-point of (0,1) is required. We now take a = 0, b = 1, without loss of generality.

In this case we have ($\delta = \frac{1}{2}$)

$$\left.\begin{aligned} u_a &= u_0 = u_1 - \delta.u1_1 + \delta_2.u2_1 - \delta_3.u3_1 + \delta_4.u4_1 - \delta_5.u5_1 + O(\delta^6) \\ u_b &= u_{2N} = u_1 + \delta.u1_1 + \delta_2.u2_1 + \delta_3.u3_1 + \delta_4.u4_1 - \delta_5.u5_1 + O(\delta^6) \end{aligned}\right\} . \qquad (6.1)$$

Introduce notation so that (6.1) can be written

$$\begin{aligned} u_a &= A_1^- + \delta_4.\alpha_1 - \delta_5.\beta_1 + O(\delta^6) \\ u_b &= A_1^+ + \delta_4.\alpha_1 + \delta_5.\beta_1 + O(\delta^6) \end{aligned} \qquad (6.2)$$

where

$$A_1^- = u_1 - \delta.u1_1 \;\ldots\ldots - \delta_3.u3_1, \qquad A_1^+ = u_1 + \delta_1.u1_1 \;\ldots\ldots + \delta_3.u3_1$$

are known, whereas $\alpha_1 \equiv u4_1$ and $\beta_1 \equiv u5_1$ are not. If now we neglect the error terms of $O(\delta^6)$ equations (6.2) are 2 linear algebraic equations for α_1, β_1. In other words (6.2) allow us to calculate the two extra 4th and 5th x-derivatives of u required for the Taylor expansions (4). We have thus found a way around the closure difficulty. Returning now to the equations (5^i), i = 1,3,.... we can write these, dropping the error terms as

$$\begin{aligned} u_{2n} &= A_{2n+1}^- + \delta_4.\alpha_{2n+1} - \delta_5.\beta_{2n+1} \\ u_{2n} &= B_{2n+1}^- - \delta_3.\alpha_{2n+1} + \delta_4.\beta_{2n+1} \\ u_{2n+2} &= A_{2n+1}^+ + \delta_4.\alpha_{2n+1} + \delta_5.\beta_{2n+1} \\ u_{2n+2} &= B_{2n+1}^+ + \delta_3.\alpha_{2n+1} + \delta_4.\beta_{2n+1} \end{aligned} \qquad (7)$$

for n = 0,1,...... N-1.

In system (7) the $A^{\pm}$, $B^{\pm}$ are known, being obtained by integrating equations (2). So are the δ_n. Finally u_0, u_{2N} are given (or the equivalent). By now requiring that u and $\partial u/\partial x$ be continuous at the matching points $x_2, x_4 \dots$ we get at x_2 for example, and writing $a_i = \delta_4 . \alpha_i$, $a_{i+1} = \delta_5 . \beta_i$ (i odd).

$$\begin{aligned} a_1 + a_2 - a_3 + a_4 &= A_3^- - A_1^+ \\ \frac{4}{5} a_1 + a_2 + \frac{4}{5} a_3 - a_4 &= (B_3^- - B_1^+)\delta/5 \end{aligned} \tag{8}$$

and hence eliminating a_4, a_1 respectively

$$\begin{aligned} 1.8\, a_1 + 2a_2 - .2a_3 &= (A_3^- - A_1^+) - \frac{\delta}{5}(B_3^- - B_1^+) \\ .2a_2 + 1.6a_3 - 1.8a_4 &= .8(A_3^- - A_1^+) + \frac{\delta}{5}(B_3^- - B_1^+) \end{aligned} \tag{9}$$

and so on. Writing $\underset{\sim}{a} = (a_1, a_2 \dots a_{2N})^T$ and $\underset{\sim}{b}$ for the right hand side of equations (9) we can write

$$A \underset{\sim}{a} = \underset{\sim}{b} \tag{10}$$

where A is a constant matrix which can easily be read off from system (9). This point deserves emphasis, that the actual numerical values of A_{ij} depend only on the truncation level used. Further if the highest space derivative in L is of order 2 then A is tridiagonal, if of order 4, A would be quindiagonal. In other words, for the class of operators L_n containing up to and including nth derivatives and, for truncation at level i, we have a unique matrix $A_i^{(n)}$ with constant coefficients. Whether or not the D.E. has variable coefficients or is linear or not makes no difference. We can therefore compute the inverse of whichever of these matrices $A_i^{(n)}$, we require once and for all and provide this as data of a computation. Thus $\underset{\sim}{a}$ can be calculated directly as $\underset{\sim}{a} = A^{-1}\underset{\sim}{b}$. This feature is quite unusual in the context of nonlinear partial differential equations. It provides in some degree that element of universality available in linear equations (namely eigenfunction expansions).

Another feature of the Parode method (partial to ordinary differential equations) is the ease with which more complicated boundary conditions can be treated. For example $u + \alpha\ \partial u/\partial x = \beta$ becomes $u + \alpha . u1 = \beta$ just another linear relation between 2 of the unknowns and thus easily fitted into the general system of equations. It is probably in 2 or more space dimensions that this simplicity becomes most attractive. Thus the condition $u + f(s)\ \partial u/\partial n = g(s)$ on the boundary ∂R of arc lengths s, f and g being given, is again in the Parode approach just another relation between three of the unknowns u, u_x, u_y.

Again, although much of our analysis has taken δ constant this is of course not absolutely necessary. Thus the interface between contiguous neighbourhoods can be chosen for convenience. In particular if shocks should occur they could be chosen as interfaces. Now, instead of requiring continuity one would obviously apply whatever jump conditions are physically necessary.

The reader will have noticed that we have not discussed the treatment of elliptic equations. This is mainly because it seems unlikely that one could be successful by a direct application so we have so far concentrated on hyperbolic and parabolic systems. By direct application we mean, in connection with the equation

$$\frac{\partial^2 u}{\partial x^2} + \frac{\partial^2 u}{\partial y^2} = 0 \qquad \text{in } R = [x \in (0,1),\ y \in (0,1)] \qquad (14)$$

with u = given function on ∂R, treating the equation as $\ddot{u} = -u2$ with $u2 = \partial^2 u/\partial x^2$ and dot respresenting differentiation with respect to y. This is because as a consequence of the non-analytic dependence of u(x,y) on boundary values we would expect the "x-sweep", mentioned on page 336 and implicit in applying the matching conditions, to be an unstable phase of the calculation for the elliptic case. However, it is clear that one can proceed in exactly the same way as is done in finite differences namely to introduce the artifact of time to replace system (14) by the associated parabolic system

$$\frac{\partial u}{\partial t} = \frac{\partial^2 u}{\partial x^2} + \frac{\partial^2 u}{\partial y^2}$$

and seeking to calculate the time-asymptotic solution. This of course immediately raises the questions whether the Parode approach can be applied in the case where two or more space dimensions are involved.

The reader will have noticed that the present method can be thought of as belonging to the class called <u>finite element</u> methods by which we mean those methods are based on some sort of explicit representation of the spatial or x-dependence of the unknown u(x,t). Thus, the Parode method being of finite element type, one can expect to be able to apply it in the case of higher dimensions. Unfortunately, the number of auxiliary unknowns increases rather rapidly. Thus a function u(x,y) of 2 independent variables (t is not germain to our present considerations) has 2 first derivatives, 3 second derivetaives and so on. Clearly the matter of determining the equivalents of u4 and u5 of the 1D case by application of boundary and continuity conditions becomes rather unpleasant. What is perhaps more important, the "naturalness" of the requirement of continuity in u and $\partial u/\partial x$ in one dimension no longer applied in two dimensions. Nor is it obvious at which points on the interface between contiguous "intervals" to demand continuity. Further, having decided on the points it is not clear which of the various possibilities, functions or its derivatives, to require to be continuous. Clearly there is here a need for further research*.

<u>Application</u>

In order to do preliminary tests of the method we have applied it to a set of prototype problems where either explicit solutions could be found against which to compare our numerical solutions, where special interest in the solution exists. These are

1. heat equation $\partial u/\partial t = \partial^2 u/\partial x^2$
2. wave equation $\partial^2 u/\partial t^2 = \partial^2 u/\partial x^2$
3. advection $\partial u/\partial t = -u\partial u/\partial x$
4. Burger's equation $\partial u/\partial t = -u\partial u/\partial x + \sigma\partial^2 u/\partial x^2$.

---

* It has come to our attention through Dr. H. R. Lewis that quite similar questions apparently arise in relation to multi-dimensional spline fitting.

In cases 1, 2, and 4 the auxiliary conditions were

initial conditions $u(x,0) = \sin \pi x$ in (0,1)

$\dot{u}(x,0) = 0$ (wave equation)

boundary conditions $u(0,t) = 0 = u(1,t)$ $t \geq 0$.

In case 3 only a rather trivial problem was considered with

$$u(x,0) = x \quad \text{in } (0,1)$$
$$u(0,t) = 0 \qquad t \geq 0 .$$

The explicit solutions to these problems are:

1. $u = \sin \pi x \ \exp(-\pi^2 t)$

2. $u = \sin \pi x \ \cos \pi t$

3. $u = \frac{x}{1+t}$

4. see Cole [2].

Results

Detailed results will not be given here for lack of space. They can however be obtained in an internal report (Department of Natural Philosophy, University of Glasgow) of the same name as this paper, which has also been sent for publication.

In general terms when applied to the 4 examples given above the method proved quite successful even with quite rough predictor-correctors and without using the explicitly known A^{-1} feature. Only Burger's example caused any difficulty and this only for small $\sigma(\sim 10^{-5})$. This is of course typical for this equation.

References

1. AMES, W.F., "Numerical methods for Partial Differential Equations" pp. 87-89, Nelson, 1969.

2. COLE, J.D., "On a quasi-linear parabolic equation occurring in aerodynamics" Quat. App. M. 9 (3) 335-236 (1951).

3. MILNE, W.E. and REYNOLDS, R.R., "Fifth-order methods for the numerical solution of ordinary differential equations" J. ACM 9, 64-70 (1962).

4. MITCHELL, A.R., "Computational methods in partial differential equations". p. 202, Wiley, 1969.

Acknowledgment

We would like to thank Professor L. Fox and A.R. Mitchell for the interest they have shown in this work.

ON DERIVING EXPLICIT RUNGE-KUTTA METHODS

J.H. Verner

Summary

Explicit Runge-Kutta methods of arbitrarily high order for linear autonomous systems of differential equations are derived using integration of certain interpolation formulae. A relationship of these methods to methods of the same order for general systems leads to easily applied algorithms for deriving methods of the latter type, and, in particular, a class of nine-stage methods of order seven is obtained.

1. Introduction

Despite considerable investigation on the derivation of high order Runge-Kutta methods, the analysis remains difficult. Even in a recent articel (Cooper and Verner [4]) which establishes sets of methods of arbitrary orders, a number of intricate arguments are required.

Here, a less direct approach is taken. Although the results are not as significant as those in [4], they are more general, and the simplicity of both the arguments and the algorithms for obtaining methods appears promising.

For the initial value problem for an autonomous system

$$x' = f[x] , \quad x(0) = z(0) , \quad x, f \in R_n, \tag{1}$$

an explicit single step method is defined for real parameters λ_{ij} independent of a stepsize h by

$$\begin{aligned} k_i &= f[z(0) + h \sum_{j=0}^{i-1} \lambda_{ij} k_j] , \qquad i = 0(1)s-1 \\ z(h) &= z(0) + h \sum_{j=0}^{s-1} \lambda_{sj} k_j . \end{aligned} \tag{2}$$

By defining abscissae $u_i = \sum_j \lambda_{ij}$, $i = 0(1)s$, it follows that there exists a vector of positive integers, $(\eta_0, \eta_1, \ldots, \eta_s)$ such that

$$\lambda_i\,[\tau] \triangleq (\tau+1) \sum_j \lambda_{ij} u_j^{\tau} - u_i^{\tau+1} = 0 \,, \quad \tau = 0(1)\eta_i - 1 \,.$$

By assuming the existence of sufficient derivatives of f, it is shown in [4] that a method (2) is of order p for (1) (denoted as an (s,p) method) if

$$\sum_{i_0} \lambda_{si_0} u_{i_0}^{\tau_0} \sum_{i_1} \lambda_{i_0 i_1} u_{i_1}^{\tau_1} \cdots \sum_{i_{k-1}} \lambda_{i_{k-2} i_{k-1}} u_{i_{k-1}}^{\tau_{k-1}} \lambda_{i_{k-1}} [\tau_k] = 0 \,, \qquad (3)$$

whenever $\tau_0 + \tau_1 + \ldots + \tau_k + k < p$, and

$$\sum_{i_0} \lambda_{si_0} u_{i_0}^{\tau_0} \sum_{i_1} \lambda_{i_0 i_1} u_{i_1}^{\tau_1} \cdots \sum_{i_{k-1}} \lambda_{i_{k-2} i_{k-1}} u_{i_{k-1}}^{\tau_{k-1}} \lambda_{i_{k-1} i_k} u_{i_k}^{\tau_k} = 0 \,, \qquad (4)$$

whenever $\tau_0 + \ldots + \tau_k + k < p - 2\eta_{i_k} - 2$.

Similar arguments imply methods are of order p for linear autonomous systems (denoted as (s,p) methods) if

$$\sum_{i_0} \lambda_{si_0} \sum_{i_1} \lambda_{i_0 i_1} \cdots \sum_{i_{k-1}} \lambda_{i_{k-2} i_{k-1}} \lambda_{i_{k-1}} [\tau] = 0 \,, \quad \tau + k < p \,. \qquad (5)$$

2. Methods for Linear Systems

Lemma 1: Conditions (5) for $k = 0(1)m$ are equivalent to

$$\sum_{i_0} \lambda_{si_0} \sum_{i_1} \lambda_{i_0 i_1} \cdots \sum_{i_k} \lambda_{i_{k-1} i_k} u_{i_k}^{\tau} = \frac{1}{(\tau+k+1)(\tau+k)\dots(\tau+1)} \tag{6}$$

$$\equiv \int_0^1 \int_0^{v_1} \cdots \int_0^{v_k} v^{\tau} dv dv_k \dots dv_1 , \quad \tau + k < p ,$$

for $k = 0(1)m$.

Proof: As (5) and (6) are identical for $k = 0$, the result is easily proved by induction on k.

For distinct abscissae $u_0, u_1, \dots, u_s$, define

$$L_{kj} \overset{\Delta}{=} \int_0^1 \int_0^{v_1} \cdots \int_0^{v_{s-k}} l_j^{k-1}(v) dv dv_{s-k} \dots dv_1 \tag{7}$$

$$\equiv \int_0^1 l_j^{k-1}(v) \frac{(1-v)^{s-k}}{(s-k)!} dv, \quad j = 0(1)k-1 , \quad k = 1(1)s ,$$

where $l_j^{k-1}(v)$ is the unique polynomial of degree $k-1$ satisfying

$$l_j^{k-1}(u_i) = \delta_{ij} , \quad i = 0(1)k-1 .$$

Lemma 2: For distinct abscissae, conditions (6), and hence (5), are valid for $p = 2$ if and only if

$$L_{s-kj} = \sum_{i_0} \lambda_{si_0} \sum_{i_1} \lambda_{i_0 i_1} \cdots \sum_{i_{k-1}} \lambda_{i_{k-2} i_{k-1}} \lambda_{i_{k-1} j} , \tag{8}$$

$$j = 0(1)k-1, \quad k = 1(1)s .$$

Proof: Lagrangian interpolation with (7) gives (6), and conversely (7) is proved using the identity

$$\sum_{i_0} \lambda_{si_0} \sum_{i_1} \lambda_{i_0 i_1} \cdots \sum_{i_k} \lambda_{i_{k-1} i_k} (u_{i_k} - u_0)\cdots(u_{i_k} - u_{j-1})(u_{i_k} - u_{j+1})\cdots$$

$$\cdots(u_{i_k} - u_{s-k-1}) = \int_0^1 \int_0^{v_1} \cdots \int_0^{v_k} (v-u_0)\cdots(v-u_{j-1})(v-u_{j+1})\cdots$$

$$\cdots(v-u_{s-k-1})dv\,dv_k\cdots dv_1 \quad . \qquad (9)$$

Theorem 1: For any s, $\{s,s\}$ methods exist.

Proof: For $s \leqslant 4$, (s,s) methods exist, and these are $\{s,s\}$ methods. For $s > 4$, choose distinct abscissae so that $L_{k,k-1} \neq 0$, $k = 1(1)s$, and for L_{ij} defined by (7), (8) is valid by choosing

$$\lambda_{si} = L_{si} , \qquad i = 0(1)s-1 ,$$

$$\lambda_{i\,i-k} = \Big(L_{i\,i-k} - \sum_{j=i-k+1}^{i-1} L_{i+1\,j}\lambda_{j\,i-k}\Big)/L_{i+1\,i} , \qquad (10)$$

$$i = s-1(-1)k, \qquad k = 1(1)s-1 .$$

Lemma 3: If $u_{s-1} = 1$, parameters defined by (10) satisfy

$$\sum_{i=j+1}^{s-1} \lambda_{si}\lambda_{ij} - \lambda_{sj}(1-u_j) = 0 , \quad j = 0(1)s-1 . \qquad (11)$$

Proof: The final expression in (7) implies $L_{s-1\,j} = L_{sj}(1-u_j)$.

Theorem 2: For $s \geqslant 3$, an $\{s,s\}$ method generated by (10) is and $(s,3)$ method. For $u_{s-1} = 1$, and $s \geqslant 4$, such a method is an $(s,4)$ method.

Proof: The definition of u_i leads to the first result, and lemma 3 is used to establish the second.

3. Methods for General Systems

Butcher [2] has shown that for (s,p) methods with $p > 4$, it is necessary that $s > p$. To obtain such methods L_{ij} are obtained using (7) for abscissae $u_0, u_{s-p+1}, \ldots, u_{s-1}$. For the remaining s-p abscissae, the choice $L_{ij} = 0$ still admits the results of section 2, and, in addition, establishes certain of conditions (4). These comments are clarified by another definition.

Definition: For $i \geq 4$, define

$$\bar{L}_{i1} \triangleq 0 ,$$

$$\bar{L}_{ij} \triangleq \int_0^1 \bar{I}_j^{i-2}(v) \frac{(1-v)^{s-i}}{(s-i)!} dv, \quad j = 0, 2(1)i-1, \tag{12}$$

where $\bar{I}_j^{i-2}(v)$ is the unique polynomial of degree i-2 satisfying

$$\bar{I}_j^{i-2}(u_k) = \delta_{jk}, \qquad k = 0, 2(1)i-1 .$$

Theorem 3: If, for $s = 6$, the parameters λ_{ij} for $i < 4$ are chosen so that $\lambda_i[0] = 0$, $i = 0(1)3$, and $\lambda_i[1] = 0$, $i = 0,2,3$, and for $i \geq 4$, λ_{ij} is obtained from (10) with $\bar{L}_{ij}$ replacing L_{ij}, then a (6,5) method is obtained if the parameters also satisfy

$$\sum_j \lambda_{sj}u_j\lambda_j[2] = \sum_i \lambda_{si}u_i \sum \lambda_{ij}\lambda_j[1] = 0 . \tag{13}$$

Proof: Conditions (10) establish (5) for $k < 3$, and then (3) and (4) excepting (13) are proved after showing that $\lambda_i[0] = \lambda_i[1] = 1$, $i = 4,5$.

Corollary: A (6,5) method is obtained if $u_5 = 1$, or if both

$$R \triangleq \frac{\lambda_2[2]}{\lambda_{21}} = \frac{\lambda_3[2]}{\lambda_{31}} , \tag{14}$$

and the abscissae satisfy

$$10u_2^2 u_3 - 8u_2 u_3 - u_2 + 2u_3 = 0 . \tag{15}$$

Proof: If $u_5 = 1$, (11) is valid and leads to (13).

Otherwise, (14) and (5) are used to show that $R = \frac{\lambda_4[2]}{\lambda_{41}} = \frac{\lambda_5[2]}{\lambda_{51}}$ implying that the two expressions of (13) are identical. Now (14) and the statement of the theorem uniquely define the parameters in terms of the abscissae. A tedious computation and (15) imply that the first expression of (13) is equal to zero.

This yields respectively a five parameter and a four parameter class of methods which have been derived previously by a more direct approach [3].

Similar arguments lead to a four parameter class of (7,6) methods also derived previously [1].

Theorem 4: A (7,6) method is obtained if the parameters λ_{ij} for $i < 4$ are chosen to satisfy

$$\lambda_i[0] = 0,\ i = 0(1)3, \qquad \lambda_i[1] = 0,\ i = 0,2,3,$$

$$R \triangleq \frac{\lambda_2[2]}{\lambda_{21}} = \frac{\lambda_3[2]}{\lambda_{31}} ,$$

for $i \geq 4$, λ_{ij} is obtained from (10) with $\overline{L}_{ij}$ replacing L_{ij}, and the abscissae satisfy $u_6 = 1$, and

$$15u_2^2 u_3 - 10u_2 u_3 - u_2 + 2u_3 = 0 .$$

The approach may be easily applied to obtain classes of higher order methods, and in particular, a four parameter class of (9,7) methods is obtained.

<u>Definition</u>: For $i \geq 6$, define

$$\bar{\bar{L}}_{i1} = \bar{\bar{L}}_{i2} = 0 ,$$

$$\bar{\bar{L}}_{ij} = \int_0^1 \bar{\bar{l}}_j^{i-3}(v) \frac{(1-v)^{s-i}}{(s-i)!} dv , \qquad j = 0, 3(1)i-1 ,$$

where $\bar{\bar{l}}_j^{i-3}(v)$ is the unique polynomial of degree i-3 satisfying

$$\bar{\bar{l}}_j^{i-3}(u_k) = \delta_{jk} , \quad k = 0, 3(1)i-1 .$$

<u>Theorem 5</u>: A (9,7) method is obtained if the parameters λ_{ij} for $i < 6$ are chosen so that

$$\lambda_i[0] = 0, \quad i = 0(1)5 , \qquad \lambda_i[\tau] = 0, \quad \tau = 1,2, \quad i = 0, 2(1)5,$$

$$\lambda_{31} = \lambda_{41} = \lambda_{51} = 0, \qquad \sum_j \lambda_{si}(u_i - u_8)(u_i - u_7)\lambda_{i2} = 0 ,$$

for $i \geq 6$, λ_{ij} is obtained from (10) with $\bar{\bar{L}}_{ij}$ replacing L_{ij}, and abscissae satisfy $u_8 = 1$, and

$$u_5 = \frac{v-12w+7vw}{3-12v+14v^2+24w-70vw+105w^2}$$

where $v = u_3 + u_4$ and $w = u_3 u_4$.

(To satisfy the first three conditions, additional restrictions on the abscissae are required: $u_1 = \frac{2}{3} u_2 = \frac{4}{9} u_3$.)

An attempt to obtain (10,8) methods by this approach was unsuccessful. A review of the results indicates that (10,8) methods of the particular type examined may not exist. Whereas only one or the other of two sets of conditions is required for (6,5) methods, both sets of analogous conditions are required for (7,6) methods, and as well for (9,7) methods. Hence the extension of the derivation of (6,5) methods to that of (7,6) methods does not suggest a corresponding extension of the derivation of (9,7) methods which would give (10,8) methods. However, additional investigation may yield algorithms of this type for generating (s,p) methods with p arbitrary.

In conclusion, it appears that other features of explicit Runge-Kutta methods such as error bounds and regions of stability might be easily investigated using the characterization of methods given above.

References

1. BUTCHER, J.C., On Runge-Kutta processes of high order, J. Aust. Math. Soc. 4 (1964) pp. 179-194.
2. BUTCHER, J.C., On attainable order of Runge-Kutta methods, Math. Comp. 19 (1965) pp. 408-417.
3. CASSITY, C.R., The complete solution of the fifth-order Runge-Kutta equations, SIAM J. of N.A. 6 (1969) pp. 432-436.
4. COOPER, G.J. and VERNER, J.H., Some explicit Runge-Kutta methods of high order, SIAM J. of N.A., to appear.

A FINITE ELEMENT FOR THREE DIMENSIONAL FUNCTION APPROXIMATION

R. Wait

Introduction

In his earlier lecture Professor Wachspress gave a method of function approximation in two dimensions that can be used to advantage in finite element calculations. The purpose of the present paper is to indicate one method by which such an approach can be extended to problems in three dimensions.

I would like to consider for a moment the functions that Professor Wachspress constructed for each quadrilateral.

FIG i

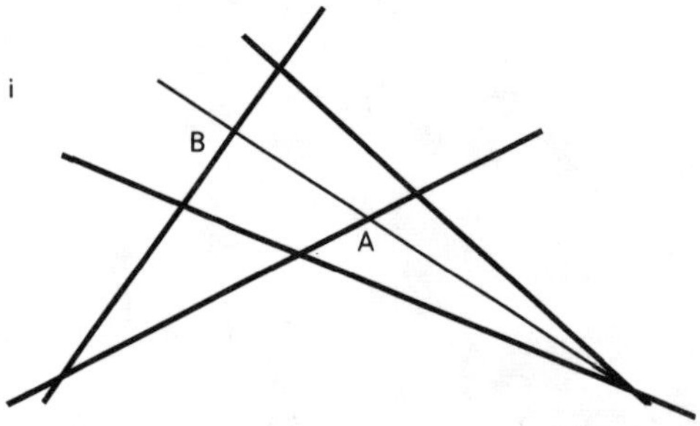

These functions had the property that the function varied linearly between pairs of points on the opposite sides. In geometric terms this means that the surface

$$z = w_i(x,y) \qquad i = 1(1)4$$

is a ruled surface. In fact if we rewrite the surface

$$z = \frac{(a_1 + b_1 x + c_1 y)(a_2 + b_2 x + c_2 y)}{\alpha + \beta x + \gamma y}$$

as $(\alpha + \beta x + \gamma y)\, z = (\quad)(\quad)$ it is clear that it is in fact a ruled quadric. An important property of such surfaces which will be used later is that quadrics can be uniquely defined, by means of two pairs of intersecting straight lines (a skew quadrilateral) or by any three skew lines.

The three dimensional element.

We take as our element an 'arbitrary' hexahedra with quadrilateral faces.

The three pairs of opposite faces meet in three lines which are in general skew, the various degenerate cases will be considered later. As stated earlier three skew lines define a quadric surface and so we can construct a unique surface that corresponds to the 'exterior diagonal' in the two-dimensional construction. We now wish to construct a function $\varphi_i(x,y,z)$ corresponding to each vertex P_i of the hexahedra, such that

1) $\varphi_i(x_j,y_j,z_j) = \delta_{ij}$

2) $\varphi_i \equiv 0$ on the faces opposite the vertex P_i

and 3) $\varphi_i(x,y,z)$ reduces to a function of the type produced by Professor Wachspress on each face containing the vertex P_i.

Such a function can be written as

$$\varphi_i(x,y,z) = k \frac{\pi_{i_1} \pi_{i_2} \pi_{i_3}}{Q}$$

where $\tilde{w}_{i_1} (\pi_{i_1} = 0)$, $\tilde{w}_{i_2} (\pi_{i_2} = 0)$ and $\tilde{w}_{i_3} (\pi_{i_3} = 0)$ are the three faces opposite the vertex P_i, $Q = 0$ is the surface which will be referred to hereafter as the E.D.Q. (exterior diagonal - quadric) and k is a scaling factor

$$k = \frac{Q(x_i,y_i,z_i)}{\pi_{i_1}(x_i,y_i,z_i) \cdot \pi_{i_2}(\cdot)\pi_{i_3}(\cdot)} ,$$

so that $\varphi_i(x_i,y_i,z_i) = 1$.

Clearly $\varphi_i(x,y,z) = 0$ on the opposite faces, the proof of the desired behaviour on the other three faces is not so instantly apparent. First we assume that the faces $\tilde{w}(\pi = 0)$ and $\tilde{w}_{i_1}$ are opposite faces, then we restrict our attention to these two planes to determine the behaviour of the function φ_i for points on the former. It should be remembered that there are two methods of defining a quadric and any given quadric can be defined by either means. In particular the E.D.Q. can be defined in either fashion.

Previously the E.D.Q. was defined by means of the three skew lines l_1, l_2 and l_3, formed by the intersections of the pairs of opposite faces.

FIG ii

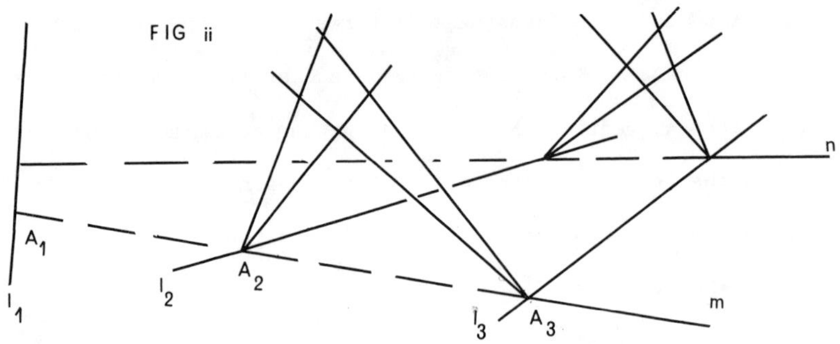

If we assume that l_1 is the intersection of $\tilde{w}$ and $\tilde{w}_{i_1}$ then l_2 and l_3 meet the plane $\tilde{w}$ in the points A_2 and A_3 which are the diagonal points of the quadrilateral formed by the intersections of the other four faces with $\tilde{w}$. Given any point on l_3 there is a unique transversal that intersects both l_2 and l_1, and this transversal also lies in the E.D.Q. Clearly the exterior diagonal m of the quadrilateral on $\tilde{w}$ is such a transversal, as one diagonal point is on l_3 and the other is on l_2 while l_1 lies wholly in $\tilde{w}$ and hence must intersect m in some point A_1. A similar argument shows that the exterior diagonal n of the quadrilateral on the opposite face also lies in the E.D.Q.

The skew quadrilateral (l_1 n l_3 m) then defines the E.D.Q. it is constructed as follows:

1) l_1 and n define the plane $\tilde{w}_{i_1}$

2) l_1 and m define the plane $\tilde{w}_1$

3) l_3 and n define a plane $\tilde{w}$ ($\pi' = 0$)

and 4) l_3 and m define a plane $\tilde{w}''(\pi'' = 0)$.

The E.D.Q. can be written as

$$\pi_{i_1} \pi'' - h \pi \pi' = 0,$$

where h is a scale factor. But on $\tilde{w}$, $\pi = 0$ and hence

$$Q = \pi_{i_1} \pi''$$

on the face $\tilde{w}$, thus

$$\varphi_i(x,y,z) = k \frac{\pi_{i_1} \pi_{i_2} \pi_{i_3}}{\pi_{i_1} \pi''}$$

$$= k \frac{\pi_{i_2} \pi_{i_3}}{\pi''}$$

on the face $\tilde{w}$.

FIG iii

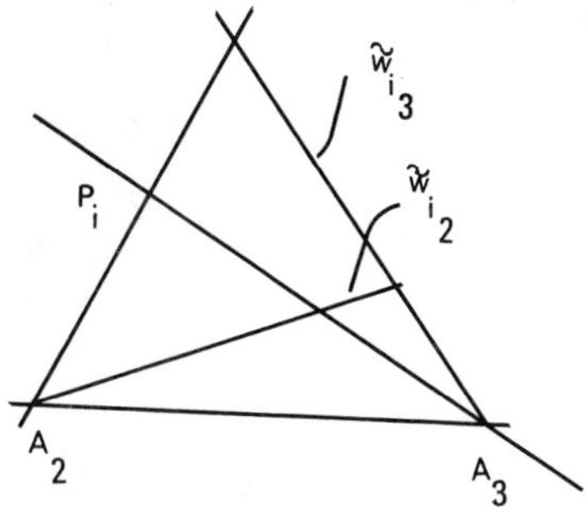

But on the face $\tilde{w}$ $\pi'' = 0$ is the equation of the exterior diagonal with $\pi_{i_2} = 0$ and $\pi_{i_3} = 0$ as the two sides opposite the vertex P_i. Hence the function has the desired properties.

Example

Consider the element defined by the planes

$$z + y - 3 = 0 \qquad \text{(top)}$$

$$z = 0 \qquad \text{(base)}$$

$$y = 0 \qquad \text{(front)}$$

$$x - y + 1 = 0 \qquad \text{(back)}$$

$$x = 0 \qquad \text{(left)}$$

$$4x + z - 4 = 0 \qquad \text{(right)}$$

Take $P_i = (0,0,0)$, then

$$\tilde{w}_{i_1} \equiv z + y - 3 = 0$$

$$\tilde{w}_{i_2} \equiv x - y + 1 = 0$$

$$\tilde{w}_{i_3} \equiv 4x + z - 4 = 0$$

$$l_1 \equiv \begin{cases} z = 0 \\ y-3 = 0 \end{cases} \qquad \text{(top \& base)}$$

$$l_2 \equiv \begin{cases} y = 0 \\ x+1 = 0 \end{cases} \qquad \text{(front \& back)}$$

$$l_3 \equiv \begin{cases} x = 0 \\ z-4 = 0 \end{cases} \qquad \text{(sides)}$$

The E.D.Q. is constructed by taking two arbitrary points on l_3 and constructing the transversals of l_2 and l_1. Thus

$$Q = (z + 8x - 4)(z + 2y - 6)-(z + 2x-4)(z + 4y - 12).$$

Hence

$$\varphi_i(x,y,z) = \frac{2(x+y-3)(y-x-1)(4x+z-4)}{(z+8x-4)(z+2y-6) - (z+2x-4)(z+4y-12)} \quad .$$

Taking w as any one of the planes x = 0, y = 0 or z = 0 will verify that the desired functions are obtained.

Degenerate cases

We assume that l_1, l_2 and l_3 are not concurrent, for if they are one of the faces has become a point. If two of the lines meet, the E.D.Q. becomes two planes. But the preceding argument still holds if for example l_2 and l_1 intersect. This point would be A_2 but very little would need to be changed. If two of the three lines are parallel the E.D.Q. again degenerates into two planes (all three parallel is impossible).

If two opposite faces are parallel then the E.D.Q. is obtained by taking the transversal of the two skew lines parallel to these two faces. If two pairs of opposite faces are parallel, the E.D.Q. is formed as the locus of the intersection of planes parallel to the two points passing through any point on the third line. All opposite faces parallel gives a parallelapiped for which the E.D.Q. becomes the point at infinity.

CONFORMAL TRANSFORMATION METHODS FOR THE NUMERICAL SOLUTION OF HARMONIC MIXED BOUNDARY VALUE PROBLEMS

J.R. Whiteman* and N. Papamichael
Department of Mathematics, Brunel University,

and

Q. Martin

Department of Mathematics, University of Texas

Introduction

The slow convergence with decreasing mesh size of finite-difference and finite element solutions to the true solutions of harmonic mixed boundary value problems containing boundary singularities has caused such problems to be much studied, and special adaptations of these standard methods have been proposed, (see Fix [4], and Whiteman and Webb [16]). In addition several numerical techniques based on analytic methods have been given for particular problems, Whiteman [10], and Fox and Sankar [5]. As Laplace's equation is invariant under conformal transformation, another attractive technique is that in which the original harmonic problem is mapped into another containing no singularities. When mapping the region in which the original problem is defined, the user of the conformal transformation has two choices. He can either apply a simple transformation and accept the, usually complicated, resulting region, as for example in [10], or he can devise a transformation which leads to a previously chosen simple region. The latter course is adopted here to solve numerically several harmonic mixed boundary problems defined on a rectangle. The method involves a new application of a well known technique (see Bowman [2], Chapter VII), and in particular is used to produce accurate solutions to a model problem which has been the subject of much interest recently, [7] - [17].

In the problems considered here the function u(x,y) satisfies

---

* The work of Dr. Whiteman was supported in part by Army Research Office (Durham) Grant DA-ARO(D) - 31 - 124 - G1050 and the National Science Foundation Grant GP-8442 awarded to the University of Texas at Austin.

$$\left.\begin{aligned}
\Delta[u(x,y)] &= 0, & (x,y) &\in R,\\
u(x,y) &= a, & (x,y) &\in S_1,\\
\frac{\partial u(x,y)}{\partial \nu} &= 0, & (x,y) &\in S_2,\\
u(x,y) &= b, & (x,y) &\in S_3,\\
\frac{\partial u(x,y)}{\partial \nu} &= 0, & (x,y) &\in S_4.
\end{aligned}\right\} \qquad (1)$$

In (1) the a and b are constants, Δ is the Laplacian operator, R is a bounded open domain with a rectangular boundary S, where $S = \bigcup S_i$, $i = 1,2,3,4$, S_i and S_{i+1} being adjacent subarcs of S, and $\partial/\partial\nu$ is the derivative in the direction of the outward normal to the boundary. The application of two Schwarz-Christoffel transformations, with an intermediate bilinear transformation, maps the region $G = R \cup S$ in the $w = x + iy$ plane into the region G' in the $w' = \xi + i\eta$ plane, where $G' = \{(\xi,\eta):\ 0 \leqq \xi \leqq 1,\ 0 \leqq \eta \leqq H\}$, and H is a known constant. The original problem is thus transformed into the problem

$$\left.\begin{aligned}
&\Delta[v(\xi,\eta)] = 0,\\
&\frac{\partial v(\xi,0)}{\partial \eta} = \frac{\partial v(\xi,H)}{\partial \eta} = 0, \quad 0 < \xi < 1,\\
&v(0,\eta) = a, \quad v(1,\eta) = b, \quad 0 \leqq \eta \leqq H,
\end{aligned}\right\} \qquad (2)$$

which has solution

$$v(\xi,\eta) = (b - a)\left(\xi + \frac{a}{b - a}\right). \qquad (3)$$

Thus if $P = (x,y) \in G$ is mapped into $P' = (\xi, \eta) \in G'$, it follows that $u(P) = v(P')$, and so from (3) the solution of (1) at P is known immediately if the real co-ordinate of the point P' is found. In practice the transformations are performed numerically thus introducing errors, and so only an approximation U(P) to u(P) is obtained.

Model Harmonic Mixed Boundary Value Problem

In the model problem the function u(x,y) satisfies $\Delta u = 0$ in a square region $-1 < x, y < 1$ with the slit $y = 0$, $0 < x < 1$, and boundary conditions

$$\frac{\partial u(x,\pm 1)}{\partial y} = 0 \qquad -1 < x < 1,$$

$$u(1,y) = \begin{cases} 1000, & 0 \leqq y \leqq 1, \\ 0, & -1 \leqq y \leqq 0, \end{cases}$$

$$\frac{\partial u(-1,y)}{\partial x} = 0, \qquad -1 < y < 1,$$

$$\frac{\partial u(x,0+)}{\partial y} = \frac{\partial u(x,0-)}{\partial y} = 0, \quad 0 < x < 1,$$

where 0+ and 0- represent respectively the upper and lower arms of the slit. Because of the antisymmetry it is sufficient to consider only the problem in the upper half region G, where $G = \{(x,y);\ -1 \leqq x \leqq 1,\ 0 \leqq y \leqq 1\}$, and to add the boundary condition $u(x,0) = 500$, $-1 \leqq x \leqq 0$; this problem is a special case of (1), and has a singularity at the origin.

Transformation For The Model Problem

The transformations used and their effects are:

(i) $G \in$ w-plane $\longrightarrow G_1 \in$ z-plane, $(z = \alpha + i\beta)$,

$$z = \mathrm{sn}(Kw,\ 1/\sqrt{2}), \tag{4}$$

where sn is the Jacobian elliptic sine, $K = K(1/\sqrt{2})$ is the complete elliptic integral of the first kind with modulus $1/\sqrt{2}$, and G_1 is the upper half z-plane.

(ii) $G_1 \in$ z-plane $\longrightarrow G_2 \in$ t-plane, $(t = g + ih)$,

$$t = \frac{2z}{1+z}, \tag{5}$$

where G_2 is the upper half t-plane.

(iii) $G_2 \in$ t-plane $\longrightarrow G' \in$ w'-plane, $(w' = \xi + i\eta)$,

$$w' = \frac{1}{K_1(m)}\ \mathrm{sn}^{-1}(t^{\frac{1}{2}}, m), \tag{6}$$

where $K_1(m)$ is the complete elliptic integral of the first kind with modulus $m = (\frac{1}{2} + 1/2\sqrt{2})^{\frac{1}{2}}$, and G' is the rectangle

$$G' = \{(\xi,\eta):\ 0 \leqq \xi \leqq 1,\ 0 \leqq \eta \leqq H\},$$

with $H = K_1((1-m^2)^{\frac{1}{2}})/K_1(m)$. The complete effect of (4), (5) and (6) is to transform the original problem into problem (2) in G', with a=500 and b=1000, which has solution

$$v(\xi,\eta) = 500\ (\xi + 1). \tag{7}$$

Using a series expansion for the elliptic sine, Copson [3], in (4) we have that

$$\alpha + i\beta = \mathrm{sn}(Kw, 1/\sqrt{2}) = \frac{2^{\frac{5}{4}} \sum_{n=0}^{\infty} \left\{ (-1)^n e^{-\pi(n+\frac{1}{2})^2} \sin[(n+\tfrac{1}{2})\pi w] \right\}}{1 + \sum_{n=1}^{\infty} \left\{ (-1)^n e^{-\pi n^2} \cos n\pi w \right\}}, \tag{8}$$

and hence (5) immediately gives the image $(g,h) \in G_2$ of $(\alpha,\beta) \in G_1$. Equation (6), when used in conjunction with equations 35-38 of Bowman [2], pp.40-41, produces, after some manipulation, the Jacobian elliptic cosine

$$\mathrm{cn}(2K_1(m)\xi, m) = \frac{A - B\{A^2m^2 + (1 - B^2m^2)(1 - m^2)\}^{\frac{1}{2}}}{\{1 - B^2m^2\}},$$

where $A = \{(1 - g)^2 + h^2\}^{\frac{1}{2}}$ and $B = (g^2 + h^2)^{\frac{1}{2}}$. Hence, adopting the notation of Abramowitz and Stegun [1], p.589, Section 17, we have that

$$\xi = F(\phi \backslash \sin^{-1} m) / 2K_1(m), \tag{9}$$

where $F(\phi \backslash \sin^{-1} m)$ is the incomplete elliptic integral of the first kind with amplitude $\phi = \cos^{-1}(\mathrm{cn}(2K_1(m)\xi, m))$, and parameter m^2. Full details of this derivation are given by Whiteman and Papamichael in [14]. Finally from (7) the solution $u(x,y) = v(\xi,\eta) = 500(\xi + 1)$. We note that more than one suitable bilinear transformation exists, and that the above can be repeated with (5) replaced, for example, by

$$t = (1 - z) \ / \ (1 - z/\sqrt{2}), \tag{10}$$

in which case the final solution is

$$v(\xi,\eta) = 500\ (2 - \xi). \tag{11}$$

Numerical Results and Discussion

Although the method produces an analytic solution to the model problem, the transformations (4), (5) and (6) are in practice performed numerically, so that rounding error is introduced. In addition the series in (8) are truncated after N terms thus introducing truncation error. The first ten figures in the calculated approximations to $\mathrm{sn}(Kw, 1/\sqrt{2})$ remain constant for $N \geq 3$, and so N is taken as 3. An algorithm of Hofsommer and Van de Riet [6] is used to calculate the incomplete elliptic integral $F(\phi \backslash \sin^{-1} m)$ in (9). The numerical results for the transformation method show everywhere four figure agreement with those of Whiteman [13]. In addition approximations calculated using the bilinear transformation (10) and the solution (11) show agreement to five figures. In the neighbourhood

$|x| \leq 1/7$, $0 \leq y \leq 1/7$ of the singularity, the results show five figure agreement with those of Whiteman [12]. Full details of all numerical results and computation times can be found in [14].

An advantage of this mapping technique is that the actual problem, rather than some approximating problem, is solved. Moreover the technique provides for smaller computation times to achieve comparable accuracy than any method previously known to us. The need for a definite error bound is recognized.

Other Harmonic Problems

More general problems can be solved with the mapping technique by the choice of an appropriate bilinear transformation in the place of (5), and a suitable value for the modulus m in (6). For example we consider a problem of type (1) where G = BCDE is the rectangle $|x| \leq 1$, $0 \leq y \leq 1$, as in the figure, and where the four

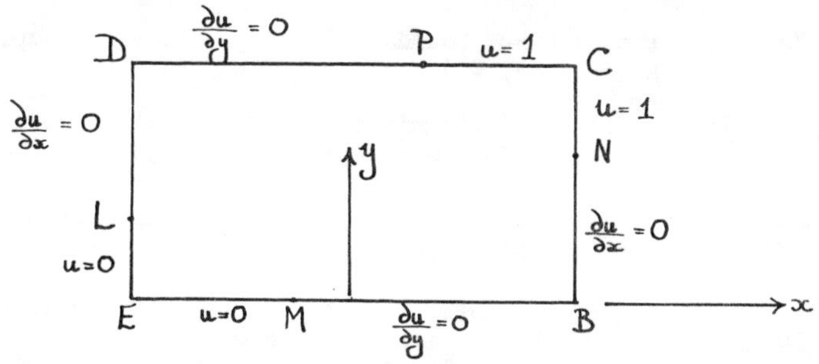

distinct end points L,M,N,P of the subarcs S_i can be taken anywhere on S. We note again that the boundary conditions on adjacent S_i's must be one Dirichlet and the other Neumann, that $G \in$ w-plane, and we let L,M,N,P be respectively the points w_1, w_2, w_3 and w_4. Application of (4) again maps G onto the upper half z-plane, with $w_i \longrightarrow (\alpha_i, 0)$, where $\alpha_i = \mathrm{sn}(Kw_i, 1/\sqrt{2})$. The bilinear transformation (5) is replaced by

$$t = \left(\frac{\alpha_4 - \alpha_2}{\alpha_4 - \alpha_1}\right) \cdot \left(\frac{z - \alpha_1}{z - \alpha_2}\right),$$

which maps $(\alpha_1, 0)$ into $(0,0)$, $(\alpha_2, 0)$ into $(\infty, 0)$ and $(\alpha_4, 0)$ into $(1,0)$. The image of $(\alpha_3, 0)$ will in the t-plane be the point $(g, 0)$, where

$$g = \left(\frac{\alpha_4 - \alpha_2}{\alpha_4 - \alpha_1}\right) \circ \left(\frac{\alpha_3 - \alpha_1}{\alpha_3 - \alpha_2}\right).$$

When $m^2 = 1/g$, the transformation (6) maps this upper half t-plane onto G' as before, so that the original problem is mapped into (2) which now has solution $v(\xi, \eta) = \xi$.

Acknowledgment

We wish to express our thanks to Drs. E. L. Wachspress and W. B. Jordan for suggesting this approach to elliptic problems containing boundary singularities.

REFERENCES

1. M. Abramowitz and I.A.Stegun, Handbook of Mathematical Functions, Dover, New York, (1965).

2. F. Bowman, Introduction to Elliptic Functions, English Universities Press, London, (1953).

3. E.T. Copson, An Introduction to the Theory of Functions of a Complex Variable, Oxford University Press, London, (1935).

4. G. Fix, Higher-order Rayleigh Ritz approximations, Journal of Math.and Mech., 18, 645-657, (1969).

5. L. Fox and R. Sankar, Boundary singularities in linear elliptic differential equations, J.Inst.Maths.Applics,5, 340-350, (1969).

6. D.J. Hofsommer and R.P.van de Riet, On the numerical calculations of elliptic integrals of the first and second kind and the elliptic functions of Jacobi, Numerische Mathematik,5, 291-302, (1963).

7. R.B.Kelman, Harmonic mixed boundary value problems in composite rectangular regions, Q.J.Mech.Appl.Math, XXIII, 549-566, (1970).

8. H. Motz, The treatment of singularities of partial differential equations by relaxation methods, Quart.Appl.Math.,4, 371-377 (1946).

9. R. Wait and A.R. Mitchell, Corner singularities in elliptic problems by finite element methods, to appear J. Comp. Phys. 1971.

10. J.R. Whiteman, Singularities due to re-entrant corners in harmonic boundary value problems, Mathematics Research Center, Technical Summary Report No.829, University of Wisconsin, Madison, (1967).

11. J.R. Whiteman, Treatment of singularities in a harmonic mixed boundary value problem by dual series methods, Q.J.Mech.Appl.Math., XXI, 41-50, (1968).

12. J.R. Whiteman, Numerical solution of a harmonic mixed boundary value problem by the extension of a dual series method. Q.J.Mech. Appl. Math., XXIII, 449-455 (1970).

13. J.R. Whiteman, Numerical solution of a harmonic mixed boundary value problem by linear programming, Mathematics Research Center, Technical Summary Report No.857, University of Wisconsin,Madison,(1968).

14. J.R. Whiteman and N. Papamichael, Numerical solution of two dimensional harmonic boundary problems containing singularities by conformal transformation methods. Dept.of Mathematics, Brunel University, Report TR/2, to appear.

15. J.R. Whiteman and Jerry C. Webb, Convergence of finite-difference techniques for a harmonic mixed boundary value problem, B.I.T., 10, 366-374, (1970).

16. J.R. Whiteman and Jerry C. Webb, Finite-difference solutions to a harmonic mixed boundary value problem, Technical Report, Center for Numerical Analysis, University of Texas, to appear.

17. L.C. Woods, The relaxation treatment of singular points in Poisson's equation, Q.J.Mech.Appl.Math., 6, 163-185. (1953).

Lecture Notes in Mathematics

Comprehensive leaflet on request

Vol. 38: R. Berger, R. Kiehl, E. Kunz und H.-J. Nastold, Differentialrechnung in der analytischen Geometrie IV, 134 Seiten. 1967 DM 12,-

Vol. 39: Séminaire de Probabilités I. II, 189 pages. 1967. DM 14,-

Vol. 40: J. Tits, Tabellen zu den einfachen Lie Gruppen und ihren Darstellungen. VI, 53 Seiten. 1967. DM 6.80

Vol. 41: A. Grothendieck, Local Cohomology. VI, 106 pages. 1967. DM 10,-

Vol. 42: J. F. Berglund and K. H. Hofmann, Compact Semitopological Semigroups and Weakly Almost Periodic Functions. VI, 160 pages. 1967. DM 12,-

Vol. 43: D. G. Quillen, Homotopical Algebra. VI, 157 pages. 1967. DM 14,-

Vol. 44: K. Urbanik, Lectures on Prediction Theory. IV, 50 pages. 1967. DM 5,80

Vol. 45: A. Wilansky, Topics in Functional Analysis. VI, 102 pages. 1967. DM 9,60

Vol. 46: P. E. Conner, Seminar on Periodic Maps. IV, 116 pages. 1967. DM 10,60

Vol. 47: Reports of the Midwest Category Seminar I. IV, 181 pages. 1967. DM 14,80

Vol. 48: G. de Rham, S. Maumary et M. A. Kervaire, Torsion et Type Simple d'Homotopie. IV, 101 pages. 1967. DM 9,60

Vol. 49: C. Faith, Lectures on Injective Modules and Quotient Rings. XVI, 140 pages. 1967. DM 12,80

Vol. 50: L. Zalcman, Analytic Capacity and Rational Approximation. VI, 155 pages. 1968. DM 13.20

Vol. 51: Séminaire de Probabilités II. IV, 199 pages. 1968. DM 14,-

Vol. 52: D. J. Simms, Lie Groups and Quantum Mechanics. IV, 90 pages. 1968. DM 8,-

Vol. 53: J. Cerf, Sur les difféomorphismes de la sphère de dimension trois (Γ_4= O). XII, 133 pages. 1968. DM 12,-

Vol. 54: G. Shimura, Automorphic Functions and Number Theory. VI, 69 pages. 1968. DM 8,-

Vol. 55: D. Gromoll, W. Klingenberg und W. Meyer, Riemannsche Geometrie im Großen. VI, 287 Seiten. 1968. DM 20,-

Vol. 56: K. Floret und J. Wloka, Einführung in die Theorie der lokalkonvexen Räume. VIII, 194 Seiten. 1968. DM 16,-

Vol. 57: F. Hirzebruch und K. H. Mayer, O (n)-Mannigfaltigkeiten, exotische Sphären und Singularitäten. IV, 132 Seiten. 1968. DM 10,80

Vol. 58: Kuramochi Boundaries of Riemann Surfaces. IV, 102 pages. 1968. DM 9,60

Vol. 59: K. Jänich, Differenzierbare G-Mannigfaltigkeiten. VI, 89 Seiten. 1968. DM 8,-

Vol. 60: Seminar on Differential Equations and Dynamical Systems. Edited by G. S. Jones. VI, 106 pages. 1968. DM 9,60

Vol. 61: Reports of the Midwest Category Seminar II. IV, 91 pages. 1968. DM 9,60

Vol. 62: Harish-Chandra, Automorphic Forms on Semisimple Lie Groups X, 138 pages. 1968. DM 14,-

Vol. 63: F. Albrecht, Topics in Control Theory. IV, 65 pages. 1968. DM 6,80

Vol. 64: H. Berens, Interpolationsmethoden zur Behandlung von Approximationsprozessen auf Banachräumen. VI, 90 Seiten. 1968. DM 8,-

Vol. 65: D. Kölzow, Differentiation von Maßen. XII, 102 Seiten. 1968. DM 8,-

Vol. 66: D. Ferus, Totale Absolutkrümmung in Differentialgeometrie und -topologie. VI, 85 Seiten. 1968. DM 8,-

Vol. 67: F. Kamber and P. Tondeur, Flat Manifolds. IV, 53 pages. 1968. DM 5,80

Vol. 68: N. Boboc et P. Mustată, Espaces harmoniques associés aux opérateurs différentiels linéaires du second ordre de type elliptique. VI, 95 pages. 1968. DM 8,60

Vol. 69: Seminar über Potentialtheorie. Herausgegeben von H. Bauer. VI, 180 Seiten. 1968. DM 14,80

Vol. 70: Proceedings of the Summer School in Logic. Edited by M. H. Löb. IV, 331 pages. 1968. DM 20,-

Vol. 71: Séminaire Pierre Lelong (Analyse), Année 1967 – 1968. VI, 190 pages. 1968. DM 14,

Vol. 72: The Syntax and Semantics of Infinitary Languages. Edited by J. Barwise. IV, 268 pages. 1968. DM 18,-

Vol. 73: P. E. Conner, Lectures on the Action of a Finite Group. IV, 123 pages. 1968. DM 10,-

Vol. 74: A. Fröhlich, Formal Groups. IV, 140 pages. 1968. DM 12,-

Vol. 75: G. Lumer, Algèbres de fonctions et espaces de Hardy. VI, 80 pages. 1968. DM 8,-

Vol. 76: R. G. Swan, Algebraic K-Theory. IV, 262 pages. 1968. DM 18,-

Vol. 77: P.-A. Meyer, Processus de Markov: la frontière de Martin. IV, 123 pages. 1968. DM 10,-

Vol. 78: H. Herrlich, Topologische Reflexionen und Coreflexionen. XVI, 166 Seiten. 1968. DM 12,-

Vol. 79: A. Grothendieck, Catégories Cofibrées Additives et Complexe Cotangent Relatif. IV, 167 pages. 1968. DM 12,-

Vol. 80: Seminar on Triples and Categorical Homology Theory. Edited by B. Eckmann. IV, 398 pages. 1969. DM 20,-

Vol. 81: J.-P. Eckmann et M. Guenin, Méthodes Algébriques en Mécanique Statistique. VI, 131 pages. 1969. DM 12,-

Vol. 82: J. Wloka, Grundräume und verallgemeinerte Funktionen. VIII, 131 Seiten. 1969. DM 12,-

Vol. 83: O. Zariski, An Introduction to the Theory of Algebraic Surfaces. IV, 100 pages. 1969. DM 8,-

Vol. 84: H. Lüneburg, Transitive Erweiterungen endlicher Permutationsgruppen. IV, 119 Seiten. 1969. DM 10.-

Vol. 85: P. Cartier et D. Foata, Problèmes combinatoires de commutation et réarrangements. IV, 88 pages. 1969. DM 8,-

Vol. 86: Category Theory, Homology Theory and their Applications I. Edited by P. Hilton. VI, 216 pages. 1969. DM 16,-

Vol. 87: M. Tierney, Categorical Constructions in Stable Homotopy Theory. IV, 65 pages. 1969. DM 6,-

Vol. 88: Séminaire de Probabilités III. IV, 229 pages. 1969. DM 18,-

Vol. 89: Probability and Information Theory. Edited by M. Behara, K. Krickeberg and J. Wolfowitz. IV, 256 pages. 1969. DM 18,-

Vol. 90: N. P. Bhatia and O. Hajek, Local Semi-Dynamical Systems. II, 157 pages. 1969. DM 14,-

Vol. 91: N. N. Janenko, Die Zwischenschrittmethode zur Lösung mehrdimensionaler Probleme der mathematischen Physik. VIII, 194 Seiten. 1969. DM 16,80

Vol. 92: Category Theory, Homology Theory and their Applications II. Edited by P. Hilton. V, 308 pages. 1969. DM 20,-

Vol. 93: K. R. Parthasarathy, Multipliers on Locally Compact Groups. III, 54 pages. 1969. DM 5,60

Vol. 94: M. Machover and J. Hirschfeld, Lectures on Non-Standard Analysis. VI, 79 pages. 1969. DM 6,-

Vol. 95: A. S. Troelstra, Principles of Intuitionism. II, 111 pages. 1969. DM 10,-

Vol. 96: H.-B. Brinkmann und D. Puppe, Abelsche und exakte Kategorien, Korrespondenzen. V, 141 Seiten. 1969. DM 10,-

Vol. 97: S. O. Chase and M. E. Sweedler, Hopf Algebras and Galois theory. II, 133 pages. 1969. DM 10,-

Vol. 98: M. Heins, Hardy Classes on Riemann Surfaces. III, 106 pages. 1969. DM 10,-

Vol. 99: Category Theory, Homology Theory and their Applications III. Edited by P. Hilton. IV, 489 pages. 1969. DM 24,-

Vol. 100: M. Artin and B. Mazur, Etale Homotopy. II, 196 Seiten. 1969. DM 12,-

Vol. 101: G. P. Szegö et G. Treccani, Semigruppi di Trasformazioni Multivoche. VI, 177 pages. 1969. DM 14,-

Vol. 102: F. Stummel, Rand- und Eigenwertaufgaben in Sobolewschen Räumen. VIII, 386 Seiten. 1969. DM 20,-

Vol. 103: Lectures in Modern Analysis and Applications I. Edited by C. T. Taam. VII, 162 pages. 1969. DM 12,-

Vol. 104: G. H. Pimbley, Jr., Eigenfunction Branches of Nonlinear Operators and their Bifurcations. II, 128 pages. 1969. DM 10,-

Vol. 105: R. Larsen, The Multiplier Problem. VII, 284 pages. 1969. DM 18,-

Vol. 106: Reports of the Midwest Category Seminar III. Edited by S. Mac Lane. III, 247 pages. 1969. DM 16,-

Vol. 107: A. Peyerimhoff, Lectures on Summability. III, 111 pages. 1969. DM 8,-

Vol. 108: Algebraic K-Theory and its Geometric Applications. Edited by R. M. F. Moss and C. B. Thomas. IV, 86 pages. 1969. DM 6,-

Vol. 109: Conference on the Numerical Solution of Differential Equations. Edited by J. Ll. Morris. VI, 275 pages. 1969. DM 18,-

Vol. 110: The Many Facets of Graph Theory. Edited by G. Chartrand and S. F. Kapoor. VIII, 290 pages. 1969. DM 18,-

Vol. 111: K. H. Mayer, Relationen zwischen charakteristischen Zahlen. III, 99 Seiten. 1969. DM 8,–

Vol. 112: Colloquium on Methods of Optimization. Edited by N. N. Moiseev. IV, 293 pages. 1970. DM 18,–

Vol. 113: R. Wille, Kongruenzklassengeometrien. III, 99 Seiten. 1970. DM 8,–

Vol. 114: H. Jacquet and R. P. Langlands, Automorphic Forms on GL (2). VII, 548 pages. 1970. DM 24,–

Vol. 115: K. H. Roggenkamp and V. Huber-Dyson, Lattices over Orders I. XIX, 290 pages. 1970. DM 18,–

Vol. 116: Sèminaire Pierre Lelong (Analyse) Année 1969. IV, 195 pages. 1970. DM 14,–

Vol. 117: Y. Meyer, Nombres de Pisot, Nombres de Salem et Analyse Harmonique. 63 pages. 1970. DM 6.–

Vol. 118: Proceedings of the 15th Scandinavian Congress, Oslo 1968. Edited by K. E. Aubert and W. Ljunggren. IV, 162 pages. 1970. DM 12,–

Vol. 119: M. Raynaud, Faisceaux amples sur les schémas en groupes et les espaces homogènes. III, 219 pages. 1970. DM 14,–

Vol. 120: D. Siefkes, Büchi's Monadic Second Order Successor Arithmetic. XII, 130 Seiten. 1970. DM 12,–

Vol. 121: H. S. Bear, Lectures on Gleason Parts. III, 47 pages. 1970. DM 6,–

Vol. 122: H. Zieschang, E. Vogt und H.-D. Coldewey, Flächen und ebene diskontinuierliche Gruppen. VIII, 203 Seiten. 1970. DM 16,–

Vol. 123: A. V. Jategaonkar, Left Principal Ideal Rings. VI, 145 pages. 1970. DM 12,–

Vol. 124: Séminare de Probabilités IV. Edited by P. A. Meyer. IV, 282 pages. 1970. DM 20,–

Vol. 125: Symposium on Automatic Demonstration. V, 310 pages. 1970. DM 20,–

Vol. 126: P. Schapira, Théorie des Hyperfonctions. XI, 157 pages. 1970. DM 14,–

Vol. 127: I. Stewart, Lie Algebras. IV, 97 pages. 1970. DM 10,–

Vol. 128: M. Takesaki, Tomita's Theory of Modular Hilbert Algebras and its Applications. II, 123 pages. 1970. DM 10,–

Vol. 129: K. H. Hofmann, The Duality of Compact Semigroups and C*- Bigebras. XII, 142 pages. 1970. DM 14,–

Vol. 130: F. Lorenz, Quadratische Formen über Körpern. II, 77 Seiten. 1970. DM 8,–

Vol. 131: A Borel et al., Seminar on Algebraic Groups and Related Finite Groups. VII, 321 pages. 1970. DM 22,–

Vol. 132: Symposium on Optimization. III, 348 pages. 1970. DM 22,–

Vol. 133: F. Topsøe, Topology and Measure. XIV, 79 pages. 1970. DM 8,–

Vol. 134: L. Smith, Lectures on the Eilenberg-Moore Spectral Sequence. VII, 142 pages. 1970. DM 14,–

Vol. 135: W. Stoll, Value Distribution of Holomorphic Maps into Compact Complex Manifolds. II, 267 pages. 1970. DM 18,–

Vol. 136: M. Karoubi et al., Séminaire Heidelberg-Saarbrücken-Strasbuorg sur la K-Théorie. IV, 264 pages. 1970. DM 18,–

Vol. 137: Reports of the Midwest Category Seminar IV. Edited by S. MacLane. III, 139 pages. 1970. DM 12,–

Vol. 138: D. Foata et M. Schützenberger, Théorie Géométrique des Polynômes Eulériens. V, 94 pages. 1970. DM 10,–

Vol. 139: A. Badrikian, Séminaire sur les Fonctions Aléatoires Linéaires et les Mesures Cylindriques. VII, 221 pages. 1970. DM 18,–

Vol. 140: Lectures in Modern Analysis and Applications II. Edited by C. T. Taam. VI, 119 pages. 1970. DM 10,–

Vol. 141: G. Jameson, Ordered Linear Spaces. XV, 194 pages. 1970. DM 16,–

Vol. 142: K. W. Roggenkamp, Lattices over Orders II. V, 388 pages. 1970. DM 22,–

Vol. 143: K. W. Gruenberg, Cohomological Topics in Group Theory. XIV, 275 pages. 1970. DM 20,–

Vol. 144: Seminar on Differential Equations and Dynamical Systems, II. Edited by J. A. Yorke. VIII, 268 pages. 1970. DM 20,–

Vol. 145: E. J. Dubuc, Kan Extensions in Enriched Category Theory. XVI, 173 pages. 1970. DM 16,–

Vol. 146: A. B. Altman and S. Kleiman, Introduction to Grothendieck Duality Theory. II, 192 pages. 1970. DM 18,–

Vol. 147: D. E. Dobbs, Cech Cohomological Dimensions for Commutative Rings. VI, 176 pages. 1970. DM 16,–

Vol. 148: R. Azencott, Espaces de Poisson des Groupes Localement Compacts. IX, 141 pages. 1970. DM 14,–

Vol. 149: R. G. Swan and E. G. Evans, K-Theory of Finite Groups and Orders. IV, 237 pages. 1970. DM 20,–

Vol. 150: Heyer, Dualität lokalkompakter Gruppen. XIII, 372 Seiten. 1970. DM 20,–

Vol. 151: M. Demazure et A. Grothendieck, Schémas en Groupes I. (SGA 3). XV, 562 pages. 1970. DM 24,–

Vol. 152: M. Demazure et A. Grothendieck, Schémas en Groupes II. (SGA 3). IX, 654 pages. 1970. DM 24,–

Vol. 153: M. Demazure et A. Grothendieck, Schémas en Groupes III. (SGA 3). VIII, 529 pages. 1970. DM 24,–

Vol. 154: A. Lascoux et M. Berger, Variétés Kähleriennes Compactes. VII, 83 pages. 1970. DM 8,–

Vol. 155: Several Complex Variables I, Maryland 1970. Edited by J. Horváth. IV, 214 pages. 1970. DM 18,–

Vol. 156: R. Hartshorne, Ample Subvarieties of Algebraic Varieties. XIV, 256 pages. 1970. DM 20,–

Vol. 157: T. tom Dieck, K. H. Kamps und D. Puppe, Homotopietheorie. VI, 265 Seiten. 1970. DM 20,–

Vol. 158: T. G. Ostrom, Finite Translation Planes. IV. 112 pages. 1970. DM 10,–

Vol. 159: R. Ansorge und R. Hass. Konvergenz von Differenzenverfahren für lineare und nichtlineare Anfangswertaufgaben. VIII, 145 Seiten. 1970. DM 14,–

Vol. 160: L. Sucheston, Constributions to Ergodic Theory and Probability. VII, 277 pages. 1970. DM 20,–

Vol. 161: J. Stasheff, H-Spaces from a Homotopy Point of View. VI, 95 pages. 1970. DM 10,–

Vol. 162: Harish-Chandra and van Dijk, Harmonic Analysis on Reductive p-adic Groups. IV, 125 pages. 1970. DM 12,–

Vol. 163: P. Deligne, Equations Différentielles à Points Singuliers Reguliers. III, 133 pages. 1970. DM 12,–

Vol. 164: J. P. Ferrier, Seminaire sur les Algebres Complètes. II, 69 pages. 1970. DM 8,–

Vol. 165: J. M. Cohen, Stable Homotopy. V, 194 pages. 1970. DM 16,–

Vol. 166: A. J. Silberger, PGL_2 over the p-adics: its Representations, Spherical Functions, and Fourier Analysis. VII, 202 pages. 1970. DM 18,–

Vol. 167: Lavrentiev, Romanov and Vasiliev, Multidimensional Inverse Problems for Differential Equations. V, 59 pages. 1970. DM 10,–

Vol. 168: F. P. Peterson, The Steenrod Algebra and its Applications: A conference to Celebrate N. E. Steenrod's Sixtieth Birthday. VII, 317 pages. 1970. DM 22,–

Vol. 169: M. Raynaud, Anneaux Locaux Henséliens. V, 129 pages. 1970. DM 12,–

Vol. 170: Lectures in Modern Analysis and Applications III. Edited by C. T. Taam. VI, 213 pages. 1970. DM 18,–

Vol. 171: Set-Valued Mappings, Selections and Topological Properties of 2^X. Edited by W. M. Fleischman. X, 110 pages. 1970. DM 12,–

Vol. 172: Y.-T. Siu and G. Trautmann, Gap-Sheaves and Extension of Coherent Analytic Subsheaves. V, 172 pages. 1971. DM 16,–

Vol. 173: J. N. Mordeson and B. Vinograde, Structure of Arbitrary Purely Inseparable Extension Fields. IV, 138 pages. 1970. DM 14,–

Vol. 174: B. Iversen, Linear Determinants with Applications to the Picard Scheme of a Family of Algebraic Curves. VI, 69 pages. 1970. DM 8,–

Vol. 175: M. Brelot, On Topologies and Boundaries in Potential Theory. VI, 176 pages. 1971. DM 18,–

Vol. 176: H. Popp, Fundamentalgruppen algebraischer Mannigfaltigkeiten. IV, 154 Seiten. 1970. DM 16,–

Vol. 177: J. Lambek, Torsion Theories, Additive Semantics and Rings of Quotients. VI, 94 pages. 1971. DM 12,–

Vol. 178: Th. Bröcker und T. tom Dieck, Kobordismentheorie. XVI, 191 Seiten. 1970. DM 18,–

Vol. 179: Seminaire Bourbaki – vol. 1968/69. Exposés 347-363. IV. 295 pages. 1971. DM 22,–

Vol. 180: Séminaire Bourbaki – vol. 1969/70. Exposés 364-381. IV, 310 pages. 1971. DM 22,–

Vol. 181: F. DeMeyer and E. Ingraham, Separable Algebras over Commutative Rings. V, 157 pages. 1971. DM 16.–

Vol. 182: L. D. Baumert. Cyclic Difference Sets. VI, 166 pages. 1971. DM 16,–

Vol. 183: Analytic Theory of Differential Equations. Edited by P. F. Hsieh and A. W. J. Stoddart. VI, 225 pages. 1971. DM 20,–

Vol. 184: Symposium on Several Complex Variables, Park City, Utah, 1970. Edited by R. M. Brooks. V, 234 pages. 1971. DM 20,–

Vol. 185: Several Complex Variables II, Maryland 1970. Edited by J. Horváth. III, 287 pages. 1971. DM 24,–

Vol. 186: Recent Trends in Graph Theory. Edited by M. Capobianco/ J. B. Frechen/M. Krolik. VI, 219 pages. 1971. DM 18.–

Vol. 187: H. S. Shapiro, Topics in Approximation Theory. VIII, 275 pages. 1971. DM 22,–

Vol. 188: Symposium on Semantics of Algorithmic Languages. Edited by E. Engeler. VI, 372 pages. 1971. DM 26,–

Vol. 189: A. Weil, Dirichlet Series and Automorphic Forms. V, 164 pages. 1971. DM 16,–

Vol. 190: Martingales. A Report on a Meeting at Oberwolfach, May 17-23, 1970. Edited by H. Dinges. V, 75 pages. 1971. DM 12,–

Vol. 191: Séminaire de Probabilités V. Edited by P. A. Meyer. IV, 372 pages. 1971. DM 26,–

Vol. 192: Proceedings of Liverpool Singularities – Symposium I. Edited by C. T. C. Wall. V, 319 pages. 1971. DM 24,–

Vol. 193: Symposium on the Theory of Numerical Analysis. Edited by J. Ll. Morris. VI, 152 pages. 1971. DM 16,–

Vol. 194: M. Berger, P. Gauduchon et E. Mazet. Le Spectre d'une Variété Riemannienne. VII, 251 pages. 1971. DM 22,–

Vol. 195: Reports of the Midwest Category Seminar V. Edited by J.W. Gray and S. Mac Lane. III, 255 pages. 1971. DM 22,–

Vol. 196: H-spaces – Neuchâtel (Suisse)- Août 1970. Edited by F. Sigrist, V, 156 pages. 1971. DM 16,–

Vol. 197: Manifolds – Amsterdam 1970. Edited by N. H. Kuiper. V, 231 pages. 1971. DM 20,–

Vol. 198: M. Hervé, Analytic and Plurisubharmonic Functions in Finite and Infinite Dimensional Spaces. VI, 90 pages. 1971. DM 16.–

Vol. 199: Ch. J. Mozzochi, On the Pointwise Convergence of Fourier Series. VII, 87 pages. 1971. DM 16,–

Vol. 200: U. Neri, Singular Integrals. VII, 272 pages. 1971. DM 22,–

Vol. 201: J. H. van Lint, Coding Theory. VII, 136 pages. 1971. DM 16,–

Vol. 202: J. Benedetto, Harmonic Analysis on Totally Disconnected Sets. VIII, 261 pages. 1971. DM 22,–

Vol. 203: D. Knutson, Algebraic Spaces. VI, 261 pages. 1971. DM 22,–

Vol. 204: A. Zygmund, Intégrales Singulières. IV, 53 pages. 1971. DM 12,–

Vol. 205: Séminaire Pierre Lelong (Analyse) Année 1970. VI, 243 pages. 1971. DM 20,–

Vol. 206: Symposium on Differential Equations and Dynamical Systems. Edited by D. Chillingworth. XI, 173 pages. 1971. DM 16,–

Vol. 207: L. Bernstein, The Jacobi-Perron Algorithm – Its Theory and Application. IV, 161 pages. 1971. DM 16,–

Vol. 208: A. Grothendieck and J. P. Murre, The Tame Fundamental Group of a Formal Neighbourhood of a Divisor with Normal Crossings on a Scheme. VIII, 133 pages. 1971. DM 16,–

Vol. 209: Proceedings of Liverpool Singularities Symposium II. Edited by C. T. C. Wall. V, 280 pages. 1971. DM 22,–

Vol. 210: M. Eichler, Projective Varieties and Modular Forms. III, 118 pages. 1971. DM 16,–

Vol. 211: Théorie des Matroïdes. Edité par C. P. Bruter. III, 108 pages. 1971. DM 16,–

Vol. 212: B. Scarpellini, Proof Theory and Intuitionistic Systems. VII, 291 pages. 1971. DM 24,–

Vol. 213: H. Hogbe-Nlend, Théorie des Bornologies et Applications. V, 168 pages. 1971. DM 18,–

Vol. 214: M. Smorodinsky, Ergodic Theory, Entropy. V, 64 pages. 1971. DM 12,–

Vol. 215: P. Antonelli, D. Burghelea and P. J. Kahn, The Concordance-Homotopy Groups of Geometric Automorphism Groups. X, 140 pages. 1971. DM 16,–

Vol. 216: H. Maaß, Siegel's Modular Forms and Dirichlet Series. VII, 328 pages. 1971. DM 20,–

Vol. 217: T. J. Jech, Lectures in Set Theory with Particular Emphasis on the Method of Forcing. V, 137 pages. 1971. DM 16,–

Vol. 218: C. P. Schnorr, Zufälligkeit und Wahrscheinlichkeit. IV, 212 Seiten 1971. DM 20,–

Vol. 219: N. L. Alling and N. Greenleaf, Foundations of the Theory of Klein Surfaces. IX, 117 pages. 1971. DM 16,–

Vol. 220: W. A. Coppel, Disconjugacy. V, 148 pages. 1971. DM 16,–

Vol. 221: P. Gabriel und F. Ulmer, Lokal präsentierbare Kategorien. V, 200 Seiten. 1971. DM 18,–

Vol. 222: C. Meghea, Compactification des Espaces Harmoniques. III, 108 pages. 1971. DM 16,–

Vol. 223: U. Felgner, Models of ZF-Set Theory. VI, 173 pages. 1971. DM 16,–

Vol. 224: Revêtements Etales et Groupe Fondamental. (SGA 1). Dirigé par A. Grothendieck XXII, 447 pages. 1971. DM 30,–

Vol. 225: Théorie des Intersections et Théorème de Riemann-Roch. (SGA 6). Dirigé par P. Berthelot, A. Grothendieck et L. Illusie. XII, 700 pages. 1971. DM 40,–

Vol. 226: Seminar on Potential Theory, II. Edited by H. Bauer. IV, 170 pages. 1971. DM 18,–

Vol. 227: H. L. Montgomery, Topics in Multiplicative Number Theory. IX, 178 pages. 1971. DM 18,–

Vol. 228: Conference on Applications of Numerical Analysis. Edited by J. Ll. Morris. X, 358 pages. 1971. DM 26,–

ISBN 3-540-05656-4
ISBN 0-387-05656-4